The Role of Renewable Resources for Ecology and Human Health

The Role of Renewable Resources for Ecology and Human Health

Editor

Milka Mileva

MDPI

Basel • Beijing • Wuhan • Barcelona • Belgrade • Novi Sad • Cluj • Manchester

Editor
Milka Mileva
Department of Virology
Institute of Microbiology
"Stephan Angeloff"
Bulgarian Academy of Sciences
Sofia
Bulgaria

Editorial Office
MDPI
St. Alban-Anlage 66
4052 Basel, Switzerland

This is a reprint of articles from the Special Issue published online in the open access journal *Life* (ISSN 2075-1729) (available at: www.mdpi.com/journal/life/special_issues/renewable_resources_life).

For citation purposes, cite each article independently as indicated on the article page online and as indicated below:

Lastname, A.A.; Lastname, B.B. Article Title. *Journal Name* **Year**, *Volume Number*, Page Range.

ISBN 978-3-7258-0108-4 (Hbk)
ISBN 978-3-7258-0107-7 (PDF)
doi.org/10.3390/books978-3-7258-0107-7

Contents

About the Editor

Milka Mileva

Milka Mileva graduated from the Plovdiv University "Paisii Hilendarski", Faculty of Chemistry, with a specialization in organic chemistry in 1993. She received her PhD degree in pharmacology in 2002 at the Medical University of Sofia. In 2011, she became an associate professor at the Institute of Microbiology "Stephan Angeloff" at the Bulgarian Academy of Sciences. Since 2015, Milka Mileva has been the Head of the Laboratory of Biological Response Modifiers and Pathogenesis of Viral Infections in the Department of Virology. Since 2023, Milka Mileva has been a professor of pharmacology. Her research interests are directed toward searching for new biologically active compounds from unexplored biological subjects, natural compounds from microalgae, aromatic plants, and essential oils, especially their redox modulating capacities, in the field of oxidative stress damage during influenza viral infection, herpes virus infection, drug delivery systems, etc. She has multidisciplinary and international achievements by executing many projects in Europe, Japan, and Bulgaria.

Preface

Dear Colleagues,

Recently, numerous studies have focused their attention on natural bioactive compounds, which have been proven to play a potentially beneficial role in the fight against a wide range of human diseases, from "modern" metabolic disorders, chronic diseases, neurodegenerative diseases, cancer, and cardiovascular disease, to resistant pathogens, bacteria, viruses, and fungi, whose higher resistance to conventional drugs is a severe problem for human health.

It turns out that the source of valuable molecules is often products considered to be waste in technological production or agricultural byproducts discarded without being recovered. Thus, a large number of these natural functional ingredients remain unused, and a clear understanding of the mechanisms of their beneficial properties needs innovative and original research.

The topics in this Special Issue included the following points:

- Useful and original approaches for the utilization of residual and waste products from technological production and agricultural activity;
- Preparation, isolation, identification, and examination of natural extracts or synthetic drugs that are modified analogs of natural biomolecules with valuable biological properties;
- Formulation and delivery of anti-inflammatory, antiviral, and antifungal compounds with high stability and good bioavailability to potentiate their activity;
- Exploration of the putative activity in the cascade of events controlling the development of disease;
- In vivo and in vitro screening involves specific compounds to identify them as new, leading compounds with the potential to cause relief of symptoms and influence the progression of some severe diseases.

In this Special Issue, we invited authors working in this field but selected and published only the best articles with the latest research results on the biological effects of waste products, studies of natural compounds, and examination of the mechanisms underlying their biological activity. We selected articles describing innovative experimental models related to the extraction and purification of natural functional components and their application, the presence of hazardous substances, and the beneficial properties of these extracts or compounds in various diseases. Investigations of complex mixtures of natural products and newly synthesized substances were carried out using modern analytical methodologies and approaches.

Milka Mileva
Editor

life

Editorial

The Role of Renewable Resources for Ecology and Human Health

Milka Mileva

Department of Virology, The Stephan Angeloff Institute of Microbiology, Bulgarian Academy of Sciences, 26, Acad. Georgi Bonchev Str., 1113 Sofia, Bulgaria; milkamileva@gmail.com

Scientists are increasingly asking very serious and topical questions: what do we throw away as waste from industrial production? Are there valuable ingredients in what we throw away? What are we not using up from the waste from agriculture or the food industry? What are the environmental pollutants? Are there useful practices and models for converting this waste?

These questions gave us the idea to create the Special Issue "The Role of Renewable Resources for Ecology and Human Health".

Day by day, many natural resources that were abundantly available to humans are diminishing and becoming scarce. Factors such as poor agricultural practices, logging, mining and mineral exploration, industrial and technological development, overconsumption, and waste generation, etc., contribute to this. This leads to environmental pollution and the depletion of natural resources.

Colleagues from different fields of science have shown interest in these issues. We accepted and printed 10 articles with very interesting scientific facts. The pruning wastes (leaves and stems) of rose farms in Taif are rich in cardiac glycosides and flavonoids and inhibit the growth *of Bacillus subtilis*, *Escherichia coli*, *Proteus vulgaris*, *Aspergillus fumigatus*, and *Candida albicans* [1]. Steam distillation of valuable rose essential oil from Bulgarian *Rosa damascena* Mill. and *Rosa alba* L. generates large volumes of wastewater. Although such wastewaters are biopollutants, they contain valuable bioactive compounds that have cytoprotective/genoprotective effects and can reduce DNA damage [2].

An experiment conducted to evaluate the yield and economic benefits of a rice-wheat cropping system (RWCS) influenced by the co-application of sewage sludge (SSL) and fertilizer shows that partial replacement of chemical fertilizers with organic fertilizers reduces both environmental pollution and economic costs and provides greater soil health benefits [3].

Kabaivanova et al. (2022) adapted a fast, efficient process for biogas/biomethane production from natural wheat straw and corn stover. The biomethane yield in biogas reached up to 60% between the second and fifth day [4].

Many plant sources are used as well in cooking as spices and food additives, and so they are involved indirectly in human metabolism through the daily food intake manifesting their healing effect. While extracts from the leaves of *Ginkgo biloba* L. are widely studied and relatively well-known and used in the pharmaceutical industry, the therapeutic properties of the seeds are less well-studied. Petrov et al. (2022) looked for beneficial biological properties of a methanolic extract of *Ginkgo biloba* seeds. The good ability of the extract to neutralize free radicals could have a beneficial effect in pathological conditions with an etiology of oxidative stress [5].

In retrospect, herbal medicinal plants have been used for thousands of years as an alternative healing remedy, including for the treatment of various viral diseases. Natural products used to treat viral infections have some advantages over the chemotherapeutics used. Products derived from medicinal plants are more easily absorbed by the body due to their natural origin and exhibit fewer side effects as they are structurally similar to normal

Citation: Mileva, M. The Role of Renewable Resources for Ecology and Human Health. *Life* **2023**, *13*, 879. https://doi.org/10.3390/life13040879

Received: 15 March 2023
Accepted: 23 March 2023
Published: 25 March 2023

cellular components. The last three years have shown humanity a significant example of the dangers and seriousness of the prevention, treatment, and control of viral diseases through the COVID-19 pandemic. The search for new approaches to control the pathogenesis of this viral infection with extremely diverse symptoms is the basis of a study conducted by Vilhelmova-Ilieva et al. on testing the antiviral activity of a group of Bulgarian medicinal plants widely used to alleviate the symptoms characteristic of COVID-19 patients. The extracts of 13 Bulgarian medicinal plants showed excellent redox-modulating potential and could find application as antioxidants and antiviral symptomatic treatment agents in the fight against COVID-19 infection [6].

The preventive effect of another natural product—snail mucus *Cornu aspersum* (Müller, 1774)—has been investigated as an antiulcerogenic agent on the pathogenesis of ethanol-induced gastric ulcer [7]. The protection of mucus in this model of gastric damage is probably due to the complex action of many factors, including the protection of molecules with antioxidant properties, substances that provide tissue regeneration, and compounds with antimicrobial properties effects that may contribute to the beneficial effect of snail mucus in stomach ulcers and make it a suitable substance for the prevention of stomach diseases.

The complex of classical cytogenetic analyses on the cytotoxic and genotoxic activity of the hydrosols obtained from the water-vapor distillation of Bulgarian *Rosa alba* L. and *Rosa damascena* Mill. did not show significant cytotoxic and genotoxic effects [8]. With their fine aroma and low genotoxic risk, they will continue to be a desirable product in various areas of human life.

Global warming and anthropogenic pollution of natural watersheds, including arsenic, is a serious risk to human health. Omeroglu et al. [9] investigated the contamination of the world's largest Soda Lake—Lake Van, with arsenic and arsenic-resistant bacteria. The authors demonstrate that the total amount of arsenic in this water body changes seasonally and is a serial human health hazard due to the use of lake water for irrigation in agricultural areas. Identification of resistant strains of microorganisms from high-contamination environments, as well as determination of the pathways by which microbial remediation of arsenic occurs, are essential.

In the production process, when processing corn, large amounts of by-products are inevitably generated, which are rich in valuable proteins. Of utmost importance is the development of biotechnologies for the valorization of such by-products as corn-milling by-products from cereals to food/feed ingredients. The selection of optimal conditions under which to preserve the nutritional value of the corn raw material and the total content of amino acids. Thermo-mechanical processing (extrusion), with subsequent 48-h fermentation, is a promising technology that produces a food material capable of potentially reducing microbial contamination, improving its nutritional value, and improving the technological properties of albumins and globulins of corn-milling by-products, as well as the digestibility and bioactivity of prolamins [10].

All these studies show that the source of valuable molecules is often products that are considered waste in technological pollution, or agricultural by-products that are discarded without being recovered. Thus, many of these natural functional ingredients still remain untapped and a clear understanding of the mechanisms of their beneficial properties needs innovative and original research. The innovative and unconventional approaches applied by the authors who published in the first volume of the Special Issue "The Role of Renewable Resources for Ecology and Human Health" have attracted the attention of leading specialists in these fields, and most of the articles now have citations. Therefore, the Editorial Board of Life Journal has suggested that we continue to collect such valuable expertise in Volume II of this edition. My heartfelt thanks to all the authors who responded to our idea and sent us their contributions, to all of you and to the future authors who will publish with us I wish you creative success in your future research, and I look forward with particular interest to your new proposals.

Data Availability Statement: https://www.mdpi.com/journal/life/special_issues/renewable_resources_life.

Acknowledgments: The Guest Editor expresses his heartfelt gratitude to the Editorial Board of the Journal Life for the opportunity to organize and select the most interesting submissions for the Special Issue "The Role of Renewable Resources for Ecology and Human Health".

Conflicts of Interest: The author declares no conflict of interest.

References

1. Galal, T.M.; Al-Yasi, H.M.; Fawzy, M.A.; Abdelkader, T.G.; Hamza, R.Z.; Eid, E.M.; Ali, E.F. Evaluation of the Phytochemical and Pharmacological Potential of Taif's Rose (*Rosa damascena* Mill var. *trigintipetala*) for Possible Recycling of Pruning Wastes. *Life* **2022**, *12*, 273. [CrossRef] [PubMed]
2. Gateva, S.; Jovtchev, G.; Angelova, T.; Dobreva, A.; Mileva, M. The Anti-Genotoxic Activity of Wastewaters Produced after Water-Steam Distillation of Bulgarian Rosa damascena Mill. and Rosa alba L. Essential Oils. *Life* **2022**, *12*, 455. [CrossRef] [PubMed]
3. Jatav, S.S.; Singh, S.K.; Parihar, M.; Alsuhaibani, A.M.; Gaber, A.; Hossain, A. Application of Sewage Sludge in a Rice (*Oryza sativa* L.)-Wheat (*Triticum aestivum* L.) System Influences the Growth, Yield, Quality and Heavy Metals Accumulation of Rice and Wheat in the Northern Gangetic Alluvial Plain. *Life* **2022**, *12*, 484. [CrossRef] [PubMed]
4. Kabaivanova, L.; Petrova, P.; Hubenov, V.; Simeonov, I. Biogas Production Potential of Thermophilic Anaerobic Biodegradation of Organic Waste by a Microbial Consortium Identified with Metagenomics. *Life* **2022**, *12*, 702. [CrossRef] [PubMed]
5. Petrov, L.; Alexandrova, A.; Argirova, M.; Tomova, T.; Georgieva, A.; Tsvetanova, E.; Mileva, M. Chromatographic Profile and Redox-Modulating Capacity of Methanol Extract from Seeds of *Ginkgo biloba* L. Originating from Plovdiv Region in Bulgaria. *Life* **2022**, *12*, 878. [CrossRef] [PubMed]
6. Vilhelmova-Ilieva, N.; Petrova, Z.; Georgieva, A.; Tzvetanova, E.; Trepechova, M.; Mileva, M. Anti-Coronavirus Efficiency and Redox-Modulating Capacity of Polyphenol-Rich Extracts from Traditional Bulgarian Medicinal Plants. *Life* **2022**, *12*, 1088. [CrossRef] [PubMed]
7. Petrov, L.; Kachaunov, M.; Alexandrova, A.; Tsvetanova, E.; Georgieva, A.; Dolashki, A.; Velkova, L.; Dolashka, P. Snail Mucus Protective Effect on Ethanol-Induced Gastric Ulcers in Mice. *Life* **2022**, *12*, 1106. [CrossRef] [PubMed]
8. Gerasimova, T.; Jovtchev, G.; Gateva, S.; Topashka-Ancheva, M.; Stankov, A.; Angelova, T.; Dobreva, A.; Mileva, M. Study on Cytotoxic and Genotoxic Potential of Bulgarian *Rosa damascena* Mill. and *Rosa alba* L. Hydrosols—In Vivo and In Vitro. *Life* **2022**, *12*, 1452. [CrossRef] [PubMed]
9. Ersoy Omeroglu, E.; Sudagidan, M.; Ogun, E. Arsenic Pollution and Anaerobic Arsenic Metabolizing Bacteria in Lake Van, the World's Largest Soda Lake. *Life* **2022**, *12*, 1900. [CrossRef] [PubMed]
10. Zadeike, D.; Gaizauskaite, Z.; Svazas, M.; Gruzauskas, R.; Gruzauskas, V.; Damasius, J.; Juodeikiene, G. Application of Solid-State Fermentation for the Improving of Extruded Corn Dry-Milling By-Products and Their Protein Functional Properties. *Life* **2022**, *12*, 1909. [CrossRef] [PubMed]

life

Article

Application of Solid-State Fermentation for the Improving of Extruded Corn Dry-Milling By-Products and Their Protein Functional Properties

Daiva Zadeike [1,*], Zydrune Gaizauskaite [2], Mantas Svazas [1], Romas Gruzauskas [1], Valentas Gruzauskas [3], Jonas Damasius [1] and Grazina Juodeikiene [1]

1 Department of Food Science and Technology, Faculty of Chemical Technology, Kaunas University of Technology, 50254 Kaunas, Lithuania
2 Food Institute, Kaunas University of Technology, 50254 Kaunas, Lithuania
3 Institute of Computer Science, Vilnius University, 08303 Vilnius, Lithuania
* Correspondence: daiva.zadeike@ktu.lt; Tel.: +370-37-300188

Abstract: In this study, the effect of solid-state fermentation (SSF) with *Lactobacillus sakei* MI401 and *Pediococcus acidilactici* PA-2 strains on functional properties of extruded (130 °C; 25 rpm) corn-milling by-products (CMB) and their albumin, globulin, and prolamin fractions was evaluated in order to produce stabilized and functionalized food/feed stock. Extrusion resulted in a considerable reduction of microbial contamination of CMB by five log cycles, increased damaged starch, water-absorption capacity, and lowered protein and fat contents by 12.4% and 37%, respectively. The application of SSF for the extruded CMB have been shown to improve the water absorption, foaming, and emulsifying capacity of albumins and globulins and also increased the digestibility and free radical scavenging activity of prolamins. The essential amino acid content (EAA) in CMB and antioxidant activity of prolamins was lowered after extrusion but significantly increased after SSF. The combination of the abovementioned treatments can be confirmed as a prospective functionalization of CMB, capable of potentially enhancing its safety and improving nutritional, biochemical, and technological properties of proteins.

Keywords: corn-milling by-products; protein fractions; extrusion; solid state fermentation; protein modification; functional properties; digestibility; radical scavenging activity

Citation: Zadeike, D.; Gaizauskaite, Z.; Svazas, M.; Gruzauskas, R.; Gruzauskas, V.; Damasius, J.; Juodeikiene, G. Application of Solid-State Fermentation for the Improving of Extruded Corn Dry-Milling By-Products and Their Protein Functional Properties. *Life* **2022**, *12*, 1909. https://doi.org/10.3390/life12111909

Academic Editor: Milka Mileva

Received: 19 October 2022
Accepted: 13 November 2022
Published: 16 November 2022

1. Introduction

With the increasing demand for a sustainable environment and healthy food, there is a rapidly growing interest in the industry to provide innovative and sustainable solutions, ensuring the safety and nutritional quality of cereal-based raw materials and food products. In this case, the agroindustry by-products can be valorized into functional components by various technological means and biotechnological methods, providing economic and environmental advantages for the development of new food products and feedstocks.

Corn (*Zea mays* L.), being the third primary cereal crop in the world, contains valuable proteins, fats, and dietary fiber. The application of corn-processing by-products in the food or feed industry can decrease the product's cost [1]. Corn and its milling by-products are mainly used as a lower-nutritional-value animal feedstock even though corn contains high amounts of phospholipids, carbohydrates, proteins rich in carotenoids, and a fiber fraction rich in phytosterols that can be used to improve the nutrition value of food and feed products [2–4]. Recently, corn proteins have been identified as a source of peptides indicating specific bioactivity that can be released during hydrolysis induced by proteolytic enzymes or microbial fermentation [5]. These peptides can be used as bioactive components in food/feed formulations because of their health benefits [6].

Consequently, to reduce microbial contamination and to extend the shelf life of corn milling by-products (CMB), it is important to stabilize raw material by applying thermal technologies, ensuring the safety of end-products [7]. Extrusion is a major technology among other processing techniques used for the production of texture-stable and microbiologically safe corn-based food/feed products. However, when submitted to high-temperature treatment, the feeding value of corn grain may change significantly. Thus, it is necessary to assess the effect of thermal treatment on the functional and technological properties of CMB nutritional components. For these reasons, an appropriate technological approach for the valorization and functionalization of CMB would make possible the improvement of its safety characteristics and functional properties.

One of the ways to increase the nutritional value of cereal-based raw material and to improve its functional properties is lactic acid fermentation. Fermentation is one of the most economical methods of producing and preserving foods and is easily employed for cereal processing. Based on the literature, fermentation with lactic acid bacteria (LAB) can modify starch and protein digestibility in cereal-based products and increase their nutrient availability [8,9]. Most of the research so far has focused on the corn prolamin hydrolysis by microbial enzyme, showing its potential to improve protein functional properties, such as solubility, foaming, and emulsifying capacity, allowing such protein products to apply as functional ingredients with an increased antioxidant effect in both food and non-food applications [10]. Evaluating the impact of the extrusion process on nutritional, functional, and antioxidant properties is mainly focused on corn extrudates [11] or studies on the microstructure, bioavailability, and other functional properties of zein proteins [12].

In this study, the effect of extrusion on the chemical composition and functionality of corn-milling by-products (CMB) was analyzed with emphasis on the amino acid profile, albumin, globulin, and prolamin functional properties. Moreover, the impact of solid-state fermentation (SSF) with *Lactobacillus sakei* and *Pediococcus acidilactici* strains on the hydration properties and the foaming and emulsifying capacity of untreated and extruded CMB albumins and globulins as well as digestibility and bioactivity of prolamins were analyzed.

2. Materials and Methods

2.1. Raw Material

Corn-grits-milling by-products (CMB), mainly consisting of bran and endosperm particles (moisture 11.21%, protein 11.52%, carbohydrates 70.65%, crude fiber 2.72%, fat 2.63%, ash 1.27%) were obtained from the local mill company (Pasvalys, Lithuania). The batch of corn material (50 kg) was stabilized using extrusion cooking at an industrial scale with a one-screw extruder (Parallal Twin Screw Extruder DKM-EII75x28A, Guangdong, China): temperature in the three extrusion zones was 70/90/130 °C, moisture content of raw material was 16%, and feeding rate was 8.2 kg/h. After extrusion, the material was dried at 80 °C to aprox. 12% moisture.

2.2. Microorganisms

The lactic acid bacteria (LAB) strains of *Lactobacillus sakei* (MI401) and *Pediococcus acidilactici* (PA-2), previously isolated from spontaneous rye sourdoughs [13], were used for the fermentation of untreated and extruded CMB material. Before the experiment, all LAB strains were multiplied in a MRS broth (Man-Rogosa-Sharpe, CM 0359, Oxoid Ltd., Hampshire, UK) for 48 h at 30 °C temperature.

2.3. Experimental Design

The effect of extrusion on the nutritional quality and functionality of corn-milling by-products (CMB) was analyzed. The untreated and extruded CMB material was analyzed for chemical composition and functional properties, such as water absorption, damaged starch, and starch gelatinization degree, and also was used for the isolation of albumin, globulin, and prolamin fractions. Further, the effect of solid-state fermentation (SSF) on untreated and extruded CMB amino acid profile, protein fraction yields, and functional properties of albumins and globulins, such as hydration ability, foaming, and emulsifying capacity, as well as digestibility and bioactivity of prolamins was evaluated.

For the fermentation of CMB at solid0state conditions (SSF), the untreated (UN) and extruded (E) corn material (50 g) and 60 mL of sterile water were mixed with the pure suspension of each LAB strain (2%, *w/w*), containing on average 9.1 $\log_{10}$ CFU/mL. The samples were incubated for 48 h at 30 °C under anaerobic conditions. Finally, different batches of fermented (F) CMB samples were prepared: UN_{Ls}, UN_{Pa} and E_{Ls}, and E_{Pa}—untreated-fermented and extruded-fermented with *L. sakei* (Ls) and *P. acidilactici* (Pa), respectively. The pH of samples was measured directly using a pH electrode (PP-15; Sartorius, Goettingen, Germany). For the analysis of the impact of SSF on amino acid profile, the 48 h SSF processing with *L. sakei* strain was applied. Each fermentation procedure was performed twice, followed by analysis of three sub-samples. Fermented samples of untreated and extruded CMB were analyzed for amino acid profile, protein fraction yields, albumin and globulin, water absorption and solubility, foaming and emulsifying capacity, as well as digestibility and bioactivity of prolamins.

2.4. Microbiological Analysis

The total number of aerobic microorganisms in CMB samples was evaluated under standard serial dilution method on plate count agar (PCA) (CM0325, Oxoid, Ltd., Hampshire, UK) and expressed as a $\log_{10}$ of colony-forming units (CFU) per gram of material [14]. Each sample (10 g) was homogenized with the 90 mL of NaCl (9 g/L solution). Serial dilutions of 10^{-4}–10^{-8} were used for the preparation of final sample. The sample solution was spread on the surface of agar in Petri plates that were incubated at 30 °C for 72 h under anaerobic conditions. For the cell number calculation, the plates with more than >300 CFU were reported as unsatisfactory as well as those of less than 30 CFU, with best results in the range of 50–250 CFU/plate. The results were expressed as the mean of three determinations. The limit of detection (LOD) is 1 CFU, and LOQ is 25 CFU.

2.5. Determination of Amino Acids Profile

Amino acids were determined by ultrafast liquid chromatography (UFLC) with automated o-phthalaldehyde (OPA)/9-fluorenylmethyl chloroformate (FMOC). The sample preparation and the UFLC analysis was performed according to Jukonyte et al. [14]. The amino acid standards (A9781 Sigma-Aldrich, Darmstadt, Germany) of 0.5 µmol/mL concentration except for L-cystine at 0.25 µmol/mL in 0.2 M sodium citrate, pH 2.2 were analyzed. A five-level calibration set was used, covering a concentration range of 0.006–0.20 µmol/mL except for alanine and cysteine, each covering a concentration range of 0.06–1.00 µmol/mL. All test samples were analyzed twice.

2.6. Chemical Analyzes

Raw material was tested for protein, fiber, fat, and ash contents according to the AOAC Official Methods [15]. The crude protein content was determined by Kjeldahl nitrogen (method 920.152), and the percentage of protein was estimated by multiplying the total nitrogen content by a factor of 5.7. Crude fiber content was determined according to the method 978.10. Ash was determined by combustion of the sample in a muffle oven at 550 °C for 24 h (method 942.05). The fat content was determined by 3 h Soxhlet extraction with hexane (method 996.01).

2.7. Determination of Xylanase and Protease Activities

For the xylanase and protease activity determination, the CMB sample (5 g) was mixed with 20 mL of 0.1 M acetate buffer (pH 4.5) or 50 mM potassium phosphate buffer (pH 7.5), respectively, and centrifuged (4500× *g*, 20 min). The obtained supernatants were used for the activity assays. Endoxylanase activity was determined spectrophotometrically by the reducing sugar assay [16] using birchwood xylan (0.5%) as substrate. One unit of xylanase activity was defined as the amount of enzyme required to release 1 µmole of xylose equivalents per minute from the birchwood xylan under the assay conditions (37 °C, pH 4.5). The mode of action of corn protease was determined by the Sigma-Aldrich non-specific protease assay using casein (0.65%) as substrate. One unit of protease activity was defined as the amount of enzyme that liberates the equivalent of 1 μg of tyrosine per minute from the substrate under the conditions of the assay (37 °C, pH 7.5). Enzyme activity was reported in terms of enzyme activity units per 100 g of corn material.

2.8. Protein Fractionation

Osborn fractionation of CMB proteins was performed according to Malumba et al. with some modifications [17]. For albumin extraction, a defatted CMB sample (10 g) was extracted with distilled water at a ratio of 1:10 (*w*/*v*) by intensive mixing for 1 h. After, the sample was centrifuged for 20 min (8000× *g*, 4 °C) to obtain albumin extract. For globulin extraction, the residue was extracted with 0.5 M NaCl at a 1:10 (*w*/*v*) ratio following the procedure described above. For the prolamin extraction, the residue was mixed with 70% ethanol, stirred at 60 °C for 90 min, and centrifuged. Albumins, globulins, and prolamins were precipitated with 1 M HCl by adjusting the pH to their isoelectric points of 4.1, 4.3, and 4.8, respectively, and kept at 4 °C overnight. The precipitates were centrifuged at 5500× *g* for 15 min and washed twice with distilled water by centrifugation. Protein sediments were neutralized to pH 7 with 0.1 M NaOH and lyophilized at −40 °C for 48 h (condenser temperature −85 °C, pressure 2×10^{-6} mPa; Zirbus Technology, Bad Grund/Harz, Germany). The protein powders were stored in a freezer at −18 °C until analysis.

2.9. Determination of Albumin and Globulin Functional Properties

The water-absorption capacity and solubility, emulsifying, and foaming properties of albumins and globulins were analyzed according to Silva-Sanchez [18] with some modifications. All analyzes were carried out in triplicate.

2.9.1. Water-Absorption Capacity and Solubility

For water-absorption capacity (WAC) and water-solubility (WS) determination, a 0.5 g (W_0) of sample in a graduated centrifuge tube was thoroughly mixed with 5 mL of distilled water, and the pH of the suspension was adjusted to 4, 7, or 9 with 0.1 M HCl or 0.1 M NaOH. Obtained dispersions were incubated for 30 min at 30 °C in a water bath with continuous shaking. After incubation, the liquid fraction was carefully removed by centrifugation (5500× *g*, 20 min). The wet residue was weighted and the WAC was expressed as grams of water absorbed by the gram of sample (g/g). The supernatant was transferred to a glass tube and dried in an oven at 105 °C to constant weight (S). WS was calculated according to the following equation:

$$WS\ (\%) = S \times 100/W_0 \quad (1)$$

2.9.2. Emulsifying Capacity and Emulsion Stability

The protein sample (0.5 g) was mixed with 10 mL of rapeseed oil in a graduated tube; further, the pH was adjusted to 4, 7, or 9 with 0.1 M HCl or 0.1 M NaOH. Protein dispersions were homogenized (IKA T5, Ultra-Turrax, Staufen, Germany) for 1 min and stored for 30 min at room temperature. After storage, the total volume (V_t) and the volume of emulsified layer (V_{em}) were fixed. Emulsifying capacity (EC) was defined as follows: EC (%) = $V_{em}/V_t \times 100$. Emulsion stability (ES) was measured by sample centrifugation at 3000× *g* for 5 min, following heating at 80 °C for 30 min, and was calculated according to the equation:

$$ES\ (\%) = V_{remaining\ em\ layer}/V_{em} \times 100 \quad (2)$$

2.9.3. Foam-Forming Capacity and Foam Stability

Protein suspensions (10 mL, 4%, *w/v*) in distilled water after adjusting the pH to 4, 7, or 9 with 0.1 M HCl or 0.1 M NaOH were homogenized (IKA T5, Ultra-Turrax, Staufen, Germany) for 1 min in the graduated conical cylinders. The volume of starting liquid phase (V_L) and the volume of foam formed immediately after mixing (V_F) were fixed. The foamed samples were held for 30 min at room temperature to evaluate foam stability. The foam-forming capacity (FFC) and foam stability (FS) after 30 min storage were defined as follows:

$$\text{FFC } (\%) = V_F/V_L \times 100, \quad (3)$$

$$\text{FS } (\%) = V_{F30min}/V_F \times 100. \quad (4)$$

2.10. In Vitro Protein Digestibility

Digestibility in vitro was determined for the CMB prolamins according to the procedure described by Almeida et al. [19]. For analysis, the protein sample (0.5 g) was suspended in 20 mL of 0.1 M HCl, containing 1.5 mg/mL pepsin, and then incubated for 3 h at 37 °C in a water bath. After incubation, 10 mL of 0.5 M NaOH was added. Subsequently, 10 mL of 0.2 M phosphate buffer (pH 8.0) containing 10 mg of pancreatin was added. The protein solutions were incubated for 24 h at 37 °C. After the pancreatic hydrolysis, 1 mL of 10% TCA (trichloracetic acid) solution was added. Further, the protein solutions were centrifuged (8000× *g* for 20 min), and the nitrogen contents in the supernatant (N_S) and the sample (N_t) were measured. Protein digestibility (PD) was calculated:

$$\text{PD } (\%) = (N_t - N_S)/N_T \times 100, \quad (5)$$

where N_t and N_S represent the nitrogen content in the sample before and after digestion, respectively. All measurements were performed at least in triplicate.

2.11. Degree of Hydrolysis

The degree of hydrolysis (DH) of prolamins was measured according to Adler-Nissenn [20]. DH was expressed as a percentage ratio between the number of peptide bonds cleaved (h) and the total number of bonds available for proteolytic hydrolysis (h_{total}):

$$\text{DH} = (B \times N_b/\alpha \times M_p \times h_{tot}) \times 100\%, \quad (6)$$

where B: volume of NaOH solution (mL); α = 10pH-pK/1 + 10pH-pK; Mp: protein mass (g); N_b: NaOH solution concentration (1 M); h_{tot} = 9.2 (for zein). All measurements were performed at least in triplicate.

2.12. Total Phenolic Content

Total phenolic content was determined in the extracts by the method of Singleton and Rossi [21] with slight modification. Lyophilized protein samples were dissolved in a 70% ethanol at a protein concentration of 1 mg/mL, and 0.5 mL of solution was added to 1.5 mL of freshly diluted (1:10) Folin–Ciocalteu reagent (2 N). The mixture was allowed to stand for 5 min, then 1.5 mL of sodium hydrocarbonate (75 g/L) was added. Afterwards, the mixture was incubated for 45 min in the dark and the absorbance read at 765 nm. The standard gallic acid solutions (0.01–1.0 mg/mL) were used for the construction of calibration curve. The results were expressed as mg gallic acid equivalent (GAE) per 1 g protein.

2.13. Determination of Antioxidant Activity

The scavenging activity of prolamins of untreated and pre-treated CMB samples was measured on 1,1-diphenyl-2-picrylhydrazyl (DPPH) free radicals according to the method of Tang and Zhuang [22]. Lyophilized protein samples were dissolved in a 70% ethanol at a protein concentration of 1 mg/mL, and 1 mL of sample solution was mixed with 1 mL of 0.1 mM DPPH solution in 95% ethanol. After shaking, the mixture was stored for 30 min at room temperature, and the absorbance of the sample solution was measured at 517 nm. Ethanol was used as a blank. The reference solution was prepared by mixing ethanol and DPPH solution. The antioxidant activity (AA) of prolamins was expressed as a percentage of DPPH radicals scavenged under the experimental conditions.

2.14. Statistical Analysis

All analyses except those of amino acids were performed at least in triplicate. The results are presented as mean values and standard deviations. The significant differences between means were assessed by analysis of variance (ANOVA) by Duncan test using the IBM SPSS Statistics 27.0 statistical package (SPSS Inc., Chicago, IL, USA). Data means were recognized as significantly different at $p < 0.05$.

3. Results and Discussion

3.1. Characterization of Stabilized CMB Material

The results of the comparative evaluation of chemical composition and physical and functional properties of CMB before and after extrusion are presented in Tables 1 and 2, respectively. The fat, protein, crude fiber, and carbohydrate contents of CMB were 4.36, 12.14, 1.24, and 77.37 g/100 g dw, respectively. Results clearly showed that extrusion processing significantly ($p < 0.05$) affected the nutritional value of corn raw material: the extruded CMB contained 12.4% and 37% lower ($p < 0.05$) protein and fat contents, respectively (Table 1). On the other hand, the crude fiber, ash, and carbohydrate contents of the extruded CMB did not show a significant reduction compared to the raw material.

Table 1. Chemical composition (g/100 g dw) of untreated and extruded corn-milling by-products (CMB).

CMB	Protein	Carbohydrates	Free Sugars	Crude Fiber	Fat	Ash
Untreated	12.14 ± 0.16^{a}	77.37 ± 0.36^{a}	1.13 ± 0.07^{b}	1.24 ± 0.11^{a}	4.36 ± 0.31^{a}	4.89 ± 0.08^{a}
Extruded	10.64 ± 0.11^{b}	79.65 ± 0.64^{a}	1.69 ± 0.06^{a}	1.31 ± 0.08^{a}	2.96 ± 0.17^{b}	5.12 ± 0.06^{a}

Results are the means ± standard deviation (n = 3). Data with different superscript letters within the column represent significant differences ($p < 0.05$).

Table 2. Physicochemical properties of untreated and extruded corn-milling by-products (CMB).

CMB Samples	TCM, $\log_{10}$ CFU/g	Mass Density, g/cm^3	WAC, g/g	Damaged Starch, %	DG, %	Xylanase, XU/100 g dw	Protease, PU/100 g dw
Control	6.89 ± 0.21^{a}	0.547^{a}	2.24 ± 0.02^{b}	33.6 ± 0.1^{b}	48.8 ± 0.7^{b}	77.8 ± 2.7^{a}	33.4 ± 0.9^{a}
Extruded	1.56 ± 0.19^{b}	0.481^{b}	3.82 ± 0.01^{a}	44.2 ± 0.1^{a}	59.9 ± 0.8^{a}	34.7 ± 1.2^{b}	23.2 ± 1.1^{b}

Results are the means ± standard deviation (n = 3). Data with different superscript letters within the column represent significant differences ($p < 0.05$). TCM, total count of anaerobic microorganisms; DG, degree of gelatinization; WAC, water-absorption capacity.

Our results regarding the effect of extrusion on the changes in cereal biopolymers during extrusion cooking are compatible with Hegazy et al. [11], who reported that the extrusion process caused a significant decrease in protein and fat contents of the corn-chickpea extrudates, while fiber, ash, and carbohydrates were not affected compared to the untreated material. A significant reduction of protein content can be due to the reaction occurring between amino groups of amino acids and carbonyl groups of reducing sugars in the presence of high temperature and pressure [23,24].

In the case of functional and safety properties of extruded CMB, the extrusion cooking (Table 2) caused a 5.3-log cycle decrease in the total microbial count, herewith significantly reducing the activity of endogenous xylanases and proteases (by 55 and 30.5%, respectively). The endogenous enzymes, such as xylanase and protease, play an important role in the digestibility of cereal nutrients. Xylanase disrupts the plant cell walls by hydrolyzing insoluble carbohydrates and simultaneously allows exogenous and endogenous enzymes to access proteins and other nutrients [25].

The obtained results indicated the increase in water-absorption capacity (WAC) as well as a degree of starch gelatinization of extruded CMB due to the significantly reduced ($p < 0.05$) particle size (mass density) and increased content of damaged starch as compared to CMB_{UN} (Table 2). According to the literature, extrusion cooking can cause the gelatinization of starch, protein denaturation, lipid separation [23], the complete or partial inactivation of microorganisms and enzymes, and an increase in soluble dietary fiber [24].

Extrusion, as a prevalent physical method causing starch pre-gelatinization, enhances its absorption and swelling power. During the extrusion process, starch is subjected to mechanical shearing at a high temperature and relatively low moisture. This process causes breakage of the covalent bonds between starch components, resulting in a strong structural destruction and partial depolymerization, which promotes the change of its functional properties [23].

The extent of structural and physicochemical changes of starch and proteinaceous components, such as endogenous enzymes and enzyme inhibitors, during extrusion primarily depends on the intensity of the extrusion process parameters. However, thermo-mechanical treatment even at 140 °C did not completely denature the proteins in wheat flour, which might be attributed to a lesser residence time of raw material within the extruder [23]. For example, extrusion temperature 143 °C led to partial inactivation (57%) of trypsin inhibitor in foods [26]. Thus, determining the optimal extrusion conditions for various parameters that will result in cereal products with higher nutritional value can be recommended.

3.2. *The Influence of the Extrusion and Solid-State Fermentation on Protein Yield and the Retention of Amino Acids in CMB*

3.2.1. The Effect of Extrusion on the Protein Extraction Yields and the Amino Acid Profile

The yield of each protein fraction was expressed (Figure 1) as the percentage of the crude CMB protein content. After the isolation of three CMB protein fractions, the protein recovery was approximately 66.76% of crude protein content (12.14 g/100 g of CMB dw).

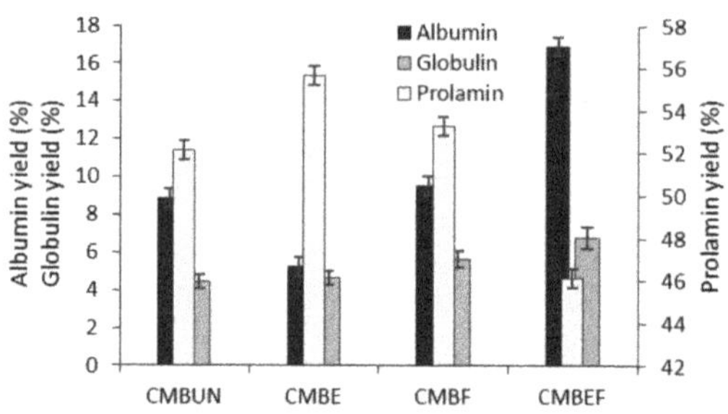

Figure 1. The yields of protein fractions isolated from the untreated (UN), extruded (E), and 48 h fermented (F) corn-milling by-products (CMB).

The Osborne solubility-based protein fractionation indicated that water-soluble albumins and salt-soluble globulins consist of about 14% of total proteins in CMB, while alcohol-soluble prolamin fraction showed the highest yield (52.17%) compared to albumins and globulins (8.87 and 4.47%, respectively) (Figure 1). Alkali-soluble glutelin fraction (not tested in this study) in corn consist by about 34% of crude proteins [27].

In our study, the most effective extrusion was noticed on the albumin and prolamin fractions as a result of heat and mechanical action. Total protein content (10.64 g/100 g dw) in extruded CMB was found significantly ($p < 0.05$) lower, and as was expected, prolamin yield (55.65%) was found significantly ($p < 0.05$) higher due to the reduced content of water-soluble albumins (5.23%), while the globulin proteins were found to be less sensitive to heat treatment during extrusion (yield 4.62%) (Figure 1).

As shown in Table 3, the most abundant amino acids in corn-milling by-products are GLU, LEU, PHE, VAL, ILE, THR, ASP, and PRO. Lack of some EAA, especially LYS, HIS, and TRP, and excess LEU can indicate the poor quality of proteins [27].

Table 3. Amino acids (mg/g protein) in untreated (UN), extruded (E), and fermented (F) 48 h CMB.

Amino Acids	UN	E	F	EF
	EAA			
Valine (VAL)	38.8 ^{a}	33.6 ^{b}	39.4 ^{a}	30.4 ^{c}
Isoleucine (ILE)	35.0 ^{a}	32.2 ^{b}	34.5 ^{a}	32.8 ^{b}
Leucine (LEU)	73.5 ^{b}	66.2 ^{c}	72.0 ^{b}	81.4 ^{a}
Tryptophan (TRP)	3.6 ^{b}	3.2 ^{c}	4.6 ^{a}	4.3 ^{a}
Lysine (LYS)	25.0 ^{b}	21.4 ^{c}	33.5 ^{a}	33.6 ^{a}
Methionine (MET)	16.3 ^{c}	15.4 ^{d}	18.2 ^{b}	22.4 ^{a}
Phenylalanine (PHE)	52.6 ^{a}	48.2 ^{b}	53.5 ^{a}	46.7 ^{b}
Threonine (THR)	33.2 ^{a}	29.9 ^{b}	33.4 ^{a}	31.2 ^{b}
Histidine (HIS)	13.7 ^{b}	12.7 ^{c}	16.3 ^{a}	12.2 ^{c}
Total EAA	291.7 ^{b}	262.8 ^{c}	305.4 ^{a}	295.0 ^{b}
	NEAA and CEAA			
Alanine (ALA)	52.9 ^{ab}	49.5 ^{c}	54.4 ^{a}	57.0 ^{a}
Asparagine (ASP)	44.2 ^{a}	40.4 ^{ab}	42.5 ^{a}	39.8 ^{b}
Serine (SER)	24.3 ^{a}	22.1 ^{b}	25.2 ^{a}	21.7 ^{b}
Glutamine (GLU)	128.4 ^{c}	116.5 ^{d}	162.2 ^{a}	146.6 ^{b}
Cysteine (CYS)	7.8 ^{b}	7.3 ^{bc}	8.6 ^{a}	7.5 ^{b}
Proline (PRO)	33.7 ^{a}	30.2 ^{b}	35.2 ^{a}	34.5 ^{a}
Glycine (GLY)	22.3 ^{b}	20.4 ^{c}	24.7 ^{a}	18.3 ^{d}
Tyrosine (TYR)	28.6 ^{b}	26.1 ^{c}	33.7 ^{a}	34.9 ^{a}
Arginine (ARG)	19.3 ^{a}	17.6 ^{b}	20.9 ^{a}	13.5 ^{c}
Total NEAA and CEAA	361.5 ^{bc}	330.1 ^{d}	397.4 ^{a}	373.8 ^{b}
Total Amino Acids	653.2 ^{b}	592.9 ^{c}	712.8 ^{a}	668.8 ^{b}

Results are the means of two determinations. Data with different superscript letters within the row represent significant differences ($p < 0.05$). EAA, essential amino acids; NEAA, non-essential amino acids; CEAA, conditionally essential amino acids.

Extrusion cooking significantly ($p < 0.05$) reduced the NEAA and CEAA and also EAA contents in extruded samples (330.1 and 262.8 mg/g protein, respectively) compared to the untreated material (361.5 and 291.7 mg/g protein, respectively) (Table 3). Among EAA, which were affected, the most reduced were LYS (14.4%) and VAL (13.4%), following LEU, THR, and TRY (average 9.8%). Most contents of amino acids decreased significantly after extrusion due to the decomposition of amino acids into molecules of ammonia under the influence of high temperature and pressure.

In the case of EAA, the tendency is in accordance with the results reported by other authors for different cereals. Paes and Maga et al. [28] reported the reduction of the contents of ILE, LEU, LYS, THR, and VAL in whole-grain corn flour extruded in a single-screw extruder (130 °C; 80 rpm) compared to the raw material. For example, the losses of LYS may be due to the loss of albumins, which are rich in lysine, during the extrusion process [28]. According to Xiao et al. [29], extrusion of barley powder (140 °C, 40 kHz) decreased the LEU, GLU, and ARG contents. Perhaps, extrusion at low moisture and high temperature leads to starch degradation, thus providing contents of reducing sugars at the same time that it modifies protein structure and favors browning reactions. Since the ε-amino group of lysine has been referred to as a major reactant in the Maillard reaction, it might explain the extrusion effect on this particular amino acid.

3.2.2. The Effect of SSF on the Improvement of Protein Yields and Amino Acid Profile

According to the results (Figure 1), SSF with *L. sakei* positively affected protein recovery: a slight increase in prolamin yield (2.1%) and significantly higher albumin and globulin contents (7.2 and 22.5%, respectively) were determined in fermented CMB compared to unfermented material (Figure 1). In extruded CMB after SSF (sample EF), the 2.2-fold higher average content of albumins and 46.9% higher level of globulins were determined, while the yield of prolamins (46.1%) was found significantly lower compared to the extruded CMB (55.65%) due to the proteolytic degradation occurred during the fermentation process (Figure 1).

This trend is in agreement with work of Cui et al. [30], demonstrating that fermentation of different maize cultivars in the presence of yeast caused a significant increase in protein content (43.5%), which was attributable to a decrease in carbon ratio in the total mass. The microorganisms utilize sugars as an energy source that causes the increased concentration of nitrogen in the fermentation medium and herewith the increase in the proportion of protein. Another possible explanation of this result might be an increase of microbial biomass during fermentation and thus an increase of total protein content. Overall, these trends were consistent with the results of protein analysis (Figure 1).

The amino acid profile is an important characteristic of evaluating the nutritional quality of protein in raw material. In our study, the quantities of most amino acids were increased in the untreated and extruded CMB after fermentation as a result of proteolysis, when peptides are broken down into amino acids by LAB-specific peptidases [30].

In the case of untreated CMB, the majority of EAA, such as LYS, TRP, HIS, VAL, ALA, and MET, contents increased after fermentation compared to unfermented CMB (Table 3). The greatest increase was fixed in LYS (34.0%), TRP (27.8%), HIS (18.9%), and MET (11.7%) contents. Moreover, the highest increase between NEAA and CEAA was determined for GLU (26.3%), TYR (17.8%), and CYS (10.2%).

For the extruded corn material, the amounts of all amino acids except ASP, CYS, SER, HIS, ARG, THR, and VAL increased after fermentation compared to unfermented extruded CMB. The contents of LYS (57%), TYR (33.7%), GLU (25.8%), and LEU (22.9%) increased the most after fermentation.

The results are consistent with the study of Thompson et al. [31], who found that the fermentation of beans and cauliflower increased the concentration of ALA, GLY, HIS, ILE, LEU, and VAL and also with Xiao et al. [29], reporting that lowered contents of GLU, GLY, ALA, and MET in barley powder after extrusion increased after fermentation due to the increased total protein content (Figure 1).

3.3. *The Influence of Extrusion and SSF on CMB Albumin and Globulin Functional Properties*

The results of the influence of extrusion on the albumin and globulin foam-formation capacity (FFC) and foam stability (FS) as well as the emulsifying capacity (EC) and emulsion stability (ES), water-absorption capacity (WAC), and solubility (WS) at different conditions (pH 4–9) are presented in Table 4.

3.3.1. Solubility and Water-Absorption Capacity

WAC is a useful parameter to indicate the possibility of proteins being incorporated into food or feed formulations. The results indicated that the WAC values of CMB albumins and globulins depended on the pH of the medium and extrusion treatment (Table 4). The CMB albumins and globulins exhibited a relatively high WAC (2.12–2.64 g/g) comparable with the study of Gao et al. [32], demonstrating that extrusion improved the solubility of rice proteins and their water-holding capacity up to 37.7% after extrusion at 130 °C. Under the mechanical action, the protein molecular space structure increases due to the degradation of large molecules, making the water molecules easy to penetrate [32].

Table 4. Water-absorption capacity (g/g), water solubility, foaming and emulsifying capacity, and foam and emulsion stability (%) of untreated and extruded CBM albumins and globulins at different pH.

Samples	Albumins			Globulins		
	pH 4	pH 7	pH 9	pH 4	pH 7	pH 9
			Untreated			
WAC	2.42 ± 0.12 ^{bc}	2.52 ± 0.11 ^{b}	2.64 ± 0.08 ^{a}	2.12 ± 0.07 ^{d}	2.27 ± 0.10 ^{c}	2.47 ± 0.06 ^{b}
WS	34.9 ± 0.1 ^{c}	64.8 ± 0.2 ^{b}	73.7 ± 0.3 ^{a}	39.4 ± 0.1 d	63.5 ± 0.2 ^{b}	72.3 ± 0.2 ^{a}
FFC	249 ± 4 ^{c}	266 ± 3 ^{b}	324 ± 6 ^{a}	178 ± 3 ^{e}	186 ± 2 ^{e}	209 ± 3 ^{d}
FS	56.4 ± 0.5 ^{d}	69.6 ± 0.7 ^{a}	63.2 ± 0.4 ^{b}	55.7 ± 0.6 ^{d}	64.8 ± 0.3 ^{b}	58.3 ± 0.2 ^{c}
EC	46.5 ± 0.2 ^{c}	52.2 ± 0.1 ^{b}	59.6 ± 0.5 ^{a}	39.4 ± 0.4 ^{e}	43.8 ± 0.6 ^{d}	49.5 ± 0.3 ^{a}
ES	45.4 ± 0.8 ^{c}	50.5 ± 0.7 ^{b}	56.8 ± 0.2 ^{a}	38.3 ± 0.1 ^{de}	39.2 ± 0.4 ^{d}	42.9 ± 0.5 ^{d}
			Extruded			
WAC	2.54 ± 0.13 ^{b}	2.72 ± 0.14 ^{a}	2.68 ± 0.11 ^{a}	2.38 ± 0.08 ^{c}	2.61 ± 0.12 ^{b}	2.59 ± 0.10 ^{b}
WS	37.6 ± 0.6 ^{d}	78.2 ± 0.4 ^{b}	88.6 ± 1.2 ^{a}	38.7 ± 0.9 ^{e}	76.7 ± 0.8 ^{b}	87.4 ± 1.1 ^{a}
FFC	215 ± 5 ^{c}	238 ± 4 ^{b}	256 ± 6 ^{a}	141 ± 3 ^{f}	152 ± 3 ^{e}	179± 2 ^{d}
FS	53.8 ± 0.2 ^{c}	65.1 ± 0.6 ^{a}	59.2 ± 0.8 ^{b}	52.1 ± 0.7 ^{d}	60.3 ± 0.4 ^{b}	54.8 ± 0.3 ^{c}
EC	40.7 ± 0.2 ^{b}	42.8 ± 0.3 ^{b}	44.3 ± 0.3 ^{a}	33.4 ± 0.4 ^{e}	35.5 ± 0.3 ^{e}	38.9 ± 0.2 ^{d}
ES	39.3 ± 0.1 ^{b}	41.2 ± 0.2 ^{b}	43. 6 ± 0.4 ^{a}	32.7 ± 0.2 ^{d}	34.2 ± 0.1 ^{d}	36.7 ± 0.2 ^{c}

Results are the means ± standard deviation (n = 3). Data with different superscript letters within the raw represent significant differences ($p < 0.05$). WAC, water-absorption capacity; WS, water solubility; FFC, foam-forming capacity; FS, foam stability; EC, emulsifying capacity; ES, emulsion stability.

In our study, WAC for CMB albumins and globulins reached the maximum at pH 9, with values of 2.64 and 2.47 g/g, respectively. Singh et al. [33] reported the water-absorption of lyophilized corn proteins ranging from 2.8 to 3.13 g/g at pH 6.5. According to Pedroche et al. [34], the alkalinization of the protein solution has a positive effect on water absorption. In our study, the alkaline conditions resulted in a slightly lower WAC for globulins of extruded CMB without influencing albumin functionality (Table 4).

The albumin- and globulin-solubility profiles indicated most solubilization at alkali and neutral conditions than at acidic pH due to the low protein solubility close to their pI (4.1 and 4.3, respectively) [17]. There was not a significant difference at neutral and alkali conditions between the solubility of albumin and globulin fractions (64.8 and 63.5% (pH 7), and 73.7 and 72.3% (pH 9), respectively) as well as between solubility at pH 4 for both protein fractions (34.9 and 39.4%, respectively).

The obtained results suggest that CMB albumins and globulins have good solubility under basic conditions, which agrees with the results for the protein isolate obtained from jackfruit seeds [35]. Analogously, Lawal et al. [36] reported low solubility of African locust bean albumins (56.7%) at pH 5, while for globulins, the lowest solubility was observed at pH 4. According to the literature, at a pH > 6.5, the solubility of most plant proteins was reported to be higher than 70% [37,38].

The protein solubility at different pH values may serve as a useful indicator of the protein performance in the food systems in addition to the extent of protein denaturation affected by heat treatment. At the isoelectric point (pH 4–5), there is no net charge on the protein, resulting in no protein–protein interactions that are disadvantageous for the solubility.

The results indicated (Table 4) that extrusion improved water absorption of both protein fractions, increasing WAC values up to 2.72 and 2.61 g/g, respectively, at neutral (pH 7) conditions. Solubility of albumins and globulins was improved significantly ($p < 0.05$) in extruded samples at pH 7 and 9 (by 17%) but was only slightly improved at pH 4 for albumins (increase by 3%), and there was no significant effect found on globulins.

3.3.2. Foaming Capacity and Foam Stability

As shown in Table 4, the foam-forming capacity (FFC) and foam stability (FS) of albumins and globulins significantly depended ($p < 0.05$) on the treatment and pH of the medium. The FFC was found to be high with values of 249–324% for albumins of untreated CMB; the lowest values were fixed at pH 4 and reached a maximum at pH 9. The FFC of the globulins was found at significantly ($p < 0.05$) lower levels (178–209%). Higher values of analyzed characteristics, appearing at alkali conditions, can be due to increased solubility of CMB proteins.

The maximum FS for corn albumins and globulins (69.6 and 64.8%, respectively) was observed at pH 7, whereas the lower FS was fixed at pH 9 (63.2 and 58.3%, respectively), and the lowest FS (~56%) occurred at pH 4. Ulloa and others [35] reported the maximum foaming capacity of 254% and FS of 164% for jackfruit seed protein at pH 10, whereas the minimum foaming capacity and FS occurred at pH 4; in addition, the FS was reduced at pH 8, compared to pH 6 and even pH 4. In contrast to these results, the foaming capacity and stability of lupin protein concentrate were found greatest at acidic pH values [39].

The foaming capacity of proteins is affected by changes in protein structure and solubility. According to Schwenzfeier and co-authors [40], the charges of soybean proteins changed with increasing pH, weakening the hydrophobic interactions and increasing protein flexibility. Thus, a faster protein diffusion to the surface of the air–water phase occurred, encapsulating air particles and increasing FFC. High foam stability of the protein in the isoelectric region is attributed to the formation of stable molecular layers in the air–water interface of the foams. At the pI of proteins, electrostatic forces reduced, lowering protein solubility and increasing surface tension; thus, the adsorption of proteins on the surface of foam bubbles become weak, resulting in a weaker foam formation and lower stability [32]. In the alkaline medium, the surface tension decreases, increasing foaming activity of proteins.

The results indicated that extrusion processing significantly ($p < 0.05$) reduced the FFC of albumin and globulin fractions to 215–256% and 141–179% and FS values to 53.8–65.1% and 52.1–60.3%, respectively (Table 4). According to Hojilla-Evangelista et al. [41], the foaming capacity and stability of dried corn germ albumin and globulin fractions at pH 7 and pH 10 were found to be of 124 and 130% and 84 and 92%, respectively. Maruatona and others [42], showing that after heat treatment (150 °C; 20 min), the foaming activity at pH 7 of bean flour decreased from 31.1 to 30.7%. The *Brassica carinata* protein isolate showed a foaming capacity of 280% at pH 10, and the foam stability after 20 min was 87% [34]. FFC is inferior to the extruded raw material proteins because the structure of the protein changes under the influence of heat and mechanical force, i.e., denaturation of proteins, causes the loss of their ability to form foam [43].

3.3.3. Emulsifying Capacity and Emulsion Stability

The non-polar and polar amino acids cause hydrophobic and hydrophilic properties of proteins, leading to them acting as emulsifiers. The EC of CMB protein fractions depended on pH of the medium (Table 4), as it was observed to be the lowest under acidic conditions (pH 4) with values of 46.5 and 39.4% for albumins and globulins, respectively. It was consistent with the low solubility of proteins and high protein–protein interactions, which led to decreasing emulsion formation. At alkali conditions (pH 9), the EC of albumins and globulins significantly increased ($p < 0.05$) up to 59.6 and 49.5%, respectively, which were close to the Physic nut seed cake protein isolate [43]. Deb and others [44] reported the highest emulsifying capacity of 59.27% for the waste banana peel albumin fraction. According to Jayasena et al. [45], by increasing the pH values of lupine protein extracts from 4 to 8, the emulsifying capacity increased from 51 to 53.4%. Liu et al. [46] reported a pH-dependent increase in the emulsifying activity of soybean proteins, with the lowest value at pH 5.8 and the highest at pH 8. The high EC values of the CMB protein fractions accompanied its high solubility and thus rapid diffusion and adsorption at the interface. According to the literature data [47], emulsifying properties of Bambara bean protein isolate depended on the treatment temperature and pH.

As our study indicated, the EC of CMB albumin/globulin was the highest at pH 9, while emulsion stability was the highest at pH 4. According to Singh et al. [33], the emulsifying capacity and stability at pH 7 ranged from 42 to 53% and from 34 to 41% for corn albumins and globulins. Perilla seed proteins showed the lowest emulsifying activity at pH 4 but tended to increase at pH > 6 [48]. Emulsifying capacity of soluble proteins depends upon the hydrophilic–lipophilic balance, which is affected by pH. At acidic pH and close to the isoelectric point of proteins, the reduced net charges initiate the formation of stronger interfacial membranes; however, at pH 7, the presence of negative charges on the polypeptide chains leads to the formation of weak interfacial membranes, which would have the enhanced oil droplet coalescence and, consequently, reduced stability [42].

Extrusion of CMB significantly reduced the EC of soluble protein fractions (Table 4). The obtained trend is in line with Obatolu and co-authors [49], reporting that the formation of emulsions was highly weakened during thermomechanical processing. With increasing pH, the EC values were reduced by 12.5–25.7% for albumins and by 15.2–21.1% for globulins. The ES values of albumins and globulins after extrusion were the lowest at acidic conditions (39.3 and 32.7%). In the alkaline medium (pH 9), the ES of extruded albumin and globulin fractions were reduced by 23.4 and 14.4% compared to untreated samples (56.8 and 42.9%), respectively (Table 4).

3.3.4. The Influence of SSF on the Changes of Extruded Albumins and Globulins Functional Properties

The results of the effect of SSF processing on the FFC and EC of albumins and globulins after extrusion are presented in Table 5. The analysis of the albumin and globulin functional properties was performed after 30 min retention at pH 9. As was expected, modification of extruded CMB by SSF positively affected the hydration properties and hydrophobicity of these protein fractions. Albumins and globulins in fermented extruded CMB showed WAC to be on average 27.9 and 13.8% higher, respectively, compared to proteins of unfermented extruded CMB (Table 4). The observed improvement of FFC and EC (on average 13.6 and 8.6%, respectively) for both protein fractions indicates the higher effect of fermentation on protein functional properties (Table 5).

Table 5. Water-absorption capacity (g/g), foaming and emulsifying capacity, and foam and emulsion stability (%) of extruded CBM albumins and globulins after fermentation with different LAB.

Samples	Albumins				Globulins			
	L. sakei		*P. acidilactici*		*L. sakei*		*P. acidilactici*	
	24 h	48 h	24 h	48 h	24 h	48 h	24 h	48 h
WAC	3.40 ± 0.12 [a]	3.11 ± 0.09 [b]	3.30 ± 0.10 [a]	3.22 ± 0.07 [ab]	2.89 ± 0.04 [c]	2.92 ± 0.07 [c]	2.96 ± 0.10 [c]	3.02 ± 0.11 [bc]
FFC	287 ± 2 [a]	295 ± 2 [a]	288 ± 2 [a]	290± 2 [a]	192 ± 1 [d]	189 ± 1 [d]	223 ± 2 [b]	211 ± 2 [c]
FS	78.8 ± 0.6 [a]	68.7 ± 0.5 [c]	82.4 ± 0.7 [a]	69.2 ± 0.3 [c]	72.9 ± 0.5 [b]	56.8 ± 0.6 [d]	73.5 ± 0.5 [b]	56.1 ± 0.2 [d]
EC	47.6 ± 0.3 [a]	48.7 ± 0.5 [a]	48.1 ± 0.6 [a]	49.6 ± 0.4 [a]	41.8 ± 0.2 [b]	42.7 ± 0.5 [b]	40.6 ± 0.7 [b]	42.3 ± 0.6 [b]
ES	46.5 ± 0.2 [a]	47.0 ± 0.4 [a]	46.0 ± 0.6 [a]	47.5 ± 0.7 [a]	37.6 ± 0.3 [c]	39.8 ± 0.5 [b]	38.9 ± 0.5 [bc]	40.9 ± 0.7 [b]

Results are the means ± standard deviation (n = 3). Data with different superscript letters within the raw represent significant differences ($p < 0.05$). WAC, water-absorption capacity; FFC, foam-forming capacity; FS, foam stability; EC, emulsifying capacity; ES, emulsion stability.

Twenty-four-hour fermentation with *L. sakei* and *P. acidilactici* strains slightly improved ES (on average by 4.8%) and significantly increased FS (on average by 25.9%) of soluble proteins compared to unfermented samples. However, it should be noted that 48 h fermentation reduced FS of albumins and globulins by 14.5 and 22.8%, respectively, compared to 24 h fermentation. Further, the results showed that the hydration, emulsifying, and foaming capacity of albumins and globulins were slightly affected by the LAB strain used for fermentation, showing a stronger effect ($p < 0.05$) of *P. acidilactici* on globulins (Table 5). There was no significant difference ($p \geq 0.05$) found between the different LAB used on albumin technological properties analyzed.

Worse functional properties in the case of extruded samples are a consequence of forming new intermolecular bonds, structures, and insoluble protein complexes. However, higher water-absorption ability should be connected with more protein released from protein complexes during fermentation [50]. Based on the literature, heat treatment by extrusion negatively affects the formation of protein emulsions, but it can be improved by applying fermentation. Obatolu et al. [49] demonstrated that lupin protein emulsifying activity after fermentation was increased by 23%, and similar results were obtained by Lampart-Szczapa [50], who showed higher lupin protein emulsion stability after solid-state fermentation with appropriate LABs.

Fermentation is a safe and green technology that uses the ability of LAB to produce organic acids and affects the structure of proteins. According to the literature, fermentation treatment resulted in an enhanced oil–water binding of sorghum protein by exposing hydrophobic groups inside the protein, thus improving its emulsifying properties [51]. The study by Tian et al. [52] showed that fermentation enhanced the surface electrostatic charge and solubility of egg yolk proteins, significantly improving its emulsifying activity.

3.4. The Effect of Extrusion and SSF on the Functional Properties and Bioactivity of CMB Prolamins

3.4.1. Digestibility and Degree of Hydrolysis

The digestibility of prolamins after SSF of untreated and extruded CMB with *L. sakei* and *P. acidilactici* varied from 83.56 to 86.69% and from 81.63 to 84.93%, respectively, compared to unfermented controls (71.61% and 76.32%, respectively) (Figure 2A), depending slightly on the LAB strain used for fermentation.

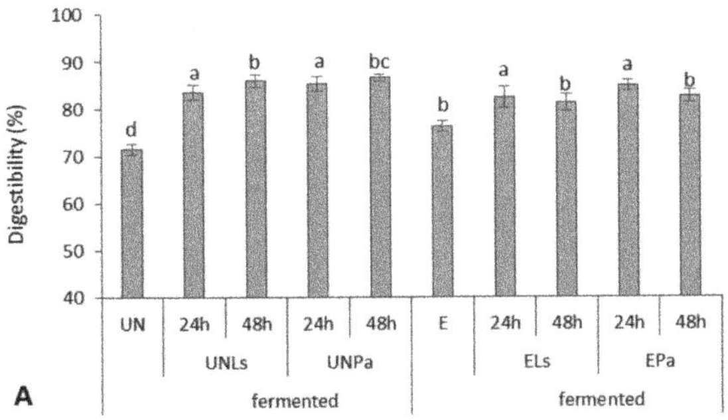

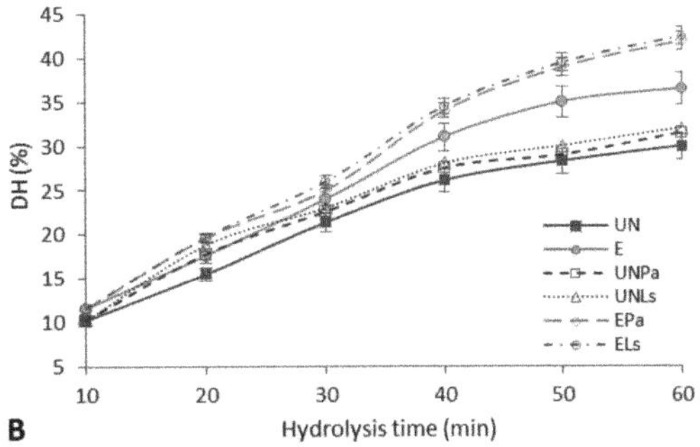

Figure 2. Digestibility (**A**) and degree of hydrolysis (DH) with trypsin (**B**) of prolamins isolated from untreated (UN) and extruded (E) CMB before and after fermentation (F) with *P. acidilactici* (Pa) or *L. sakei* (Ls). Different lowercase letters represent significant differences between data values ($p < 0.05$).

Extruded samples showed a significantly higher (by 20.4%) degree of hydrolysis (DH) of prolamins during 60 min hydrolysis compared to untreated CMB (Figure 2B). With increasing the time of hydrolysis with trypsin, the DH of extruded CMB prolamins was increased from 13.2% to 36.5% and for untreated CMB prolamins from 10.5 to 30%. The obtained trend was in line with Tang and Zhuang [22], reporting that the DH of corn zein fraction reached 24.5% using alkaline proteases and 15% using trypsin. According to the authors, the degree of hydrolysis tended to slightly decrease when increasing the hydrolysis time to 100 min [22].

The increase in the DH of extruded CMB proteins can be explained by the fact that structural changes in proteins occur due to the high temperature of 130–180 °C, pressure, and the screw of the extruder. During the extrusion of corn meal, the partially denatured proteins are stabilized by weak hydrogen bonding, electrostatic, and hydrophobic interactions [53], and such partial protein hydrolysis can improve protein digestibility [54]. As already mentioned, changes in the functional properties of protein fractions can be explained by the action of proteolytic enzymes and the formation of low-molecular-weight proteins and peptides.

The application of the SSF process additionally improved the DH of untreated CMB prolamins (5.8%) and significantly ($p < 0.05$) increased the DH of extruded CMB prolamins (on average by 13.6%). The results of the impact of fermentation on the digestibility of CMB proteins are close to the results of other authors. Krungleviciute et al. [55] reported an increase from 73% to 85% in the digestibility of lupine proteins after fermentation with *P. acidilactici* and *L. sakei* strains. Millet protein digestibility also was increased from 60.5 to 86.0% after 24 h fermentation [56]. It is believed that during fermentation, microorganisms produce different proteolytic enzymes [57] depending on the medium and process conditions, which can lead to an increase in protein digestibility.

3.4.2. Antioxidant Activity of Prolamins

The results of the influence of the extrusion and SSF on the antioxidant activity (AA) of prolamins (Figure 3A) showed that DPPH radical scavenging activity of extruded CMB prolamin fraction was considerably lower (14.7%) compared to untreated CMB prolamins. A slight increase in AA was found for the prolamins of untreated and extruded CMB after 24 h fermentation (6.5 and 8.3%, respectively). The 48 h SSF caused the significant increase (10.0% for CMB_{UN} and 20.2% for CMB_E) in DPPH scavenging activity, indicating that the AA of CMB proteins partially lost during the extrusion cooking can be enhanced by the application of LAB fermentation process. Results also show that in the case of AA, there was not a significant difference ($p \geq 0.05$) between LAB strains. In this case, more effective conditions to enhance the DPPH radical scavenging activity of extruded CMB, namely a 48 h fermentation with *L. sakei*, can be recommended (Figure 3A).

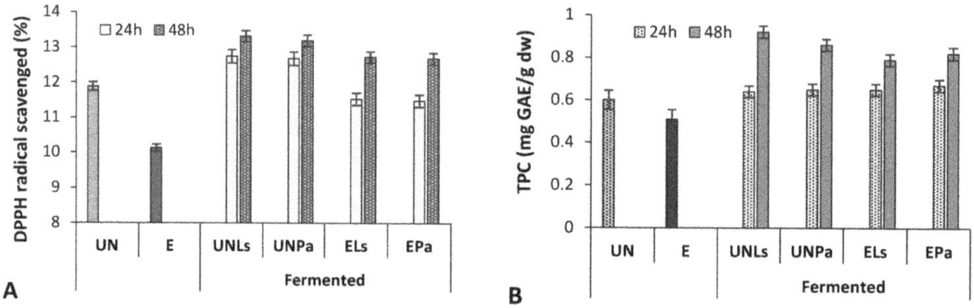

Figure 3. Antioxidant activity (DPPH) (**A**) and total phenolic content (TPC) (**B**) in prolamin fraction of untreated (UN) and extruded (E) CMB before and after fermentation with *L. sakei* (Ls) and *P. acidilactici* (Pa) strains.

The AA values of hydrolyzed by SSF prolamins were comparable to the results reported in the literature. Yang et al. [10] reported a 31.60–56.98% scavenging ability of fermented by *Cordyceps militaris* 202 zein against DPPH. Tang and Zhuang [22] found zein alkaline protease hydrolysates exhibiting high antioxidant activity against DPPH (12.05%).

According to our study, LAB fermentation improved the AA of CMB prolamins mainly due to the release of bioactive compounds caused by bioacidification and LAB hydrolytic enzyme activity. This statement is consistent with the Cui et al. [30], demonstrating that SSF significantly increased total phenolic content (23.4%) of different maize cultivars. Moreover, free and bound phenolic compounds have been isolated from the wheat albumin, glutelin, prolamin, and globulin protein fractions [58].

Regression analysis of data obtained in our study confirmed that the increased antioxidant activity of prolamins correlated with the increased content of phenolics compounds (R^2 = 0.6834) (Figure 3). Extrusion probably caused the breakdown of phenolics, while fermentation helped to degrade cell walls of dietary fiber as well as improve the extraction efficiency of phenolic compounds.

Stanisavljević et al. [59] reported that LAB fermentation with appropriate strains can exhibit strong proteolytic activity on pea proteins, producing high antioxidant active peptides. This indicates that protein hydrolysates contain components that are electron donors and could convert free radicals to stable form, depending on the molecular weight and hydrophobicity of the constituting peptides [60].

Overall, the study demonstrates that SSF with *L. sakei* and *P. acidilactici* is a potential method for the improvement of functional properties of CMB; moreover, such material can be used as a functional ingredient for food/feed applications with increased bioavailability and antioxidant potential of proteins.

4. Conclusions

Since corn is widely used for grits and snack production, large amounts of by-products rich in valuable proteins are inevitably generated in the production process; therefore, it is relevant to develop biotechnologies for the valorization of such cereal by-products to the food/feed ingredients. However, high-temperature extrusion changes significantly the feeding value of corn raw material. Thermo-mechanical treatment (extrusion) lowered fat and protein in CMB and also reduced protein fraction extraction yields, significantly reducing total amino acid content (9.2%). Among EAA, the most reduced were LYS and VAL, followed by LEU, THR, and TRY. However, it was demonstrated that SSF can partially improve the nutritional value of extruded CMB, positively affecting the recovery of protein fractions and increasing the EAA as well as total amino acids content on average by 12.5%, and improving protein functional properties. The study showed that 24 h fermentation significantly increased the hydration foaming and emulsifying capacity of corn albumins and globulins. The 48 h fermentation with tested LAB caused the significant increase in digestibility and DPPH scavenging activity of prolamins, indicating that the antioxidant activity partially lost during the extrusion cooking can be enhanced by the application of fermentation process.

The study demonstrates that the combination of extrusion-fermentation treatment can be confirmed as a prospective functionalization of CMB as feed/food stock capable of potentially reducing microbial contamination, enhancing its nutritional value, and improving technological properties of CMB albumins and globulins and digestibility and bioactivity of prolamins.

Author Contributions: Methodology and formal analysis, Z.G.; data curation, writing—original draft, writing—review and editing, and validation, D.Z.; methodology, formal analysis, and software, M.S. and V.G.; data curation and methodology, R.G.; data curation, writing—review and editing, and resources, J.D.; conceptualization, supervision, and project administration, G.J. All authors have read and agreed to the published version of the manuscript.

Funding: This research received no external funding.

Institutional Review Board Statement: Not applicable.

Informed Consent Statement: Not applicable.

Data Availability Statement: Not applicable.

Acknowledgments: The authors gratefully acknowledge the EUREKA Network Project E!13309 "SUSFEETECH" (Nr. 01.2.2-MITA-K-702-05-0001) and COST Action CA18101 "SOURDOugh biotechnology network towards novel, healthier and sustainable food and bIoproCesseS" (SOURDOmICS).

Conflicts of Interest: The authors declare no conflict of interest.

References

1. Revilla, P.; Alves, M.L.; Anđelković, V.; Balconi, C.; Dinis, I.; Mendes-Moreira, P.; Redaelli, R.; Ruiz de Galarreta, J.I.; Vaz Patto, M.C.; Žilić, S.; et al. Traditional foods from maize (*Zea mays* L.) in Europe. *Front. Nutr.* **2022**, *8*, 683399. [CrossRef] [PubMed]
2. Deepak, T.S.; Jayadeep, P.A. Prospects of maize (corn) wet milling by-products as a source of functional food ingredients and nutraceuticals. *Food Technol. Biotechnol.* **2022**, *60*, 109–120. [CrossRef] [PubMed]
3. Hussain, M.; Saeed, F.; Niaz, B.; Afzaal, M.; Ikram, A.; Hussain, S.; Mohamed, A.A.; Alamri, M.S.; Anjum, F.M. Biochemical and nutritional profile of maize bran-enriched flour in relation to its end-use quality. *Food Sci. Nutr.* **2021**, *9*, 3336–3345. [CrossRef] [PubMed]
4. Loy, D.D.; Lundy, E.L. Nutritional properties and feeding value of corn and its coproducts. In *Corn: Chemistry and Technology*; Serna-Saldivar, S.O., Ed.; Elsevier Inc.: Amsterdam, The Netherlands, 2019; pp. 633–659.
5. Trinidad-Calderón, P.A.; Acosta-Cruz, E.; Rivero-Masante, M.N.; Díaz-Gómez, J.L.; García-Lara, S.; López-Castillo, L.M. Maize bioactive peptides: From structure to human health. *J. Cereal. Sci.* **2021**, *100*, 103232. [CrossRef]
6. Li, G.; Liu, W.; Wang, Y.; Jia, F.; Wang, Y.; Ma, Y.; Gu, R.; Lu, J. Chapter one—Functions and applications of bioactive peptides from corn gluten meal. *Adv. Food Nutr. Res.* **2019**, *87*, 1–41.
7. Juodeikiene, G.; Zadeike, D.; Vidziunaite, I.; Bartkiene, E.; Bartkevics, V.; Pugajeva, I. Effect of heating method on the microbial levels and acrylamide in corn grits and subsequent use as functional ingredient for bread making. *Food Bioprod. Process.* **2018**, *112*, 22–30. [CrossRef]
8. Rodríguez-España, M.; Figueroa-Hernández, C.Y.; de Dios Figueroa-Cárdenas, J.; Rayas-Duarte, P.; Josué Hernández-Estrada, Z. Effects of germination and lactic acid fermentation on nutritional and rheological properties of sorghum: A graphical review. *Curr. Res. Food Sci.* **2022**, *5*, 807–812. [CrossRef]
9. Velikova, P.; Stoyanov, A.; Blagoeva, G.; Popova, L.; Petrov, K.; Gotcheva, V.; Angelov, A.; Petrova, P. Starch utilization routes in lactic acid bacteria: New insight by gene expression assay. *Starch-Stärke* **2016**, *68*, 953–960. [CrossRef]
10. Yang, S.; Zheng, M.; Cao, Y.; Dong, Y.; Yaqoob, S.; Liu, J. Optimization of liquid fermentation conditions for biotransformation zein by *Cordyceps militaris* 202 and characterization of physicochemical and functional properties of fermentative hydrolysates. *Braz. J. Microbiol.* **2018**, *49*, 621–631. [CrossRef]
11. Hegazy, H.S.; El-Bedawey, A.E.-F.A.; El-Sayed, H.R.; Gaafar, A.M. Effect of extrusion processs on nutritional, functional properties and antioxidant activity of germinated chickpea incorporated corn extrudates. *Am. J. Food Sci. Nutr. Res.* **2017**, *4*, 59–66.

12. Wang, X.; Jin, S.; Gou, C.; Hu, L.; Zhang, J.; Li, F.; Zhai, D.; Zhao, Y.; Jihong Huang, J.; Hui, M. Extraction optimization and functional properties of corn germ meal albumin protein as a potential source of novel food ingredients. *J. Food Process. Preserv.* **2022**, *46*, e16218. [CrossRef]
13. Digaitiene, A.; Hansen, A.S.; Juodeikiene, G.; Eidukonyte, D.; Josephsen, J. Lactic acid bacteria isolated from rye sourdoughs produce bacteriocin-like inhibitory substances active against *Bacillus subtilis* and fungi. *J. Appl. Microbiol.* **2012**, *112*, 732–742. [CrossRef]
14. Jukonyte, R.; Zadeike, D.; Bartkiene, E.; Lele, V.; Juodeikiene, G.; Cernauskas, D.; Suproniene, S. A potential of brown rice polish as a substrate for the lactic acid and bioactive compounds production by the lactic acid bacteria newly isolated from cereal-based fermented products. *LWT Food Sci. Technol.* **2018**, *97*, 323–331. [CrossRef]
15. AOAC. *Official Methods of Analysis*, 17th ed.; Approved Methods of Analysis 920.152, 978.10, 942.05, 996.01; The Association of Official Analytical Chemists: Gaithersburg, MD, USA, 2018.
16. Miller, G.L. Use of dinitrosalicylic acid reagent for determination of reducing sugar. *Anal. Chem.* **1959**, *31*, 426–428. [CrossRef]
17. Malumba, P.; Vanderghem, C.; Deroanne, C.; Béra, F. Influence of drying temperature on the solubility, the purity of isolates and the electrophoretic patterns of corn proteins. *Food Chem.* **2008**, *111*, 564–572. [CrossRef]
18. Silva-Sanchez, C.; Gonzalez-Castaneda, J.; de Leon-Rodriguez, A.; De La Rosa, B.A.P. Functional and rheological properties of amaranth albumins extracted from two Mexican varieties. *Plant Food Hum. Nutr.* **2004**, *59*, 169–174. [CrossRef]
19. Almeida, C.C.; Monteiro, M.L.G.; Costa-Lima, B.R.C.D.; Alvares, T.S.; Conte-Junior, C.A. In vitro digestibility of commercial whey protein supplements. *LWT Food Sci. Technol.* **2015**, *61*, 7–11. [CrossRef]
20. Adler-Nissen, J. Determination of the degree of hydrolysis of food protein hydrolysate by trinitrobenzensulfonic acid. *J. Agric. Food Chem.* **1979**, *27*, 1256–1262. [CrossRef]
21. Singleton, V.L.; Rossi, J.A. Colorimetry of total phenolics with phosphomolybdic-phosphofungistic acid reagents. *Am. J. Enol. Vitic.* **1965**, *16*, 144–158.
22. Tang, N.; Zhuang, H. Evaluation of antioxidant activities of zein protein fractions. *J. Food Sci.* **2014**, *79*, C2174–C2184. [CrossRef]
23. Moisio, T.; Forssell, P.; Partanen, R.; Damerau, A.; Hill, S.E. Reorganisation of starch, proteins and lipids in extrusion of oats. *J. Cereal Sci.* **2015**, *64*, 48–55. [CrossRef]
24. Zhang, M.; Bai, X.; Zhang, Z. Extrusion process improves the functionality of soluble dietary fiber in oat bran. *J. Cereal Sci.* **2011**, *54*, 98–103. [CrossRef]
25. Petry, A.L.; Huntley, N.F.; Bedford, M.R.; Patience, J.F. The influence of xylanase on the fermentability, digestibility, and physicochemical properties of insoluble corn-based fiber along the gastrointestinal tract of growing pigs. *J. Anim. Sci.* **2021**, *99*, 159. [CrossRef] [PubMed]
26. Kamau, E.H.; Nkhata, S.G.; Ayua, E.O. Extrusion and nixtamalization conditions influence the magnitude of change in the nutrients and bioactive components of cereals and legumes. *Food Sci. Nutr.* **2020**, *8*, 1753–1765. [CrossRef] [PubMed]
27. Sethi, M.; Singh, A.; Kaur, H.; Phagna, R.K.; Rakshit, S.; Chaudhary, D.P. Expression profile of protein fractions in the developing kernel of normal, Opaque-2 and quality protein maize. *Sci. Rep.* **2021**, *11*, 2469. [CrossRef] [PubMed]
28. Paes, M.C.D.; Maga, J. Effect of extrusion on essential amino acids profile and color of whole-grain flours of quality protein maize (qpm) and normal maize cultivars. *Rev. Bras. Milho Sorgo* **2004**, *3*, 10–20. [CrossRef]
29. Xiao, X.; Li, J.; Xiong, H.; Tui, W.; Zhu, Y.; Zhang, J. Effect of extrusion or fermentation on physicochemical and digestive properties of barley powder. *Front. Nutr.* **2022**, *8*, 794355. [CrossRef]
30. Cui, L.; Li, D.-J.; Liu, C.-Q. Effect of fermentation on the nutritive value of maize. *Int. J. Food Sci. Technol.* **2012**, *47*, 755–760. [CrossRef]
31. Thompson, H.O.; Onning, G.; Holmgren, K.; Strandler, S.; Hultberg, M. Fermentation of cauliflower and white beans with *Lactobacillus plantarum*—Impact on levels of riboflavin, folate, vitamin B-12, and amino acid composition. *Plant Foods Hum. Nutr.* **2020**, *75*, 236–242. [CrossRef]
32. Gao, Y.; Sun, Y.; Zhang, Y.; Sun, Y.; Jin, T. Extrusion modification: Effect of extrusion on the functional properties and structure of rice protein. *Processes* **2022**, *10*, 1871. [CrossRef]
33. Singh, N.; Kaur, M.; Singh, K.S. Physicochemical and functional properties of freeze-dried and oven dried corn gluten meals. *Dry Technol.* **2005**, *23*, 975–988. [CrossRef]
34. Pedroche, J.; Yust, M.M.; Lqari, H.; Giron-Calle, J.; Alaiz, M.; Vioque, J.; Millan, F. *Brassica carinata* protein isolates: Chemical composition, protein characterization and improvement of functional properties by protein hydrolysis. *Food Chem.* **2004**, *88*, 337–346. [CrossRef]
35. Ulloa, J.A.; Villalobos Barbosa, M.K.; Resendiz Vazquez, J.A.; Ulloa, P.R.; Ramírez Ramírez, J.C.; Carrillo, Y.S.; González Torres, L. Production, physico-chemical and functional characterization of a protein isolate from jackfruit (*Artocarpus heterophyllus*) seeds. *CyTA J. Food* **2017**, *15*, 497–507.
36. Lawal, O.S.; Adebowale, K.O.; Ogunsanwo, B.M.; Sosanwo, O.A.; Bankole, S.A. On the functional properties of globulin and albumin protein fractions and flours of African locust bean (*Parkia biglobossa*). *Food Chem.* **2005**, *92*, 681–691. [CrossRef]
37. Bora, P.S. Functional properties of native and succinylated lentil (*Lens culinaris*) globulins. *Food Chem.* **2002**, *77*, 171–176. [CrossRef]
38. Liu, C.-M.; Peng, Q.; Zhong, J.-Z.; Liu, W.; Zhong, Y.-J.; Wang, F. Molecular and functional properties of protein fractions and isolate from cashew nut (*Anacardium occidentale* L.). *Molecules* **2018**, *23*, 393. [CrossRef]

39. Sathe, S.K.; Deshpande, S.S.; Salunkhe, D.K. Functional properties of winged bean (*Psophocarpus tetragonolobus* (L.) DC) proteins. *J. Food Sci.* **1982**, *47*, 503–509. [CrossRef]
40. Schwenzfeier, A.; Wierenga, A.P.; Gruppen, H. Isolation and characterization of soluble protein from the green microalgae *Tetraselmis* sp. *Bioresour. Technol.* **2011**, *102*, 9121–9127. [CrossRef]
41. Hojilla-Evangelista, M.P. Extraction and functional properties of non-zein proteins in corn germ from wet-milling. *J. Am. Oil Chem. Soc.* **2012**, *89*, 167–174. [CrossRef]
42. Maruatona, N.G.; Duodu, G.K.; Minnaar, A. Physicochemical, nutritional and functional properties of marama bean flour. *Food Chem.* **2010**, *121*, 400–405. [CrossRef]
43. Saetae, D.; Kleekayai, T.; Jayasena, V.; Suntornsuk, W. Functional properties of protein isolate obtained from physic nut (*Jatropha curcas* L.) seed cake. *Food Sci. Biotechnol.* **2011**, *20*, 29–37. [CrossRef]
44. Deb, S.; Kumar, Y.; Saxena, D.C. Functional, thermal and structural properties of fractionated protein from waste banana peel. *Food Chem. X* **2022**, *13*, 100205. [CrossRef] [PubMed]
45. Jayasena, V.; Chih, H.J.; Nasar-Abbas, S.M. Functional properties of Australian sweet lupin protein isolated and tested at various pH levels. *Res. J. Agric. Biol. Sci.* **2010**, *6*, 130–137.
46. Liu, C.; Wang, X.; Ma, H.; Zhang, Z.; Gao, W.; Xiao, L. Functional properties of protein isolates from soybeans stored under various conditions. *Food. Chem.* **2008**, *111*, 29–37. [CrossRef]
47. Ngui, S.P.; Nyobe, C.E.; Bakwo Bassogog, C.B.; Tang, E.N.; Minka, S.R.; Mune Mune, M.A. Influence of pH and temperature on the physicochemical and functional properties of Bambara bean protein isolate. *Heliyon* **2021**, *7*, e07824. [CrossRef] [PubMed]
48. Park, B.Y.; Yoon, K.Y. Functional properties of enzymatic hydrolysate and peptide fractions from perilla seed meal protein. *Pol. J. Food Nutr. Sci.* **2019**, *69*, 119–127. [CrossRef]
49. Obatolu, V.A.; Asoyiro, S.B.; Ogunsunmi, L. Processing and functional properties of yam beans (*Sphenostylis stenocarpa*). *J. Food Process. Preserv.* **2007**, *31*, 240–249. [CrossRef]
50. Lampart-Szczapa, E.; Konieczny, P.; Nogala-Kałucka, M.; Walczak, S.; Kossowska, I.; Malinowska, M. Some functional properties of lupin proteins modified by lactic fermentation and extrusion. *Food Chem.* **2006**, *96*, 290–296. [CrossRef]
51. Elkhalifa, A.E.O.; Schiffler, B.; Bernhardt, R. Effect of fermentation on the functional properties of sorghum flour. *Food Chem.* **2005**, *92*, 1–5. [CrossRef]
52. Tian, L.; Hu, S.; Jia, J.; Tan, W.; Yang, L.; Zhang, Q.; Liu, X.; Duan, X. Effects of short-term fermentation with lactic acid bacteria on the characterization, rheological and emulsifying properties of egg yolk. *Food Chem.* **2021**, *341*, 128163. [CrossRef]
53. Zheng, X.-Q.; Li, L.-T.; Liu, X.-L.; Wang, X.-J.; Lin, J.; Li, D. Production of hydrolysate with antioxidative activity by enzymatic hydrolysis of extruded corn gluten. *Appl. Microbiol. Biotechnol.* **2006**, *73*, 763–770. [CrossRef]
54. Elyas, S.H.A.; Tinay, A.H.; Yousif, E.N.; Elsheikh, E.E. Effect of natural fermentation on nutritive value and in vitro protein digestibility of pearl millet. *Food Chem.* **2002**, *78*, 75–79. [CrossRef]
55. Krungleviciute, V.; Starkute, V.; Bartkiene, E.; Bartkevics, V.; Juodeikiene, G.; Vidmantiene, D.; Maknickiene, Z. Design of lupin seeds lactic acid fermentation—Changes of digestibility, amino acid profile and antioxidant activity. *Vet. Zootech.* **2016**, *73*, 47–52.
56. Chu, J.; Zhao, H.; Lu, Z.; Lu, F.; Bie, X.; Zhang, C. Improved physicochemical and functional properties of dietary fiber from millet bran fermented by *Bacillus natto*. *Food Chem.* **2019**, *294*, 79–86. [CrossRef]
57. Soria-Hernández, C.; Serna-Saldívar, S.; Chuck-Hernández, C. Physicochemical and functional properties of vegetable and cereal proteins as potential sources of novel food ingredients. *Food Technol. Biotechnol.* **2015**, *53*, 269–277. [CrossRef]
58. Gammoh, S.; Aludatt, M.H.; Alhamad, M.N.; Rababah, T.; Ereifej, K.; Almajwal, A.; Ammar, Z.A.; Al Khateeb, W.; Hussein, N.M. Characterization of phenolic compounds extracted from wheat protein fractions using high-performance liquid chromatography/liquid chromatography mass spectrometry in relation to anti-allergenic, anti-oxidant, anti-hypertension, and anti-diabetic properties. *Int. J. Food Prop.* **2017**, *20*, 2383–2395. [CrossRef]
59. Stanisavljević, N.S.; Vukotić, G.N.; Pastor, F.T.; Sužnjević, D.; Jovanović, Ž.S.; Strahinić, I.D.; Fira, Đ.A.; Radović, S.S. Antioxidant activity of pea protein hydrolysates produced by batch fermentation with lactic acid bacteria. *Arch. Biol. Sci.* **2015**, *67*, 1033–1042. [CrossRef]
60. Tang, X.; He, Z.; Dai, Y.; Xiong, Y.L.; Xie, M.; Chen, J. Peptide fractionation and free radical scavenging activity of zein hydrolysate. *J. Agric. Food Chem.* **2010**, *58*, 587–593. [CrossRef]

life

MDPI

Article

Arsenic Pollution and Anaerobic Arsenic Metabolizing Bacteria in Lake Van, the World's Largest Soda Lake

Esra Ersoy Omeroglu [1,*], Mert Sudagidan [2] and Erdal Ogun [3]

1 Basic and Industrial Microbiology Section, Biology Department, Faculty of Science, Ege University, Bornova-Izmir 35100, Turkey
2 KIT-ARGEM R&D Center, Konya Food and Agriculture University, Meram-Konya 42080, Turkey
3 Molecular Biology and Genetic Department, Faculty of Science, Van Yuzuncu Yil University, Van 65080, Turkey
* Correspondence: esraerso@gmail.com; Tel.: +90-5352057836

Abstract: Arsenic is responsible for water pollution in many places around the world and presents a serious health risk for people. Lake Van is the world's largest soda lake, and there are no studies on seasonal arsenic pollution and arsenic-resistant bacteria. We aimed to determine the amount of arsenic in the lake water and sediment, to isolate arsenic-metabolizing anaerobic bacteria and their identification, and determination of arsenic metabolism. Sampling was done from 7.5 m to represent the four seasons. Metal contents were determined by using ICP-MS. Pure cultures were obtained using the Hungate technique. Growth characteristics of the strains were determined at different conditions as well as at arsenate and arsenite concentrations. Molecular studies were also carried out for various resistance genes. Our results showed that Lake Van's total arsenic amount changes seasonally. As a result of 16S rRNA sequencing, it was determined that the isolates were members of 8 genera with *ars*C resistance genes. In conclusion, to sustain water resources, it is necessary to prevent chemical and microorganism-based pollution. It is thought that the arsenic-resistant bacteria obtained as a result of this study will contribute to the solution of environmental arsenic pollution problems, as they are the first data and provide the necessary basic data for the bioremediation studies of arsenic from contaminated environmental habitats. At the same time, the first data that will contribute to the creation of the seasonal arsenic map of Lake Van are obtained.

Keywords: Lake Van; arsenic pollution; arsenate; arsenite; anaerobic bacteria

check for updates

Citation: Ersoy Omeroglu, E.; Sudagidan, M.; Ogun, E. Arsenic Pollution and Anaerobic Arsenic Metabolizing Bacteria in Lake Van, the World's Largest Soda Lake. *Life* **2022**, *12*, 1900. https://doi.org/10.3390/life12111900

Academic Editor: Milka Mileva

Received: 25 October 2022
Accepted: 12 November 2022
Published: 16 November 2022

1. Introduction

Microbiota members in alkaline habitats such as soda lakes involve anaerobic microorganisms capable of reducing certain metals and metalloids. During the last 20 years, respiratory reduction of selenate [Se(VI)], selenite [Se(IV)], arsenate [As(V)], Fe(III), Co(III), and Cr(VI) by various microorganisms and this process has been shown to be of geochemical, ecological and environmental importance [1].

Among these metalloids, arsenic is a common element that can be found in the environment in different forms (commonly trivalent [As(III)] or pentavalent [As(IV)] form). Inorganic arsenic is the most abundant arsenic species in nature and is commonly found in different environments [2]. It is generally found in soil as insoluble sulfo salts and sulfides such as Arsenopyrite, Orpiment, Realgar, Lollingite, and Tennantite. Although arsenic is naturally found in the earth's crust, arsenic contamination is mainly caused by various anthropogenic activities such as excessive use of arsenic in pesticides, herbicides, wood preservatives, and medicinal products [3].

Although arsenic is present in living systems at a rate of 2 mg/kg and is called an 'essential toxin' because it is required in trace amounts for growth activities [4]. This metalloid is highly toxic to living organisms at high concentrations, and its toxicity is mainly based on its chemical diversity [2]. Although As(V) and As(III) arsenic forms are the most common forms in natural environments, arsine (-III) and elemental arsenic (0) forms are also basically found in nature. Although both arsenate and arsenite are very toxic, the toxicity of arsenite is approximately 100 higher than that of arsenate [3,5]. These two forms induce various types of cellular damage in biological systems. Considering its general mechanism, it is seen that arsenate and inorganic phosphate are structurally very similar. Therefore, both are taken into the cell by the same system. Phosphate is transported across the cell membrane, and by replacing phosphate with arsenic, reactions that require phosphorylation are damaged, and adenosine triphosphate synthesis is inhibited [3]. In bacteria, yeast, and mammals, under neutral conditions, arsenite is taken into the cell by aquaglyceroporins, which are glycerol transport proteins. Arsenite toxicity is due to the ability of cysteine residues in proteins to bind to sulfhydryl groups and, as a result, to inactivate these proteins [5]. Various toxic effects on human health occur when exposure to arsenic is above the concentration allowed by standards for drinking water's recommended limit of 10 μg/L as set out in the guidelines of the World Health Organization (WHO) and the Environmental Protection Agency (EPA).

From the point of view of microorganisms, it is seen that some microorganisms can cope with arsenic toxicity by using different ways (precipitation, chelation, compartmentalization, extrusion, or biochemical transformation). Therefore, such microorganisms play an important role in the arsenic geocycle [2].

The presence of arsenic in nature is critical to the health of millions of people around the world [5]. Contamination of water and soil with arsenic is a serious threat to public health and the environment worldwide [6]. Because arsenic is highly soluble, it is very difficult to remove arsenic from contaminated water. Looking at the world, millions of people are affected by arsenic in India, Bangladesh, Nepal, Thailand, China, Taiwan, Vietnam, Chile, Hungary, and some parts of the USA [7,8]. In Bangladesh, 57 million people were exposed to arsenic due to the use of arsenic-contaminated wells [9]. When we look at the countries with an arsenic problem, we see that isolation and identification of arsenic-resistant bacteria are carried out by sampling from arsenic-rich environments to solve this problem. Because bioremediation of arsenic in contaminated areas requires arsenic reduction and oxidation using arsenic-resistant microorganisms. For this process to be successful, it is necessary to determine the arsenic metabolism of these microorganisms. For this reason, various studies have been carried out by expert researchers in different countries [9–15].

When the studies in the literature are examined, it is revealed that regions with high salt content, alkaline pH properties, volcanic and geothermal activity, and anthropogenic factors are preferred for the isolation of arsenic-resistant bacteria [8]. Lake Van has these features. Lake Van is localized on the Eastern Anatolian high plateau in southeast Turkey and has a maximum depth of 445 m. In terms of volume, it is the fourth-largest terminal lake and the largest soda lake in the world [16–18]. Due to tectonic movements, there may be a 130 km expansion in the size of the lakes. Nemrut (3050 m above sea level) and Süphan (3800 m above sea level) are semi-active volcanoes located around Lake Van. It is stated that the last eruption took place in Nemrut in 1441. The lake water is salty (21.4‰) and alkaline (155 mEq^{-1}, pH 9.81) because volcanic rocks undergo chemical erosion and go through the evaporation process [19].

There is no study in which Lake Van was selected as the sampling point for arsenic-resistant bacteria isolation. Therefore, this study was carried out mainly to achieve four main objectives: 1. to make seasonal water and sediment sampling, 2. to contribute to the creation of the arsenic map of Lake Van by revealing the degree of influence of arsenic concentration in this region by seasonal factors, 3. to isolation and identification of anaerobically arsenate-reducing and arsenite-oxidizing bacteria and 4. to determine their arsenic metabolic pathways. In this way, it is thought that it will contribute to the sustainability of Lake Van and all living creatures there.

2. Materials and Methods

2.1. Study Site

Lake Van is a volcanic barrier lake formed as a result of the explosion of the Nemrut volcanic mountain, closing the front of the tectonic depression area in the region. The lake is around 430 km from the land. The average depth of the lake is 171 m, and its deepest point is 451 m. The age of the lake is thought to be 600,000 years. The surface area of Lake Van, which has many coves, is 3713 km^2. Lake Van is an aquatic ecosystem different from both freshwater and marine ecosystems. Besides being the largest soda lake in the world, Lake Van is the fourth largest hydrologically closed lake. The eastern part of Lake Van is a region where tectonic movements are intense due to volcanic activity [20], and this type of mobility is one of the factors that cause arsenic transition to the environment. In addition, anthropogenic pollution agents also cause an increase in the amount of arsenic in the lake. A total of 70 percent of the sewage water belonging to various settlements around the lake is transferred to the lake without treatment, causing the pollution rate to increase gradually. It is estimated that 1800 L of sewage flow into the lake every second. On the other hand, tons of garbage and domestic waste feed the lake continuously in terms of pollution via 18 rivers and streams connecting to the lake. In addition, the existence of various social facilities, the use of which increases in certain seasons, is also a pollution factor. When evaluated in terms of the specified criteria, the Edremit-Gebze water sports facility was determined as the sampling point (Figure 1). The Edremit region is located on the eastern wing of Lake Van and is also a region with a high potential for anthropogenic pollution due to various water sports activities and the presence of various businesses in its vicinity. As an indicator, Aydın et al. have shown some regions are not suitable for microbiological criteria in the Van-Edremit region [21]. In addition, since the earthquake, one of the factors that increase the presence of arsenic occurred in Van on 23 October 2011, it appears as another important factor increasing arsenic pollution. When evaluated in terms of all these factors, as a preliminary study, samples were taken from Edremit, Iskele, Gevaş, Akdamar, and Ergil, and their arsenic contents were analyzed. The highest value and microbial activity were obtained in the Edremit region. For this reason, the Edremit region was chosen as the sampling area.

Figure 1. Satellite image of the sampling area.

2.2. Water and Sediment Sampling

The amount of water in terminal deep lakes such as Lake Van may vary seasonally [22,23]. Evaporation occurs with increasing temperature, and accordingly, the arsenic amount in lake water may be variable. So, water and sediment sampling representing four seasons from 7.5 m in the Edremit region was performed. Seasonal sampling was carried out from the same coordinate and depth by determining the days when the weather conditions were suitable. This sample does not represent the entire lake. However, since it was determined as the region with the highest arsenic pollution as a result of the preliminary study sampling was carried out in this region to determine how this anthropogenic pollution changes seasonally.

During sampling, the coordinates and various physical parameters, such as pH, temperature, and dissolved oxygen of the sampling point, were determined (see Appendix A Figure A1). Sampling codes and designated physical parameters are seen in Table 1.

Table 1. Various parameters of seasonal water and sediment samplings.

Season	Winter	Spring	Summer	Autumn
Sampling date	20 January 2015	12 May 2015	28 August 2015	25 November 2015
Sample code				
Water	**WS-1**	**WS-3**	**WS-4**	**WS-5**
Sediment	**S-1**	**S-3**	**S-4**	**S-5**
Coordinate *	38°25′31″ N 43°16′53″ E	38°25′31″ N 43°16′53″ E	38°25′31″ N 43°16′53″ E	38°25′31″ N 43°16′53″ E
Hour *	14:32	09:55	10:30	14:45
pH **	9.52	9.86	9.82	9.80
Temperature **	3 °C	10.8 °C	23.2 °C	11.1 °C
Dissolved O_2 **	11.31 mg/L	9.86 mg/L	6.39 mg/L	8.82 mg/L
Moisture *	87%	91%	37%	69%
Pressure *	811 hPa	834 hPa	1021 hPa	815 hPa

*: iGPS was used. **: HQ40d multi LDO101 and HQ40d multi PHC101 were used.

2.3. Chemical Analysis

Water and sediment samples collected seasonally from the sampling point at Lake Van were analyzed for total arsenic analysis using Inductively Coupled Plasma Mass Spectrometry (ICP-MS) [24]. The samples were examined in terms of 26 different heavy metals. All heavy metal analyzes were performed using the Epa Methods 3051 analysis method [25].

2.4. Isolation of Anaerobic and Alkaliphilic Arsenate-Reducing and Arsenite-Oxidizing Bacteria

All isolation and purification assays were carried out according to Widdel et al. [26]. Saltikov et al. [27], Zargar et al. [12] and Sarkar et al. [28]. Anaerobic conditions were achieved by injecting nitrogen gas into the chamber with the Hungate Technique. Inoculation was carried out from both sediment and water samples to four different media. For the first inoculation, 5 mL of Big Soda Minimal Medium (BSM) [NaCl (20.0 g/L), KH_2PO_4 (0.48 g/L), K_2HPO_4 (0.6 g/L), $(NH_4)2SO_4$ (0.45 g/L), $MgSO_4$ (0.20 g/L), 2 mg/mL vitamin B_{12} (200 µL/L), SL10 trace element solution (2 mL/L) and 1000× vitamin solution (20 mL/L)] was used in Hungate tubes. A 5% percentage of 20× carbs solution [Na_2CO_3 (5.3 g/100 mL), $NaHCO_3$ (2.1 g/100 mL)] was used to adjust 9.5 the pH of the medium. This medium was a modified one in terms of NaCl amount. Because the NaCl content of Van Lake was 20‰. So, 20‰ NaCl was added instead of 50‰. Tubes containing BSM, 10 mM sodium arsenate [$Na_2HAsO_4.7H_2O$] as an electron acceptor, and 10 mM Na-lactate (medium 1) or Na-acetate (medium 2) as electron donors were inoculated to select for arsenate [As(V)] reducing bacteria. For isolation of arsenite [As(III)] oxidizing bacteria, tubes containing BSM, 10 mM Na-nitrate as an electron donor, and 1 mM sodium arsenite ($NaAsO_2$) (medium 3) or 0.5 mM $NaAsO_2$ (medium 4) were used. Sediment sample (1 g) and water sample 8 (1 mL) were inoculated into medium 1–4 under anaerobic conditions. Samples were then incubated at 30 °C for 5 days. 1 mL of inoculum belonging to each tube was transferred to the same medium in Balch tubes and incubated at the same condition [12,26–28]. The transfer to Balch tubes was repeated three times. In the anaerobic chamber, the inoculum obtained from the last tube was transferred to BSM agar plates. In that step, from which tube the inoculum was taken, it was transferred to the same content of agar medium. After the growth was observed during the anaerobic incubation period (5 days), the colonies were patched on fresh BSM agar plates. Several morphologically different colonies were selected (see Appendix A Figure A2). For all arsenic-resistant isolates, glycerol stocks were prepared with a 25% final glycerol concentration and stored at −80 °C.

2.5. Arsenic Transformation Assay

An arsenic transformation assay was performed using a silver nitrate test [28]. Different kinds of agar plates (Luria Bertani (LB), BSM, and Chemically Defined Medium [CDM]) were supplemented with sodium arsenate (10 mM) and sodium arsenite. All isolates were streaked onto agar plates and incubated at 30 °C for 5 days. After the incubation period, all plates were flooded with a 0.1 M silver nitrate ($AgNO_3$) solution conserved in dark conditions. Color precipitations were examined to evaluate the results. A brownish and a bright yellow precipitate indicated the presence of silver arsenate (Ag_3AsO_4) (arsenite oxidizing bacteria) and silver arsenite (Ag_3AsO_3) (arsenate reducing bacteria), respectively. As positive controls, *Shewanella* ANA-3 [27] and *Halomonas* sp. BSL-1 [29] were used.

2.6. Physiological Characterization and Identification of Arsenic-Resistant Bacteria

Arsenic-resistant bacteria obtained from different water and sediment samples of Lake Van were initially characterized in terms of colony morphologies (color, shape, size) and basic microscopic observations (Gram, spore, capsule, and PHB stains) [13]. Isolates were then biochemically analyzed for the activities of nitrate reduction [30], starch hydrolysis, gelatin liquefaction test [31], production of extracellular lipase, lecithinase [32], protease [31], carbohydrate assimilation, along with fermentation tests like glucose, lactose, mannose, fructose, and galactose [30,33].

For molecular identification, the genomic DNA of arsenic-resistant isolates was obtained with PureLink® Genomic DNA Kits according to the manufacturer's instructions. For checking the integrity of DNA, gel electrophoresis was performed using 1% agarose gel. The universal primers used for the isolation of the 16S rRNA gene region were 11F (5′-AGTTGATCCTGGCTCAG-3′) and 1492R (5′-ACCTTGTTACGACTT-3′) [12]. The reaction mixtures were composed of 5 µL 10X PCR buffer, 2.5 mM dNTPs, 10 pmol each primer, 10 ng DNA template, 3 units Taq DNA polymerase, and sterile deionized water to a final volume of 50 µL. The PCR protocol for amplification of the 16S rRNA gene was comprised of denaturation at 94 °C for 5 min, followed by 35 cycles at 94 °C for 1 min, annealing at 48 °C for 45 s, and extention at 72 °C for 2 min. A final extention was carried out for 10 min at 72 °C. To reveal individual differences, amplified 16S rRNA gene regions of each isolate were cut with *Eco*RI (CutSmart Biolabs, New England Biolabs, Ipswich, MA, USA, R3101S), *Rsa*I (CutSmart Biolabs, New England Biolabs, R0167S), and *Hae*III (CutSmart Biolabs, New England Biolabs, R01108S) restriction enzymes according to manufacturer's instructions. The bidirectional sequence analysis of purified amplicons (with Sephadex® G-50 kit, Sigma Aldrich G5080, Merck KGaA, Darmstadt, Germany) was conducted with ABI 3130xl genetic analyzer (Applied Biosystems, Waltham, MA, USA) using the same primers. Chromatograms were analyzed with Finch TV, ApE, and Bioedit programs. 16S rRNA partial sequence of selected isolates was compared to reference sequences in the GenBank database using BLASTN, and accession numbers for the strains were obtained.

2.7. PCR Amplification of Arsenic-Related Marker Genes

In this study, we examined some putative arsenic resistance genes, including cytosolic arsenate reductase gene (*ars*C) [10,28], arsenite efflux gene (*ars*B) [10], and respiratory arsenate reductase gene (*arr*B) [34]. PCR amplification of *ars*C, *ars*B, and *arr*B gene regions was done by using arsC-1-F (5′-GTAATACGCTGGAGATGATCCG-3′) and *ars*C-1-R (5′-TTTTCCTGCTTCATCAACGAC-3′), *ars*B-1-F (5′-CGGTGGTGTGGAATATTGTC-3′)

and *ars*B-1-R (5′-GTCAGAATAAGAGCCGCACC-3′), and *arr*B-F (5′-AACACGAACGACG GTATTCACTGG-3′) and *arr*B-R (5′-ATACCTTGCTCTGTGGATCATCTA-3′) primers, respectively. The reaction mixtures composed of 2.5 μL 10X PCR buffer, 2.5 mM dNTPs, 10 pmol each primer, 10 ng DNA template, 2 units Taq DNA polymerase, 5% DMSO, 2 mM $MgCl_2$ and sterile deionized water to a final volume of 25 μL for *ars*C and *ars*B gene regions. The same mixture was used to amplify the *arr*B gene region, excluding the amount of template DNA and primers. A total of 25 ng DNA and 20 pmol of each primer were optimized for amplification. PCR profile was as follows: initial denaturation at 94 °C for 3 min (5 min for *arr*B), followed by 30 cycles (for *ars*C and *ars*B) or 35 cycles (for *arr*B) of denaturation (94 °C, 30 s for *ars*C and *ars*B; 40 s for *arr*B), annealing (56 °C, 30 s for *ars*C and *ars*B; 62 °C 1 min for *arr*B), extension (72 °C 30 s for *ars*C and *ars*B, 1 min for *arr*B), and final extention at 72 °C for 7 min. Obtaining 370 bp, 219 bp, and 500 bp bands as amplification products for *ars*C, *ars*B, and *arr*B was evaluated as having positive results.

2.8. Effect of Physicochemical Parameters

Hydrogen ion concentration (pH), temperature, and salinity variation were selected as physicochemical parameters for bacterial growth. All assays were performed using the microplate technique in LB broth with 2% NaCl or pH 9.5. The amount of bacterial inoculum was designated as 1 McFarland. Each arsenic-resistant strain was inoculated into three microplate wells for all different physical parameters under anaerobic conditions. A 5-day incubation period was preferred for all physicochemical parameters. 25 °C, 30 °C, 35 °C, 45 °C, 55 °C, and 65 °C were selected for incubation temperature. Depending upon the optimum temperature, they were incubated in microplates with different pHs of 4.5, 5.5, 6.5, 7.5, 8.5, 9.5, and 10.5 to determine the optimum pH. To adjust the pH, 0.1 M different buffers (acetate, phosphate, and carbonate) were used. Similarly, the effect of salinity variation on bacterial growth was studied in a microplate. For this assay, LB with pH 9.5 was amended with different concentrations of NaCl as 1%, 2%, 3%, 4%, 5%, 6%, 7%, 8%, 9%, 10%, and 11%. After an anaerobic 5 days incubation period, the optical density of the growing cultures in different conditions was determined spectrophotometrically at 620 nm [13]. An amount of 20 μL of 2,3,5 triphenyl tetrazolium chloride (TTC) with 10% concentration for each microplate well was used as the indicator of viability for clear observation of bacterial viability. After adding TTC, all microplates were incubated under anaerobic conditions for an additional day. The formation of red-colored triphenyl formazan (TPF) was evaluated as a positive result [35].

2.9. Arsenic Tolerance Assay

The resistance of the strains against As(V) [$Na_2HAsO_4.7H_2O$] and As(III) ($NaAsO_2$) was determined by minimum inhibitory concentration (MIC). To screen the strains based on arsenic resistivity, the overnight grown culture was inoculated in a microplate well containing LB with 2% NaCl and pH 9.5, As(V) or As(III) as sources of the arsenic. All concentrations of arsenic were examined in three wells at 30 °C. The MIC value was tested by growing arsenic metabolizing bacteria in different concentrations of As(V) (0–320 mM) and As(III) (0–32 mM) [28]. The bacterial growth was determined via optical density measurement at 620 nm, and vitality was corrected by using TTC. The MIC was defined as the lowest concentration of As(V) and As(III) that suppressed bacterial growth [35].

3. Results

3.1. Sample Collection

To isolate arsenic-resistant anaerobic bacteria, sediment and water samples of Lake Van were collected seasonally. For each season, the same sampling coordinate was used. After sampling, 4 different sediment and water samples were obtained. The pH of the samples was found to be between 9.52–9.86. As expected, temperature, dissolved O_2, and moisture changed for each season (Table 1).

3.2. Determination of Heavy Metal Amount in Van Lake Samples

According to EPA and WHO, 10 ppb (µg/L) has been indicated as the limit of arsenic concentration.

The results of 24 different heavy metal analyzes of sediment and lake water samples are shown in Table 2. In general terms, as expected, it was determined that the amounts changed depending on seasonal changes (Table 2).

Table 2. Concentration of some heavy metal for Lake Van samples for each season.

Heavy Metal	Season							
	Winter		Spring		Summer		Autumn	
	WS-1	S-1	WS-3	S-3	WS-4	S-4	WS-5	S-5
Na	7587.000	17,430.000	7230.000	4885.000	8471.00	6616.0	13,690.000	20,800.00
Mg	104.000	19,660.000	98.060	7028.000	107.60	7716.0	167.200	73,670.00
Al	0.253	21,100.000	0.005	7931.000	1.40	8848.0	4.032	24,830.00
K	409.200	7220.000	462.000	3401.000	459.10	2774.0	737.400	11,360.00
Ca	34.060	125,200.000	3.949	30,760.000	11.20	25,980.0	23.080	233,500.00
V	0.005	57.130	0.004	22.800	0.01	19.58	0.019	79.570
Cr	0.012	80.760	ND	30.950	0.01	30.03	0.083	131.800
Mn	0.024	509.900	ND	173.400	0.04	148.70	0.057	771.300
Fe	0.080	18,800.000	ND	7043.000	3.06	7272.0	2.491	281,800.00
Co	ND	10.520	ND	3.963	ND	3.8	ND	16.010
Ni	0.011	69.120	0.021	26.540	ND	31.4	0.036	93.030
Cu	ND	14.230	ND	7.818	ND	4.9	0.185	8.558
Zn	0.714	53.840	0.250	31.030	ND	13.13	0.222	26.100
As	0.214	9.370	0.263	2.730	0.14	2.61	0.261	26.070
Se	0.150	0.999	0.100	4.830	0.07	2.49	0.039	1.207
Mo	0.012	1.423	0.007	1.727	0.02	1.01	0.027	1.194
Ag	ND	81.610	ND	0.235	ND	ND	0.008	0.354
Cd	ND	0.186	ND	0.085	ND	0.06	ND	0.199
Sn	ND	1.711	ND	0.740	0.07	0.29	0.120	6.546
Sb	ND	0.299	ND	0.164	ND	ND	ND	0.528
Ba	0.003	133.800	0.013	56.880	0.04	38.78	0.043	297.200
Hg	ND	ND	ND	0.591	0.13	0.015	ND	ND
Tl	ND	0.259	ND	0.103	ND	ND	ND	0.181
Pb	0.045	8.456	ND	4.075	0.003	2.35	0.024	15.160

Concentration of heavy metal (mg/kg or mg/L)

ND: Not determined. Those beginning with the WS and S codes represent samples of lake water and sediment, respectively.

When evaluated in terms of arsenic, it was determined to be above the standards, especially in the sediment samples. Within the scope of the study, seasonal sampling was carried out with the thought that seasonal differences may change the amount of arsenic in the lake water and sediment and therefore change the dominant microbiota. The ICP-MS analysis showed that the total amount of arsenic in the sediment was very high in the autumn season, and there was a gradual decrease in the winter, spring, and summer seasons. It was determined that the total amount of arsenic found in the sediment, especially in the autumn season, resulted in a 10-fold decrease in the summer (see Appendix A, Figure A3).

3.3. Isolation of Arsenate and Arsenite Metabolizing Anaerobic Bacteria

A total of 81 arsenic-resistant bacteria were isolated from the samples of lake water and sediment from Lake Van. After this stage, the studies were continued with the 12 arsenic-resistant isolates selected. They were designated as 4-S-1-1 A, 4-S-1 A2, 1-WS-1, 1-WS-1-1, 1-S-1 (2), 2-WS-1-1, 3-S-1 K, 3-S-1 A, 1-S-3-1, 3-S-4-1, 1-WS-5-1, and 1-S-5 (3) and also they were isolated in different season and different media content (Table 3).

Table 3. The features of isolation source of arsenic-resistant bacteria and the results of 16S rRNA analysis of the isolates.

No.	Isolate	Sample	Season	Medium		Closest Neighbor	Percentage Similarity	Accession Number
				e^- acceptor	e^- donor			
1	4-S-1-1 A	Sediment	Winter	Na-nitrate (10 mM)	As(III) (0.5 mM)	*Alkalimonas delamerensis*	98%	KY681793
2	4-S-1 A2	Sediment				*Idiomarina* sp.	99%	KY681796
3	1-WS-1	Lake water		As(V) (10 mM)	Na-lactate (10 mM)	*Branchybacterium paraconglomeratum*	99%	KY681785
4	1-WS-1-1	Lake water				*Microbacterium schleiferi*	99%	KY681786
5	1-S-1 (2)	Sediment				*Anaerobacillus* sp.	99%	KY681780
6	2-WS-1-1	Lake water		As(V) (10 mM)	Na-acetate (10 mM)	*Nitrincola* sp.	98%	KY989222
7	3-S-1 K	Sediment		Na-nitrate (10 mM)	As(III) (1 mM)	*Halomonas* sp.	99%	KY681792
8	3-S-1 A	Sediment				*Marinobacter halophilus*	99%	KY681791
9	1-S-3-1	Sediment	Spring	As(V) (10 mM)	Na-lactate (10 mM)	*Anaerobacillus* sp.	99%	KY681781
10	3-S-4-1	Sediment	Autumn	Na-nitrate (10 mM)	As(III) (1 mM)	*Halomonas* sp.	100%	KY989223
11	1-WS-5-1	Lake water	Summer	As(V) (10 mM)	Na-lactate (10 mM)	*Halomonas campisalis*	100%	KY681788
12	1-S-5 (3)	Sediment	Summer			*Halomonas* sp.	99%	KY681784

3.4. Silver Nitrate Test

The $AgNO_3$ screening technique was used to detect the oxidation of As(III) to As(V) or the reduction of As(V) to As(III) [35]. Although different media and conditions were used in addition to the media specified in the literature within the scope of this experiment, no results were obtained for the alkaliphilic isolates.

3.5. Identification of Arsenic-Resistant Bacteria

It was determined that 12 isolates selected within the scope of the study belong to eight different genera (Table 3). According to the results of partial 16S rRNA sequence analysis, it was determined that there were similarities with various species at rates ranging from 98% to 100%. Molecular identification of 7 strains could be made at genus level and did not show similarity with any other species within the ranges suitable for the criteria. It has been determined that 4 strains (3-S-1 K, 3-S-4-1, 1-WS-5-1, and 1-S-5 (Table 3)) are members of the *Halomonas* genus, which is also found extensively in other soda lakes in the world. It was observed that 2 strains (1-S-1 (2) and 1-S-3-1) in the sediment samples had members of *Anaerobacillus*, an obligate anaerobic species [36].

When we look at the determined characteristics of arsenic-resistant strains, it is seen that they exhibit different profiles depending on the strain difference. As a common feature of all strains, it appears there is no extracellular gelatinase activity (Table 4).

Table 4. Some characteristics of arsenic-resistant bacteria from Lake Van.

Characteristics	**Strains**											
	1	**2**	**3**	**4**	**5**	**6**	**7**	**8**	**9**	**10**	**11**	**12**
Colony color	Blue	Blue	Cream	Cream	Cream	Cream	Cream	Blue	Transparent	Cream	Cream	Blue
Gram staining	Gr(-)	Gr(-)	Gr(+)	Gr(+)	Gr(+)	Gr(-)	Gr(-)	Gr(-)	Gr(+)	Gr(-)	Gr(-)	Gr(-)
Cell morphology	Bacil	Bacil	Bacil	Bacil	Bacil	Bacil	Bacil	Bacil	Bacil	Bacil	Bacil	Coccobacil
Formation of												
Capsule	-	+	+	+	+	+	+	+	+	+	+	+
Endospore	-	+	+	-	+	+	+	-	-	+	+	+
PHB	+	+	+	-	+	+	+	-	-	-	-	+
Nitrate reduction	+	-	-	-	+	+	-	+	+	+	+	+
Production of												
Lipase	-	-	-	-	-	-	-	-	+	-	-	-
Lecithinase	-	-	-	-	-	+	+	-	-	+	-	+
Amylase	+	-	+	-	+	-	-	-	+	+	+	+
Protease	-	+	-	-	+	+	+	+	-	-	-	-
Gelatinase	-	-	-	-	-	-	-	-	-	-	-	-
Assimilation of												
Glucose	+	+	+	+	+	+	+	+	+	+	+	+
Lactose	+	+	+	+	+	+	+	+	+	+	+	+
Mannose	+	+	-	+	+	+	+	+	+	+	+	+
Fructose	+	+	+	+	+	+	+	+	-	+	+	-
Galactose	+	+	+	+	+	+	-	-	-	+	+	+
Fermentation of												
Glucose	+	-	-	-	-	+	-	-	-	-	+	-
Lactose	+	-	+	+	-	+	+	+	-	+	+	+
Mannose	+	-	+	-	-	+	+	+	-	+	-	+
Fructose	+	+	-	-	-	+	-	-	-	+	+	+
Galactose	-	+	+	-	-	+	-	-	-	+	+	+

1: 4-S-1-1 A, 2: 4-S-1 A2, 3: 1-WS-1, 4: 1-WS-1-1, 5: 1-S-1 (2), 6: 2-WS-1-1, 7: 3-S-1 K, 8: 3-S-1 A, 9: 1-S-3-1, 10: 3-S-4-1, 11: 1-WS-5-1, 12: 1-S-5 (3). PHB: Poly-β-hyrdoxybutyrate.

3.6. Detection of Arsenic Marker Genes

Within the scope of the study, it was investigated whether microorganisms have certain gene regions for respiration of As(V) and are resistant to arsenic. When strains are evaluated in terms of having *ars*C, *ars*B, and *arr*B gene regions selected as target gene regions, it has been determined that all strains have shown a common profile in terms of gene regions and all species have the *ars*C gene, which is the arsenate reductase enzyme gene, which confers the ability of microorganisms to convert arsenate to arsenite prior to extrusion of the latter oxidation [37]. In the meantime, it was observed that none of the strains had the *ars*B and *arr*B gene regions, although the required products (219 bp and 500 bp products for *ars*B and *arr*B genes, respectively) were observed in the positive controls. The products of the gene regions were not observed in the isolates.

3.7. Effect of Physicochemical Parameters

As a result of the experiment carried out under anaerobic conditions, it was determined that there was no microbial activity at 55 °C and 65 °C in common, but there were differences at other temperatures. It was observed that the temperatures at which the best reproduction took place were 25 °C and 30 °C. It was determined that strain 2-WS-1-1 (*Nitrincola* sp.) was the strain that showed the best growth at these temperatures. By increasing the incubation temperature to 45 °C, it was observed that strain 1-S-5 (3) (*Halomonas* sp.) exhibited microbial activity, unlike other strains (Figure 2A).

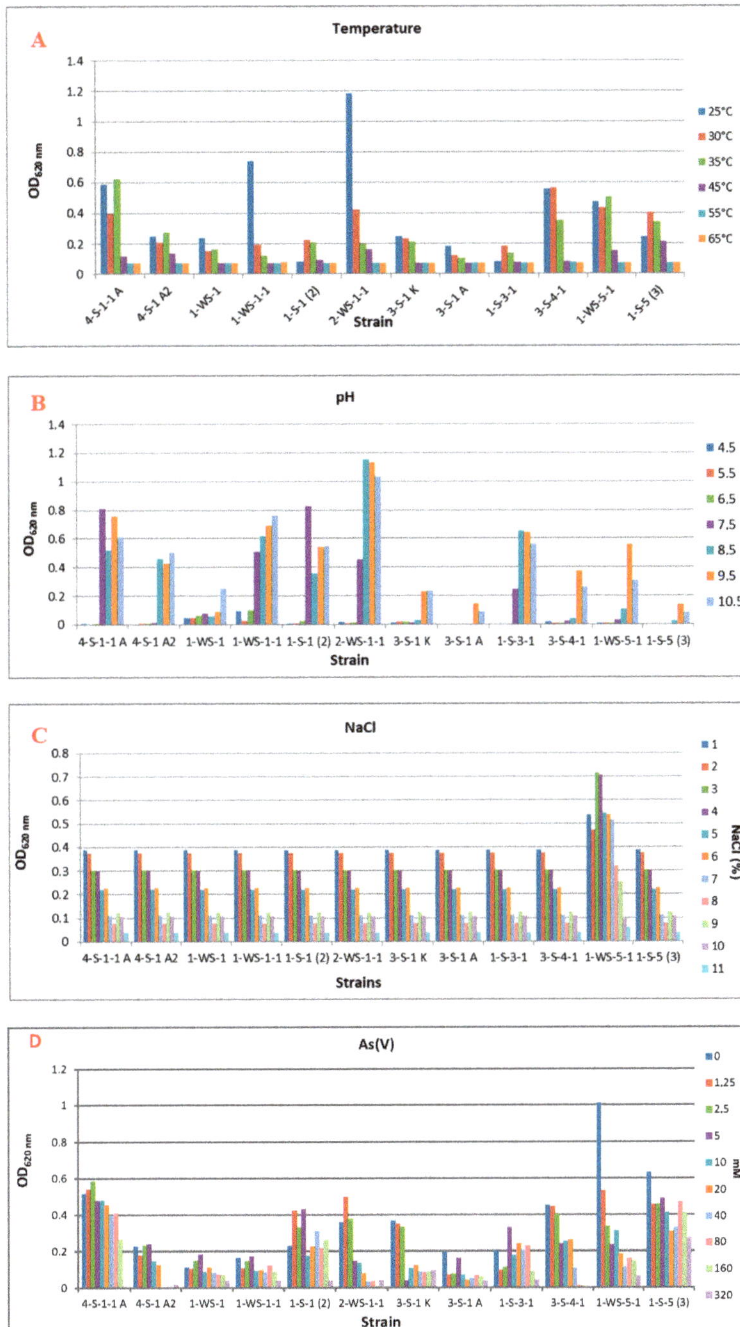

Figure 2. *Cont.*

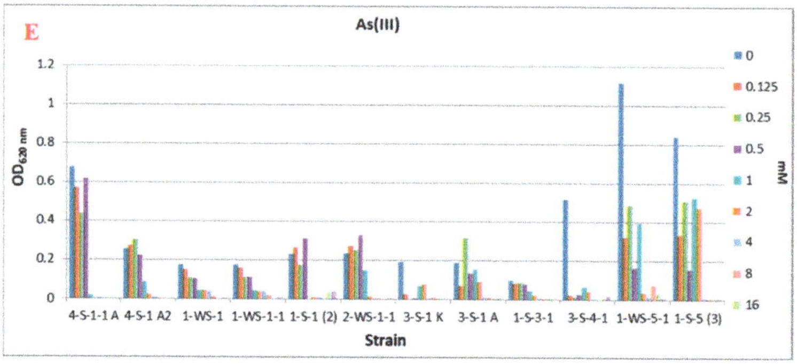

Figure 2. Effect of some physicochemical parameters and different concentrations of As(V) and As(III) on bacterial growth. (**A**) Temperature, (**B**) pH, (**C**) NaCl, (**D**) As(V), (**E**) As(III).

It was determined that there was no growth at pH 4.5, 5.5, and 6.5 as a common feature in all strains, and strain 1-S-1 (2) (*Anaerobacillus* sp.) showed intense growth, unlike the others. While it was determined that strains 3-S-1 A (*Marinobacter halophilus*), 3-S-4-1 (*Halomonas* sp.), 1-WS-5-1 (*Halomonas campisalis*), and 1-S-5 (3) (*Halomonas* sp.) at pH 9.5 showed higher microbial activity than other strains, at pH 10.5, 4-S-1 A2 (*Idiomarina* sp.), 1-WS-1 (*Branchybacterium paraconglomeratum*), and 1-WS-1-1 (*Microbacterium schleiferi*) strains showed better growth under these conditions (Figure 2B).

When the microbial activity is compared depending on the changes in the salt concentration in the medium, it has been determined that all strains can grow in environments containing 1–11% NaCl, with the concentration at which they grow best (Figure 2C).

3.8. MICs

Strains 4-S-1-1 A (*Alkalimonas delamerensis*) and 3-S-4-1 (*Halomonas* sp.) did not show any significant microbial activity in the medium containing 320 mM As(V). The strain 1-S-5 (3) (*Halomonas* sp.) showed good growth in the medium containing 160 mM and 320 mM As(V), unlike the other strains. In the medium that does not contain any As(V), it was determined that the strain that showed the best growth was strain 1-WS-5-1 (*Halomonas campisalis*). When we look at the As(V) concentrations that can be tolerated as a result of incubation under anaerobic conditions, it is seen that growth is generally high at values between 0–5 mM (Figure 2D).

In the environment that does not contain As(III), it was determined that strains 4-S-1-1 A (*Alkalimonas delamerensis*), 3-S-4-1 (*Halomonas* sp.), 1-WS-5-1 (*Halomonas campisalis*) and 1-S-5 (3) (*Halomonas* sp.) showed significantly better growth. It was observed that As(III) concentrations higher than 1 mM in the medium were generally not tolerated by the strains, and only strain 1-S-5 (3) (*Halomonas* sp.) could grow under these conditions (Figure 2E).

4. Discussion

The amount of water in deep terminal lakes such as Lake Van may change seasonally [22,23]. In this context, samplings were carried out to represent 4 seasons, with the thought that seasonal differences may change the amount of arsenic in the lake water and sediment and therefore change the dominant microbiota. While it is expected that the warming of the weather will decrease the amount of lake water and increase the amount of arsenic, it has been determined that there is approximately a similar amount of arsenic in the spring (2.730 µg/L) and summer (2.61 µg/L) seasons for sediment samples. However, in the transition to autumn (26.070 µg/L), it was determined that there was 10 times more arsenic than the amount of arsenic obtained from the spring and summer seasons. When the amounts of arsenic in the summer (2.61 µg/L) and spring (2.730 µg/L) seasons are compared with the winter data (9.370 µg/L), it has been determined that it contains three times more arsenic in winter. However, it has been revealed that there is a threefold decrease in the amount of arsenic in the transition from autumn (26.070 μg/L) to winter (9.370 μg/L). Although no such data has been encountered regarding Lake Van, as a result of the study conducted by Savarimuthu et al. in 2006, it was determined that the amount of arsenic in the water wells of West Bengal, India, changes seasonally. The lowest value is in summer (694 μg/L, 906 μg/L, and 794 μg/L for summer, monsoon, and winter, respectively) [38]. Because the sampling area is a region where anthropogenic inputs are quite high, it is thought that the increase is considered high due to the accumulation of arsenic in the sediment in the next season as a result of the intense activity in the summer months, because the sampling point is a region with a high potential for anthropogenic pollution due to various water sports activities and the presence of various businesses in its vicinity. As an indicator of this, it has been shown some regions are unsuitable for microbiological criteria in the Van-Edremit region [21]. The fact that the amount of arsenic contained in the lake water samples belonging to different seasons was approximately the same (0.214 μg/L, 0.263 μg/L, 0.14 μg/L, 0.261 μg/L for winter, spring, summer, and autumn, respectively), was considered as a data supporting this situation.

Isolation and identification studies were carried out in various lakes with characteristics similar to Lake Van. Alkaline and salty lakes, such as Lake Van, are potential isolation sources for arsenic-resistant bacteria, such as Mono Lake, Searles Lake, and Dali Lake. The obtained data is important because no studies have been carried out at Lake Van in this context, and it is the largest soda lake in the World [20]. Arsenic-resistant bacteria have been isolated from lakes such as Searles Lake and Mono Lake in the United States, which have similar characteristics to Lake Van, and have been introduced to the literature as a new species. It is seen that similar studies were carried out at Dali Lake in China [14,39,40]. When culture-dependent and metagenomic studies conducted with other lakes that have geochemically similar characteristics to Lake Van are examined, it is seen that Gammaproteobacteria members are dominant in such soda lakes [41]. As a matter of fact, as a result of the study, Gammaproteobacteria members such as *Halomonas*, *Alkalimonas*, *Idiomarina*, *Nitrincola*, and *Marinobacter* were found to be lake biota members (Table 3). Apart from Gammaproteobacteria members, it has been shown in studies based on various 16S gene sequences that Firmicutes, Actinobacteria, and Bacteriodetes members dominate in alkaliphilic lakes in general [41]. Since our study aimed to detect arsenic metabolizing, alkaliphilic, and culturable bacteria, the identified species were limited. However, it has been determined that the bacteria living in Lake Van with this feature are members of Gammaproteobacteria (*Halomonas*, *Alkalimonas*, *Idiomarina*, *Nitrincola*, and *Marinobacter*), Actinobacteria (*Branchybacterium paraconglomeratum* and *Microbacterium schleiferi*) and Firmicutes (*Anaerobacillus* sp.) in accordance with the literature (Table 3). It is important and necessary to detect microorganisms living in habitats, such as Lake Van, where extreme conditions dominate, using cultural and culture-independent techniques. Because these microorganisms are the source of many biotechnological products, they are also important in terms of understanding the various cycles prevailing in the lake ecosystem due to their important role in the food chains. As a result of the dominant species in the cycles, such as carbon, nitrogen, and phosphorus and their metabolic activities, the products released are very important in terms of the quality of the water, the determination of the species that can live in these conditions, and especially the preservation of the existence of various endemic creatures.

The most biotechnological use of arsenic-resistant bacteria is the removal of arsenic pollution, a major environmental and public health problem. As(V) reduction and As(III) oxidation processes can potentially promote arsenic removal from contaminated areas. Because both processes directly affect the mobility and bioavailability of As. As a result of this, microbial activities play a key role in biogeochemical arsenic cycling [42]. Therefore, arsenic metabolism pathways should be determined to use suitable strains in the assays. In this context, the $AgNO_3$ screening technique is highly preferred due to its low cost [28]. However, in this study, although different environments and conditions were used in addition to the experimental environments specified in the literature, no results could be obtained for alkaliphilic isolates. These assays have determined that the silver nitrate test has the disadvantages of being very toxic and, at the same time, is unable to give results in every environment and organism for the determination of arsenate reductase and arsenite oxidase enzyme activities.

Another way to identify pathways is to screen for resistance genes. In the study's results, it was observed that no strains possessed the arsB gene region, an integral membrane protein that can extrude arsenite from the cell cytoplasm, thus reducing arsenite accumulation [43]. A novel gene cluster encoding respiratory As(V) reductase has been identified for the first time in *Shewanella* ANA-3. There are two genes in this gene cluster, *arr*A and *arr*B [44]. Within the scope of our study, the *arr*B gene region was selected as the target region, but it was determined that arsenic-resistant strains did not have this gene region. However, a 500 bp product was obtained, which indicates the positive result in the *Shewanella* ANA-3 strain used as a positive control. Various microorganisms capable of metabolizing various toxic forms of arsenic exist in nature. As a result of the evolution of each species, in accordance with the habitat conditions it is adapted to, they have the appropriate gene regions and, therefore, their products. In this context, within the scope of our study, strains belonging to different genera and species, which also have different Gram reactions and show variations in oxygen requirements, were obtained from various water and sediment samples of Lake Van. However, it was determined that only the *ars*C gene region was found in all strains from the gene regions examined in general. While it is expected that the strains identified as *Anaerobacillus* sp. from the obtained strains are anaerobic in terms of oxygen requirement, it is expected to contain the *arr*B gene region, which is one of the respiratory arsenate reductase genes, but the opposite result was obtained. Further studies should be conducted to determine whether it is a new species and whether it has a new resistance gene that functions under anaerobic conditions.

Electron donors in the environment are also of great importance in the metabolism of various forms of arsenic [45]. As a result of the experiment, similar to other studies [46], it was determined that the obtained arsenic-resistant strains preferred Na-lactate as an electron donor in environments containing As(V).

Considering the As(V) and As(III) tolerances of the strains, although As(III) is a more toxic form, it has been determined that it can reproduce in environments containing As(III). In particular, in the *Halomonas* sp. 1-S-5 (3) strain obtained from the autumn season sediment sample, where the total arsenic content is the highest, growth was intense in the medium containing As(III), and the MIC value increased to 2 mM. In the presence of As(V) in the medium, it is seen that the MIC value for this strain reached >320 mM. Although this strain was isolated in a medium containing As(V), it was determined that it was resistant to both As(V) and As(III). A similar situation was observed in *Anaerobacillus* sp. 1-S-1 (2) and 1-S-3-1 strains. When the studies are examined, *Idiomarina* sp. [47], *Microbacterium* schleiferi [48], *Anaerobacillus* sp. [49], *Nitrincola* sp. [50], *Bacillus* sp., and *Halomonas* sp. [51] data on arsenic resistance and resistance genes are seen. However, no data were found for *Alkalimonas delamerensis* [52], *Barnchybacterium paraconglomeratum* [53], and *Marinobacter halophilus* [54] strains and presented as preliminary data.

5. Conclusions

As an epilog, it should be emphasized that Lake Van is the largest soda lake in the world, and there are no studies carried out in Turkey or the world regarding anaerobic arsenic-metabolizing bacteria. With the increasing global warming and anthropogenic pollution, arsenic-contaminated areas are increasing. As a result, the increase in arsenic exposure brings about serious health problems and also causes an increase in resistance to various antimicrobial agents. These results make the arsenic problem one of the most important global problems since millions of people worldwide are constantly and regularly exposed to arsenic. When evaluated in terms of Lake Van, the increase in arsenic accumulation in the lake water and, therefore, in pearl mullet, an endemic fish species consumed as food by the local people, causes public health and the continuation of this species to be endangered. Due to the use of lake water for irrigation purposes in agricultural areas, the size of the threat is increasing. Therefore, isolation and identification of resistant strains from environments with high arsenic contamination, as well as identifying the pathways they use to metabolize arsenic, have become essential. This is because the strains obtained can be used for the microbial remediation of arsenic, and at the same time, information about the microbiota in these extreme environments can be obtained, and possible new genus and species can be brought to the literature. At the same time, a world arsenic map and statistics were obtained by determining the amount of arsenic in isolation areas. The data herein will contribute to the sustainability of Lake Van and the understanding of bacterial arsenic metabolism with the contribution of different strains. The study will be evaluated as a model organism source for arsenic bioremediation studies and for being the first data on Lake Van, the world's largest soda lake.

Author Contributions: E.E.O. collected samples, performed laboratory work, and wrote the manuscript. M.S. analyzed data and prepared some figures. E.O. reviewed the manuscript. E.E.O. critically reviewed the manuscript. All authors have read and agreed to the published version of the manuscript.

Funding: This research was supported with funds provided by TUBITAK (Scientific and Technological Research Council of Turkey) with project number <114Y036> and Ege University Scientific Research Projects Coordination Department (BAP) with project number <15-BİL-005>. Financial support by TUBITAK and BAP is gratefully acknowledged.

Institutional Review Board Statement: Not applicable.

Informed Consent Statement: Not applicable.

Data Availability Statement: The data sets and materials supporting the results of this article are included within the article and also in Appendix A.

Conflicts of Interest: The authors declare no conflict of interest.

Appendix A

Figure A1. *Cont.*

Figure A1. Pressure and salt resistant ladle (**A–C**) and sampling (**D–K**).

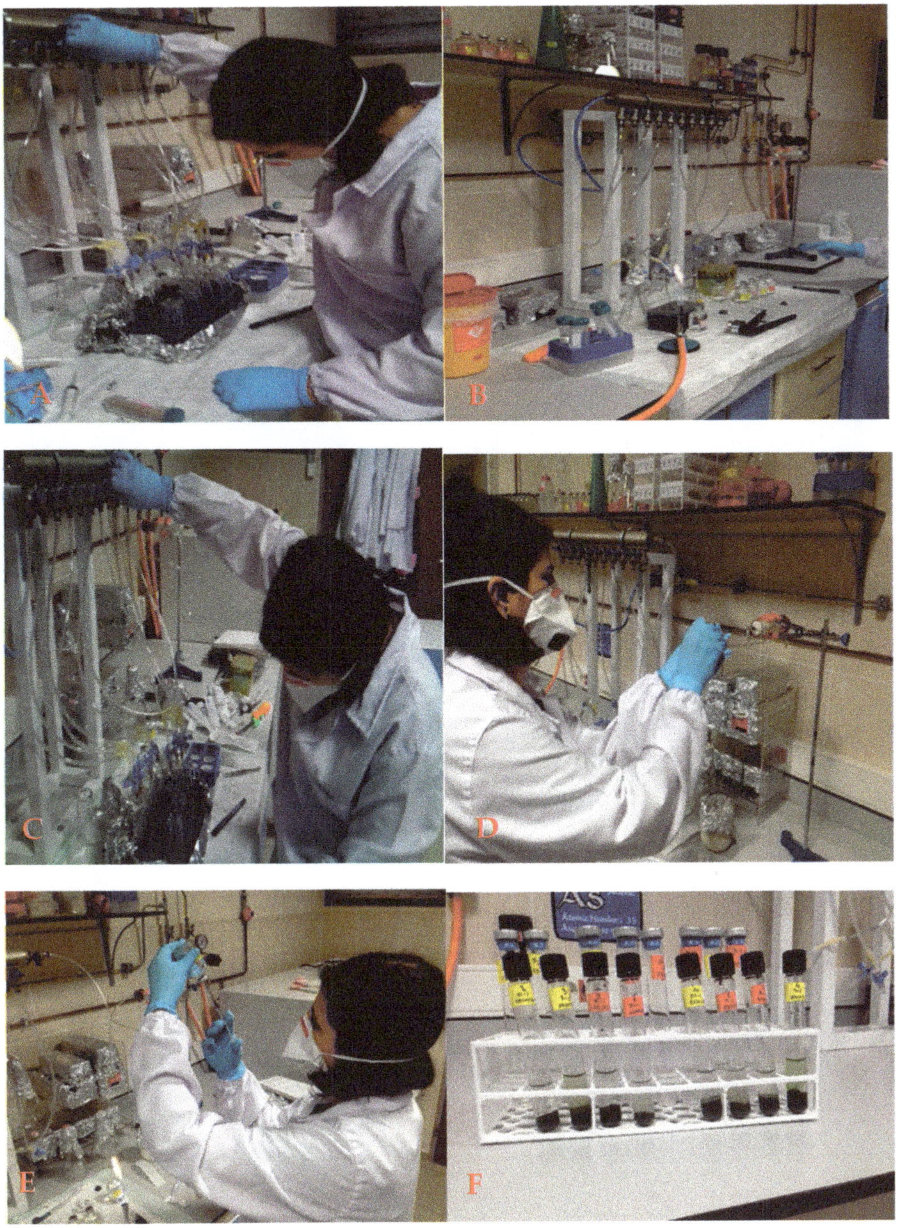

Figure A2. *Cont.*

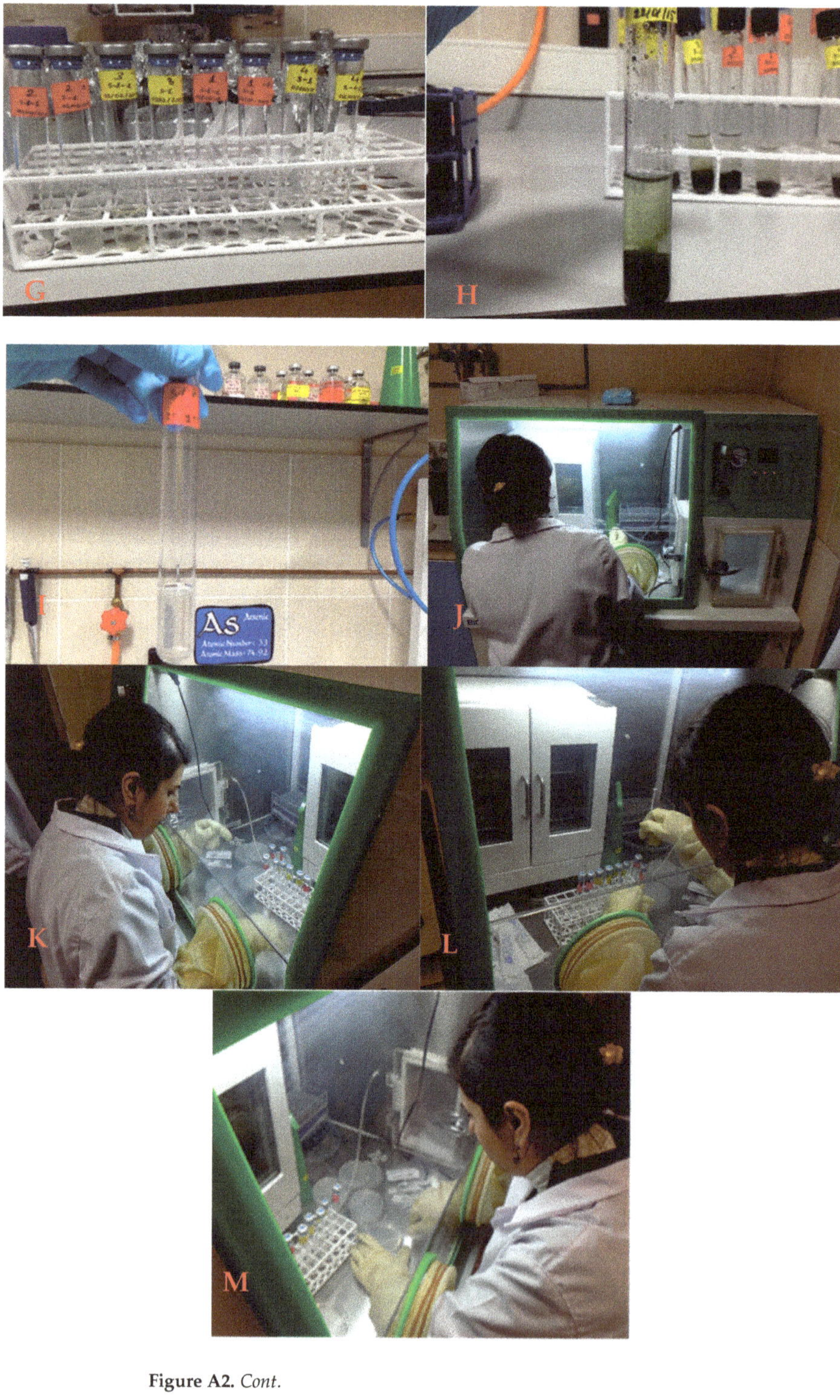

Figure A2. *Cont.*

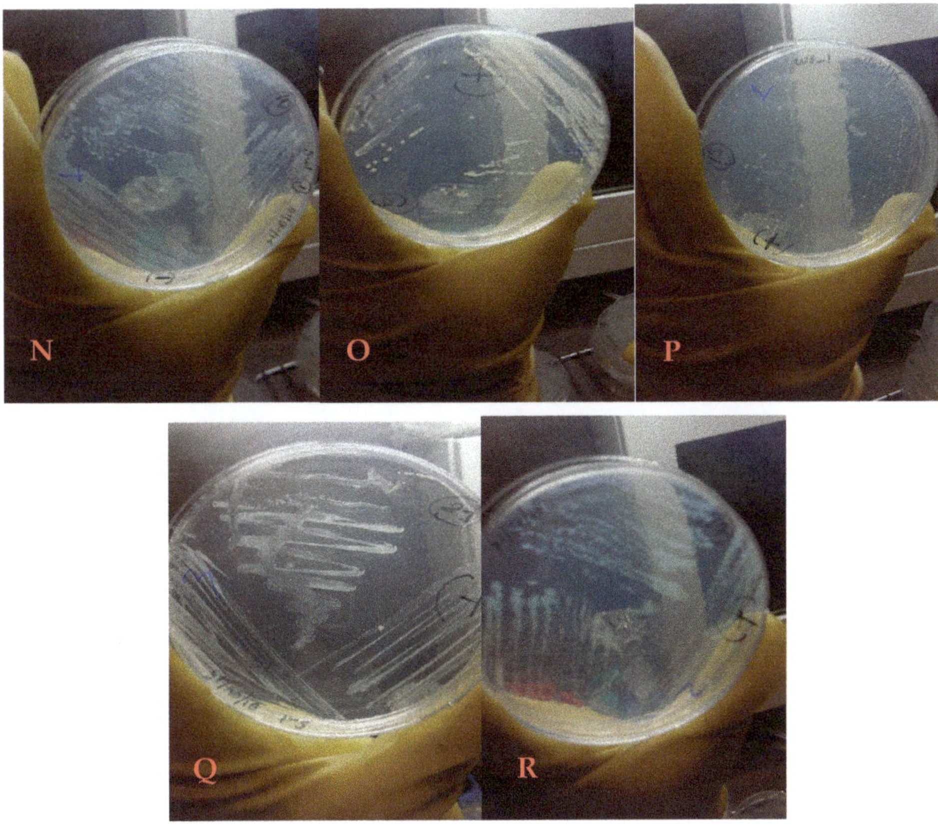

Figure A2. Isolation of anaerobic and alkalifilic arsenate-reducing and arsenite-oxidizing bacteria. Preparation of media with Hungate technique (**A–E**), Bacterial growth in Balch and Hungate tubes (**F–I**), Obtaining pure culture using anaerobic chamber (**J–M**), Different types of colonies formed on agar media (**N–R**).

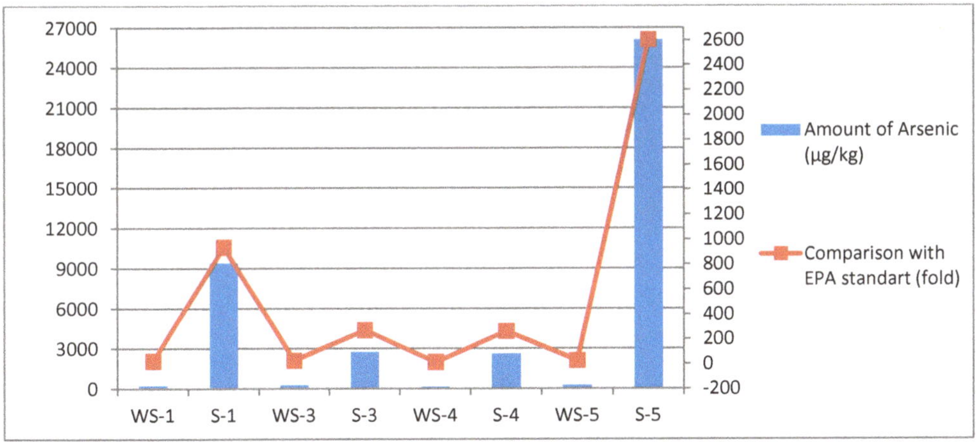

Figure A3. Comparison of arsenic amounts of Lake Van samples.

References

1. Zavarzina, D.G.; Tourova, T.P.; Kolganova, T.V.; Boulygina, E.S.; Zhilina, T.N. Description of *Anaerobacillus alkalilacustre* gen. nov., sp. nov.—strictly anaerobic diazotrophic *Bacillus* isolated from soda lake and transfer of *Bacillus arseniciselenatis*, *Bacillus macyae*, and *Bacillus alkalidiazotrophicus* to *Anaerobacillus* as the new Combinations *A. arseniciselenatis* comb. nov., *A. macyae* com. nov., and *A. alkalidiazotrophicus* com. nov. *Microbiology* **2009**, *78*, 723–731.
2. Valdés, N.; Rivera-Araya, J.; Bijman, J.; Escudero, L.; Demergasso, C.; Fernández, S.; Ferrer, A.; Chávez, R.; Levicán, G. Draft genome sequence of *Nitrincola* sp. strain A-D6, an arsenic resistant Gammaproteobacterium isolated from a salt flat. *Genome Announc.* **2014**, *2*, 1–2. [CrossRef] [PubMed]
3. Dey, U.; Chatterjee, S.; Mondal, N.K. Isolation and characterization of arsenic-resistant bacteria and possible application in bioremediation. *Biotechnol. Rep.* **2016**. [CrossRef] [PubMed]
4. Stolz, J.; Basu, P.; Oremland, R. Microbial transformation of elements: The case of arsenic and selenium. *Int. Microbiol.* **2002**, *5*, 201–207.
5. Krumova, K.; Nikolovski, M.; Groudeva, V. Isolation and identification of arsenic-transforming bacteria from arsenic contaminated sites in Bulgaria. *Biotechnol. Biotechnol. EQ* **2008**, *22*, 721–728. [CrossRef]
6. Kale, S.P.; Salaskar, D.; Ghosh, S.; Sounderajan, S. Isolation and identification of arsenic resistant *Providencia rettgeri* (KDM3) from industrial effluent contaminated soil and studies on its arsenic resistance mechanisma. *J. Microb. Biochem. Technol.* **2015**, *7*, 194–201.
7. Selim Reza, A.H.M.; Jean, J.S.; Yang, H.J.; Lee, M.K.; Woodall, B.; Liu, C.C.; Lee, J.F.; Luo, S.D. Occurence of arsenic in core sediments and groundwater in The Chapai-Nawabganj District, Northwestern Bangladesh. *Water. Res.* **2010**, *44*, 2021–2037. [CrossRef]
8. Baba, A.; Sozbilir, H. Source of arsenic based on geological and hydrogeochemical properties of geothermal system in Western Turkey. *Chem. Geol.* **2012**, *334*, 364–377. [CrossRef]
9. Salam, M.A.; Hossain, M.S.; Ali, M.E.; Asad, M.A.; Ali, M.H. Isolation and characterization of arsenic resistant bacteria from different environment in South-West Region of Bangledesh. *Res. J. Environ. Sci.* **2009**, *3*, 110–115. [CrossRef]
10. Saltikov, C.W.; Olson, B.H. Homology of *Escherichia coli* R773 *ars*A, *ars*B, and *ars*C genes in arsenic-resistant bacteria isolated from raw sewage and arsenic-enriched creek waters. *Appl. Environ. Microbiol.* **2002**, *68*, 280–288. [CrossRef]
11. Handley, K.M.; Héry, M.; Lloyd, J.R. *Marinobacter santoriniensis* sp. nov., an arsenate-respiring and arsenite-oxidazing bacterium isolated from hydrothermal sediment. *Int. J. Syst. Evol. Microbiol.* **2009**, *59*, 886–892. [CrossRef] [PubMed]
12. Zargar, K.; Hoeft, S.; Oremland, R.; Saltikov, C.W. Identification of a novel oxidase gene, *arx*A, in the haloalkaliphilic, arsenite-oxidizing bacterium *Alkalilimnicola echrlichii* Strain MLHE-1. *J. Bacteriol.* **2010**, *192*, 3755–3762. [CrossRef] [PubMed]
13. Shakya, S.; Pradhan, B.; Smith, L.; Shrestha, J.; Tuladhar, S. Isolation and characterization of aerobic culturable arsenic-resistant bacteria from surfacewater and groundwater of Rautahat District, Nepal. *J. Environ. Manag.* **2012**, *95*, 250–255. [CrossRef] [PubMed]
14. Blum, J.S.; Kulp, T.R.; Han, S.; Lanoil, B.; Saltikov, C.W.; Stolz, J.F.; Miller, L.G.; Oremland, R.S. *Desulfohalophilus alkaliarsenatis* gen. nov., sp. nov., an extremely halophilic sulfate- and arsenate-respiring bacterium from Searles Lake, California. *Extremophiles* **2012**, *16*, 727–742. [CrossRef]
15. Bandyopadhyay, S.; Schumann, P.; Das, S.K. *Pannonibacter indica* sp. nov., a highly arsenate-tolerant bacterium isolated from a hot spring in India. *Arch. Microbiol.* **2013**, *195*, 1–8. [CrossRef]
16. *The Geology of Lake Van*; PLATES, Publ. No. 169; Degens, E.T.; Kurtman, F. (Eds.) MTA: Ankara, Turkey, 1978; 158p.
17. Reimer, A.; Landmann, G.; Kempe, S. Lake Van, eastern Anatolia, hydrochemistry and history. *Aquat. Geochem.* **2009**, *15*, 195–222. [CrossRef]
18. Çağatay, M.N.; Öğretmen, N.; Damcı, E.; Stockhecke, M.; Sancar, Ü.; Eriş, K.K.; Özeren, S. Lake level and climate records of the last 90 ka from the Northern Basin of Lake Van, eastern Turkey. *Quat. Sci. Rev.* **2014**, *104*, 97–116. [CrossRef]
19. Glombitza, C.; Stockhecke, M.; Schubert, C.J.; Vetter, A.; Kallmeyer, J. Sulfate reduction controlled by organic matter availability in deep sediment cores from the saline, alkaline Lake Van (Eastern Anatolia, Turkey). *Front. Microbiol.* **2013**, *4*, 209. [CrossRef]
20. Stockhecke, M.; Anselmetti, F.S.; Meydan, A.F.; Odermatt, D.; Sturm, M. The annual particle cycle in Lake Van (Turkey). *Palaeogeogr. Palaeoclimatol. Palaeoecol.* **2012**, *333*, 148–159. [CrossRef]
21. Aydın, E.; Parlak, M.; Guducuoglu, H.; Bayram, Y. Determination of microbiological pollution level of Lake Van and Lake Erçek situated within the borders of Van province. *Türk Mikrobiyol. Cem. Derg.* **2001**, *51*, 132–142. [CrossRef]
22. Landmann, G.; Reimer, A. Climatically induced lake level changes at Lake Van, Turkey, during the pleistocene/holocene transition. *Glob. Biogeochem. Cycles* **1996**, *10*, 797–808. [CrossRef]
23. Altunkaynak, A. Forecasting surface water level fluctuation of Lake Van by artificial neural networks. *Water Resour. Manag.* **2007**, *21*, 399–408. [CrossRef]
24. Zhao, F.J.; McGrath, S.P.; Meharg, A.A. Arsenic as a food chain contaminant: Mechanisms of plant uptake and metabolism and mitigation strategies. *Annu. Rev. Plant. Biol.* **2010**, *61*, 535–559. [CrossRef] [PubMed]
25. Yang, L.; Donahoe, R.J.; Redwine, J.C. In situ chemical fixation of arsenic-contaminated soils: An experimental study. *Sci. Total Environ.* **2007**, *387*, 28–41. [CrossRef] [PubMed]

26. Widdel, F.; Pfenning, N. Studies on dissimilatory sulfate-reducing bacteria that decompose fatty acids ı. ısolation of new sulfate-reducing bacteria enriched with acetate from saline environments. Description of *Desulfobacter postgatei* gen. nov., sp., nov. *Arch. Microbiol.* **1981**, *129*, 395–400. [CrossRef]
27. Saltikov, C.W.; Wildman, R.A.; Newman, D.K. Expression dynamics of arsenic respiration and detoxification in *Shewanella* sp. atrain ANA-3. *J. Bacteriol.* **2005**, *187*, 7390–7396. [CrossRef]
28. Sarkar, A.; Kazy, S.F.; Sar, P. Characterization of arsenic resistant bacteria from arsenic rich groundwater of West Bengal, India. *Ecotoxicol.* **2013**, *22*, 363–376. [CrossRef]
29. Conrad, A. *The Arx Anaerobic Arsenite-Oxidization Pathway Is Conserved in Halomonas and Ectothiorhodospira Strains Isolated from Big Soda Lake*; University of California: Santa Cruz, NV, USA, 2014.
30. Benson, H.J. *Microbiological Applications, Laboratory Manual in General Microbiology*, 8th ed.; W. C. Brown House: Dubuque, IA, USA, 2002; p. 440.
31. Ersoy Omeroglu, E. Isolation, Fenotypic and Molecular Characterization of Bioluminescent Bacteria from Izmir Bay. Doctoral Thesis, Ege University, Bornova, Turkey, 2011.
32. Bilgehan, H. *Klinik Mikrobiyolojik Tanı*, 4th ed.; Baris Yayinlari-Fakulteler Kitabevi: İzmir, Turkey, 2004.
33. Yumoto, I.; Hirota, K.; Nodasaka, Y.; Yokota, Y.; Hoshino, T.; Nakajima, K. *Alkalibacterium psychrotolerans* sp. nov., a psychrotolerant obligate alkaliphile that reduces an indigo dye. *Int. J. Syst. Evol. Microbiol.* **2004**, *54*, 2379–2383. [CrossRef]
34. Chang, J.S.; Yoon, I.H.; Kim, K.W. Isolation and *ars* detoxification of arsenite-oxidizing bacteria from abandoned arsenic-contaminated mines. *J. Microbiol. Biotech.* **2007**, *17*, 812–821.
35. Ersoy Omeroglu, E.; Karaboz, I.; Sukatar, A.; Yasa, I.; Kocyigit, A. Determination of heavy metal susceptibilities of *Vibrio harveyi* strains by using 2, 3, 5-triphenyltetrazolium chloride (TTC). *Rapp. Comm. Int. Mer. Médit.* **2007**, *38*, 364.
36. Borsodi, A.K.; Aszalós, J.M.; Bihari, P.; Nagy, I.; Schumann, P.; Spröer, C.; Kovács, A.L.; Bóka, K.; Dobosy, P.; Óvári, M.; et al. *Anaerobacillus alkaliphilus* sp. nov., a novel alkaliphilic and moderately halophilic bacterium. *Int. J. Syst. Evol. Microbiol.* **2019**, *69*, 631–637. [CrossRef] [PubMed]
37. Ben Fekih, I.; Zhang, C.; Li, Y.P.; Zhao, Y.; Alwathnani, H.A.; Saquib, Q.; Rensing, C.; Cervantes, C. Distribution of arsenic resistance genes in prokaryotes. *Front. Microbiol.* **2018**, *9*, 2473. [CrossRef] [PubMed]
38. Savarimuthu, X.; Hira-Smith, M.M.; Yuan, Y.; von Ehrenstein, O.S.; Das, S.; Ghosh, N.; Mazumder, D.N.G.; Smith, A.H. Seasonal variation of arsenic concentration in tubewells in West Bengal, India. *J. Health Popul. Nutr.* **2006**, *24*, 277–281.
39. Boldareva, E.N.; Briantseva, I.A.; Tsapin, A.; Nelson, K.; DIu, S.; Turova, T.P.; Boĭchenko, V.A.; Boĭchenko, V.A.; Stadnichuk, I.N.; Gorlenko, V.M. The new bacteriochlorophyll a-containing bacterium *Roseinatronobacter monicus* sp. nov. from the hypersaline soda Mono Lake (California, United States). *Mikrobiologiia* **2007**, *76*, 95–106. [CrossRef] [PubMed]
40. Zhai, L.; Liao, T.; Xue, Y.; Ma, Y. *Bacillus daliensis* sp. nov., An alkaliphilic, Gram-positive bacterium isolated from a soda lake. *Int. J. Syst. Evol. Microbiol.* **2012**, *62*, 949–953. [CrossRef]
41. Ersoy Omeroglu, E.; Sudagidan, M.; Yurt, M.N.Z.; Tasbasi, B.B.; Acar, E.E.; Ozalp, V.C. Microbial community of soda Lake Van as obtained from direct and enriched water, sediment and fish samples. *Sci. Rep.* **2021**, *11*, 1–13. [CrossRef]
42. Yamamura, S.; Amachi, S. Microbiology of inorganic arsenic: From metabolism to bioremediation. *J. Biosci. Bioeng.* **2014**, *118*, 1–9. [CrossRef]
43. Yang, H.C.; Fu, H.L.; Lin, Y.F.; Rosen, B.P. Pathways of arsenic uptake and efflux. *Curr. Top. Membr.* **2012**, *69*, 325–358. [CrossRef]
44. Saltikov, C.W.; Newman, D.K. Genetic identification of a respiratory arsenate reductase. *PNAS Microbiol.* **2003**, *100*, 10983–10988. [CrossRef]
45. Kudo, K.; Yamaguchi, N.; Makino, T.; Ohtsuka, T.; Kimura, K.; Dong, D.T.; Amachi, S. Release of arsenic from soil by a novel dissimilatory arsenate-reducing bacterium, *Anaeromyxobacter* sp. Strain PSR-1. *App. Environ. Microbiol.* **2013**, *79*, 4635–4642. [CrossRef]
46. Antón, A.I. *Isolation of Bacteria That As(V) to As(III)*; Marine Biological Laboratuary: Woods Hole, MA, USA, 1998.
47. Sinha, R.K.; Krishnan, K.P.; Kurian, P.J. Draft genome sequence of *Idiomarina* sp. strain 5.13, a highly stress-resistant bacterium isolated from the Southwest Indian Ridge. *Genome Announc.* **2017**, *5*, 1–2. [CrossRef] [PubMed]
48. Macur, R.; Jackson, C.R.; Botero, L.M.; Mcdermott, T.R.; Inskeep, W.P. Bacterial populations associated with the oxidation and reduction of arsenic in an unsaturated soil. *Environ. Sci. Technol.* **2004**, *38*, 104–111. [CrossRef] [PubMed]
49. Wang, J.P.; Liu, B.; Liu, G.H.; Ge, C.B.; Chen, Q.Q.; Zhu, Y.J.; Chen, Z. Genome sequence of *Anaerobacillus macyae* JMM-4T (DSM 16346), the first genomic ınformation of the newly established genus *Anaerobacillus*. *Genome Announc.* **2015**, *3*, 1–2. [CrossRef] [PubMed]
50. Hamamura, N.; Itai, T.; Liu, Y.; Reysenbach, A.L.; Damdinsuren, N.; Inskeep, W.P. Identification of anaerobic arsenite-oxidizing and arsenate-reducing bacteria associated with an alkaline saline lake in Khovsgol, Mongolia. *Environ. Microbiol. Rep.* **2014**, *6*, 476–482. [CrossRef] [PubMed]
51. Jain, R.; Jha, S.; Mahatma, M.K.; Jha, A.; Kumar, G.N. Characterization of arsenite tolerant *Halomonas* sp. Alang-4, originated from heavy metal polluted shore of Gulf of Cambay. *J. Environ. Sci. Health A* **2016**, *51*, 478–486. [CrossRef]
52. Ma, Y.; Xue, Y.; Grant, W.D.; Collins, N.C.; Duckworth, A.W.; Van Steenbergen, R.P.; Jones, B.E. *Alkalimomas amylolytic*a gen. nov., and sp. nov., *Alkalimonas delamerensis* gen. nov., sp. nov., novel alkaliphilic bacteria from soda lakes in China and East Africa. *Extremophiles* **2004**, *8*, 193–200. [CrossRef]

53. Pham, T.H.; Lai, T.H.; Dang, P.N.; Inoue, D.; Sei, K.; Ike, M. Identification of some predominant bacteria isolated from JetA1 Fuel in Vietnam by sequence analysis of 16S rRNA gene. *OUKA* **2008**, 301–308. Available online: https://hdl.handle.net/11094/13035 (accessed on 24 October 2022).
54. Zhong, Z.P.; Liu, Y.; Liu, H.C.; Wang, F.; Zhou, Y.G.; Liu, Z.P. *Marinobacter halophilus* sp. nov., a halophilic bacterium isolated from a salt lake. *Int. J. Syst. Evol. Microbiol.* **2015**, *65*, 2838–2845. [CrossRef]

life

Article

Study on Cytotoxic and Genotoxic Potential of Bulgarian *Rosa damascena* Mill. and *Rosa alba* L. Hydrosols—*In Vivo* and *In Vitro*

Tsvetelina Gerasimova [1], Gabriele Jovtchev [1], Svetla Gateva [1], Margarita Topashka-Ancheva [1], Alexander Stankov [1], Tsveta Angelova [1], Ana Dobreva [2] and Milka Mileva [3,*]

[1] Institute of Biodiversity and Ecosystem Research, Bulgarian Academy of Sciences, 2 Gagarin Str., 1113 Sofia, Bulgaria
[2] Institute for Roses and Aromatic Plants, Agricultural Academy, 49 Osvobojdenie Blvd., 6100 Kazanlak, Bulgaria
[3] The Stephan Angeloff Institute of Microbiology, Bulgarian Academy of Sciences, 26 Acad. G. Bonchev Str., 1113 Sofia, Bulgaria
* Correspondence: milkamileva@gmail.com or mileva@microbio.bas.bg; Tel.: +359-29793185

Abstract: The *Rosa alba* L. and *Rosa damascena* Mill. growing in Bulgaria are known for their extremely fine essential oil and valuable hydrosols. Irrespectively of its wide use in human life, little research exists on the cytotoxic and genotoxic activity of the hydrosols. This set our goal to conduct cytogenetic analyses to study these effects. A complex of classical cytogenetic methods was applied in three types of experimental test systems—higher plant *in vivo*, ICR mice *in vivo*, and human lymphocytes *in vitro*. Mitotic index, PCE/(PCE + NCE) ratio, and nuclear division index were used as endpoints for cytotoxicity and for genotoxicity—induction of chromosome aberrations and micronuclei. Rose hydrosol treatments range in concentrations from 6% to 20%. It was obtained that both hydrosols did not show considerable cytotoxic and genotoxic effects. These effects depend on the type of the tested rose hydrosols, the concentrations applied in the experiments, and the sensitivity and specificity of the test systems used. Human lymphocytes *in vitro* were the most sensitive to hydrosols, followed by higher plant and animal cells. Chromosomal aberrations and micronucleus assays suggested that *R. damascena* and *R. alba* hydrosols at applied concentrations possess low genotoxic risk. Due to the overall low values in terms of cytotoxic and/or genotoxic effects in all test systems, hydrosols are promising for further use in various areas of human life.

Keywords: *Rosa alba* L. and *Rosa damascena* Mill. hydrosols; cytotoxic and genotoxic potential; test systems *in vivo* and *in vitro*; chromosome aberrations; micronuclei; mitotic index and nuclei division index

check for updates

Citation: Gerasimova, T.; Jovtchev, G.; Gateva, S.; Topashka-Ancheva, M.; Stankov, A.; Angelova, T.; Dobreva, A.; Mileva, M. Study on Cytotoxic and Genotoxic Potential of Bulgarian *Rosa damascena* Mill. and *Rosa alba* L. Hydrosols—*In Vivo* and *In Vitro*. *Life* **2022**, *12*, 1452. https://doi.org/10.3390/life12091452

Academic Editor: Othmane Merah

Received: 23 August 2022
Accepted: 15 September 2022
Published: 19 September 2022

1. Introduction

Rose hydrosol is an abundantly represented product obtained by water–steam distillation of essential rose oil, the most famous of which are *Rosa damascena* Mill. and *Rosa alba* L. These hydrosols find wide spread applications in cosmetics, perfumery, pharmacy, phytotherapy, the food industry, and many other areas of human life [1]. A couple of papers, including review papers, have reported about the physicochemical, biochemical, and pharmacological characteristics of rose hydrosols. The main information is based on *R. damascena* hydrosol data [2–10]. The hydrosol antiseptic and antispasmodic actions have been established [6]. Hydrosol prepared from the Bulgarian *Rosa damascena* shows the ability to inhibit skin inflammation caused by *Candida albicans* and methicillin-resistant *Staphylococcus aureus* [11]. It was reported that the effect depends on the extraction method and concentration applied [12]. Rose hydrosol is effective in the prevention of fresh-cut taro browning [10]. On the contrary, other authors have not obtained an antibacterial effect of *R. damascena* hydrosol on skin flora after hand rubbing was detected [13]. Less is known about the volatile constituents of white rose (*Rosa alba* L.) water and its biological activity [8,14–16].

The yield of both essential oil and hydrosol is influenced by many factors. Genotype, harvest time, geographical region [17], method of extraction, and storage are decisive [3,18,19]. During the harvesting process, hand picking must be followed [20]. The raw material for the distillation such as leaves, petals, flowers, or fruits plays a crucial role [14,15,21]. Distillation methods are also important—factory-type distillation or village-type distillation [22].

Rose hydrosols contain valuable chemical compounds that are circa 10–50% of the rose oil's constituents, including flavonoids, anthocyanins, terpenes, and glycosides. The determination and analysis of the essential oil components and their hydrosols can be performed using various methods, such as GC-MS [2,5,8,23,24] and HPLC [25,26]. For results interpretation reference samples with various origins—synthetic, conventional, and traditional samples, were used [27]. A percentage comparison between the essential oil and hydrosol of *Rosa damascena* dried petals and fresh flowers for cosmetics uses were published in a Tentative Report for Public Comment [28].

The chemical composition of essential oils and their byproducts, as well as the balance of ingredients, play a crucial role when interpreting their effects [29,30]. Information about the biological effects and genotoxic activity exists only for some chemical compounds. Geraniol, as an important component of rose essential oil, shows the potential to decrease cytotoxic and genotoxic action of MNNG in root tip meristem cells of *Hordeum vulgare* and human lymphocytes *in vitro* [31]. Geraniol exhibited a cytotoxic effect on various types of cancer cells and suppressed tumor proliferation [32]. Another component eugenol and its analogues were also tested for antitumor activity using different human cancer cell lines. Inhibitory activity of cell growth has been established [33]. It was obtained that eugenol significantly reduced the MNNG-induced gastric tumors by suppressing NF-κB activation and modulating the expression of NF-κB target genes. Thus, eugenol offers an immense potential in cancer chemoprevention and therapy [34].

There is very little evidence to date with regard to the cytotoxic and genotoxic potential of *R. alba* and *R. damascena* essential oils and the byproducts of rose oil distillation. In recent years, our research group has been working to fill this gap. In this sense, we already know much more about the cytotoxic/genotoxic activity and protective antimutagenic effect of some rose products (essential oils and wastewaters) of Bulgarian *Rosa damascena* Mill. and *Rosa alba* L. on certain groups of organisms and/or cells [30,35,36]. It is important to study whether such aromatic plant products, widely used in human practice, are safe. Since no information is available concerning the cytotoxicity and genotoxicity of rose hydrosols, we set the goal to study the cytotoxic and genotoxic/clastogenic effects of *R. alba* L. and *R. damascena* Mill. hydrosols derived during the water–steam distillation of rose oils by classical cytogenetic methods in a complex of *in vivo* and *in vitro* test systems.

2. Materials and Methods

2.1. Preparation of R. alba L. and R. damascena Mill. Hydrosols

Hydrosols were obtained from roses grown in the experimental field belonging to the Institute for Roses and Aromatic Plants, Kazanlak, harvested from 2019 and 2020. The authenticity of the rose species was confirmed by Ana Dobreva. The voucher specimens were deposited in the IBER-BAS herbarium where the number of *R. damascena* Mill. is SOM 177 768 and of *R. alba* L. is SOM 177 769, respectively. The rose flowers were subjected to water–steam distillation using semi-industrial processing line. The following parameters for distillation were used: 8 kg raw rose material for charge and hydro module 1:6, in a rate of 8–10% and duration of 150 min. at distillate temperature maintained at 28–30 °C. The hydrosol obtained from this process was redistilled at very low speed to obtain the same quantity of rose hydrosol as the charged material, using the same apparatus at the same temperature [8]. After detailed analysis, the hydrosol was stored in a refrigerator at 4 °C in dark, in sterilized containers, until further use.

2.2. *Chemicals*

The standard experimental mutagen N-methyl-N′-nitro-N-nitrosoguanidine MNNG (50 μg/mL) (CAS-Nr.: 70-25-7) used as a positive control was provided by Fluka—AG, Buchs, Switzerland. RPMI 1640 medium was from Sigma-Aldrich, (Steinheim, Germany); fetal calf serum from Sigma-Aldrich, (Sao Paulo, Brazil); phytohemagglutinin PHA and cytochalasin-B from Sigma-Aldrich, (Jerusalem, Israel); acridine orange, KCl, and acid aceticum glaciale were purchased from Sigma-Aldrich Chemie GmbH, Merck (Steinheim, Germany). Solution of 0.9% NaCl and gentamycin were provided by Sopharmacy (Sofia, Bulgaria).

2.3. *Test Systems and Experimental Design*

Three different test systems widely used in genotoxic screening were used to study cytotoxic/genotoxic effect of rose hydrosols applying tests for induction of chromosome aberrations (CA) and induction of micronuclei (MN).

2.3.1. Plant Test System *In Vivo*

The structurally reconstructed karyotype MK14/2034 of *H. vulgare* has seven easily distinguishable chromosome pairs, which allow for easier detection of the mutagen-specific features of aberration distribution [37]. Despite these special properties, this karyotype is comparable in sensitivity to its standard karyotype—spring barley.

Presoaked barley seeds (1 h in tap water) were germinated in dark for 17 h in Petri dishes on moist filter paper at 24 °C. These seeds were treated with rose hydrosols in concentrations of 6%, 14%, and 20% for 1 and 4 h. Recovery times of 18, 21, 24, 27, and 30 h were examined. For scoring chromosome aberrations, after these recovery times, seedlings were affected with 0.025% colchicine in a saturated solution of a-bromonaphthalene (2 h) and fixed in ethanol:glacial acetic acid in ratio of 3:1. The barley root tip meristems were hydrolyzed in 1N HCl at 60 °C for 9 min, stained with Schiff's reagent, macerated in 4% pectinase, and squashed onto clean slides. Untreated meristems were used as a negative untreated control.

For scoring MN, the barley root tip meristem cells were not treated with colchicine, and the meristems were fixed after 30 h recovery time [38].

2.3.2. Animal Test System *In Vivo*

Eight-week-old male and female ICR strain albino mice (21.0 ± 1.5 g b. w.) were delivered from Slivnitza animal breeding house, Bulgarian Academy of Sciences, Sofia. Animals were transported to the Animal House Facility of Institute of Biodiversity and Ecosystem Research and were kept for several days at standard laboratory conditions—temperature 20–22 °C, photoperiod 7 am to 7 pm, and unimpeded access to typical diet and fluids for rearing laboratory animals. The ICR mice were randomly allocated in four experimental groups (eight male/eight female animals each) and kept in standard cages, isolating the control and the treatment groups to avoid cross contamination. All tested substances were given as a single treatment by i.p. injection. The following experimental groups (n = 8, 4♂4♀each) were defined: Group 1. Animals injected with 11% hydrosol solution (0.01 mg/mL); Group 2. Animals injected with 20% hydrosol solution (0.01 mg/mL). Group 3: Positive control group injected with a model mutagen MNNG 50 μg/mL (0.01 mg/mL). Group 4: Untreated control group received i.p. only with 0.9% NaCl (0.01 mg/mL) under identical conditions. Throughout the experiment, experimental groups were inspected twice per day after primary i.p. compound supplementation for any kind of toxicity or unwellness indications.

The protocol for chromosome aberrations was applied for each experimental group starting at the 24th (4♂4♀) or 48th (4♂4♀) after single treatment with the respective solution [39]. To receive chromosomes in metaphase stage appropriate for cytogenetical observations, a mitotic inhibitor colchicine—0.04 mg/g b.w. was introduced i.p. an hour prior bone marrow cells insulation. For scoring of micronuclei, blood smears were prepared prior colchicine treatment. All experimental animals were humanely euthanized by diethyl ether; bone marrow cells were isolated by flushing from femoral bone and subjected to subsequent hypotonization for 15 min with 0.075 M KCl at 37 °C. The cell fixation procedure includes a solution of cold methyl alcohol: glacial acetic acid (3:1), followed by dripping on pre-cleaned and pre-cooled wet microscopic glass slides and being left to naturally dry. A solution of 5% Giemsa was used for slides staining (Sigma Diagnostic).

The rodent erythrocyte micronucleus assay, according to the regulatory requirements [40,41], was used as another method for evaluation of genotoxic or clastogenic effects of the tested hydrosols. The micronucleus assay was fulfilled following the OECD test guidance No. 474 for chemicals safety evaluation [40]. Peripheral blood samples were taken once, 48 h after initial treatment, from all dose groups. For each of the eight mice in the group, 5 μL peripheral blood was collected through the tail. The blood was diluted with 45 μL Sörensen's phosphate buffer (pH 6.8) [42], and a drop of this solution was smeared on a slide, dried, and fixed for 10 min with absolute methanol. The smeared slides were Acridine orange (AO) stained following recommendations of Hayashi et al. (1983), using some modifications. AO (50 μL of a 1 mg/mL solution) was dropped on dry blood sample slides and spread by immediately covering the slide with coverslip glass. The analysis was performed with an Axio Scope A1—Carl Zeiss Fluorescent Microscope at 400x magnification, with FITC 495 nm excitation filter.

The experiments were performed in accordance with Bulgaria's Directorate of Health Prevention and Humane Behavior toward Animals. Bulgarian Food Safety Agency (BFSA) published Certificate number 125 and standpoint 45/2015 for 5-year period, for use of animals in experiments for the Stephan Angeloff Institute. The Ethical Committee of the Stephan Angeloff Institute approved the experimental design and protocols of the work from 4 October 2020. This certificate was obtained in connection with a project application, and the experiments were carried out from 2020–2021.

2.3.3. Human lymphocytes *In Vitro*

Lymphocyte cultures were prepared from peripheral venous blood of healthy non-smoking/non-drinking donors (men and women), who do not accept any medication, aged between 33 to 40 years. Each culture contained RPMI 1640 medium, 12% fetal calf serum, 40 mg/mL gentamycin, and 0.1% mitogen phytohemagglutinin (PHA) and was cultured at 37 °C.

The method of Evans [43] was applied to study the chromosome aberrations. The lymphocyte cultures were treated with each rose hydrosol in concentrations of 6%, 11%, 14%, and 20% for 1 h and 4 h. After the treatment, the cells were washed in fresh RPMI medium and cultured at 37 °C. At the 72nd hour after PHA stimulation, the cells were affected with 0.02% colchicine, followed by hypotonization in 0.56% KCl and fixation in solution of methanol: acetic acid (3:1, v/v). After centrifugation, the pellet was dropped on clean glass slides and stained in 2% Giemsa. Untreated lymphocyte samples were used as a negative control.

For micronuclei at the 44th hour after PHA stimulation, cytochalasin-B (6 μg/mL) was added to each culture according to cytokinesis-block micronucleus (CBMN) method of Fenech [44]. After 24 h the lymphocytes were centrifuged, hypotonized with 0.56% KCl, and fixed in methanol: acetic acid (3:1). The suspension obtained after subsequent centrifugation was dropped onto clean slides and stained in 2% Giemsa.

All procedures were conducted corresponding to the Declaration of Helsinki, approved by Commission on Ethics and Academic Unity of Institute of Biodiversity and Ecosystem Research (Number: No 1 from 18 February 2022). All donors were informed about the experimental procedure and signed written informed consent forms.

2.4. *Cytogenetic Endpoints*

The cytotoxic effect was evaluated by mitotic index, PCE/(PCE + NCE) ratio, and nuclear division index (NDI). To assess the genotoxic/clastogenic activity of the tested rose hydrosols, induction of chromosome aberrations and induction of micronuclei were used as endpoints.

2.4.1. Endpoints for Cytotoxicity

Mitotic Index (MI)

The value of mitotic index, which gives information about the cytotoxic effect of the tested hydrosols, was assessed in all three test systems. The mitotic index is the number of metaphases per 1000 observed cells per each experimental variant. For animal cells, to assess the mitotic division, protocol of Darzynkiewicz [45] was used, where the MI was determined by calculating all dividing bone marrow cells among not less than 1500 counted cells per laboratory mouse.

PCE/(PCE+NCE) and Nuclear Division Index (NDI)

To obtain additional information about the cytotoxic effect of the rose hydrosols, PCE/(PCE + NCE) ratio in each treated ICR mice was calculated, where PCE were the polychromatic erythrocytes and NCE—normochromatic erythrocytes.

In human lymphocytes, the cytotoxicity of the tested substances was also assessed by nuclear division index (NDI), using test for induction of micronuclei. The following formula was used for its calculation, (N1 + 2N2 + 3N3 + 4N4)/N, where N1–N4 are the number of cells with 1–4 nuclei, and N is the total number of scored cells.

2.4.2. Endpoints for Genotoxicity

Induction of Chromosome Aberrations (CA)

To assess the genotoxic effect of rose hydrosols, the percentage of well-spread metaphases with chromosome aberrations (MwA% $\pm$ SD) was calculated in all test systems. Chromatid breaks, isochromatid breaks, chromatid translocations, intercalary deletions, centromere fusions, and telomere/telomere fusions were determined (Figure 1). More than 2000 cells were calculated for induction of chromosome aberrations in barley test system and human lymphocytes from each experimental variant. Up to 50 well-dispersed metaphases from either laboratory mouse were subjected to chromosomal aberrations analysis performed by light microscopy (Olympus CX 31) $\times$ 1000.

"Aberration hot spots" were determined in barley chromosomes (reconstructed barley karyotype MK14/2034) to receive detailed information about the specific clastogenic effects. This karyotype gave the opportunity to study defined chromosome segments in different chromosomal positions and their involvement in induced aberrations. The "aberration hot spots" in the plant chromosomes were determined to obtain information about the DNA segments with higher susceptibility to the tested hydrosol sample and/or to the mutagen. To analyze the regional specificity of aberration induction the metaphase chromosomes of *H. vulgare* were subdivided into 48 segments of almost equal sizes. The segments are numbered with respect to their position in the standard karyotype [37,46] (see Figure 1A).

Induction of Micronuclei (MN)

Generally, the micronucleus test is fast and easy to perform. Between 3000–6000 cells for the presence of MN and spindle defects, which later forms MN too for each treatment variant, were inspected for barley and human lymphocytes *in vitro* (Figure 1A(c),C(d)). Percentage of micronuclei (MN% $\pm$ SD) was calculated.

The criteria used for the identification of micronuclei and to distinguish them from artifacts (Figure 1B(c–f)) in the cytoplasm of animal cells were described by Schmid [47]. Micronuclei were easily distinguishable in polychromatic erythrocytes (PCE), which emit red fluorescence [48]. From each slide, at least 2000 PCE or at least 4000 PCE per animal were screened for the incidence of micronucleated immature erythrocytes. The ratio of PCE to total (PCE + NCE) erythrocytes was defined for each laboratory mouse separately, by counting at least 2000 normochromatic erythrocytes per animal. Only monolayers without overlapping cells were targeted for each slide.

In the *in vivo* animal test system, the relationship between PCE and NCE as well as the MN rates of the exposed and control samples (negative, positive) were examined.

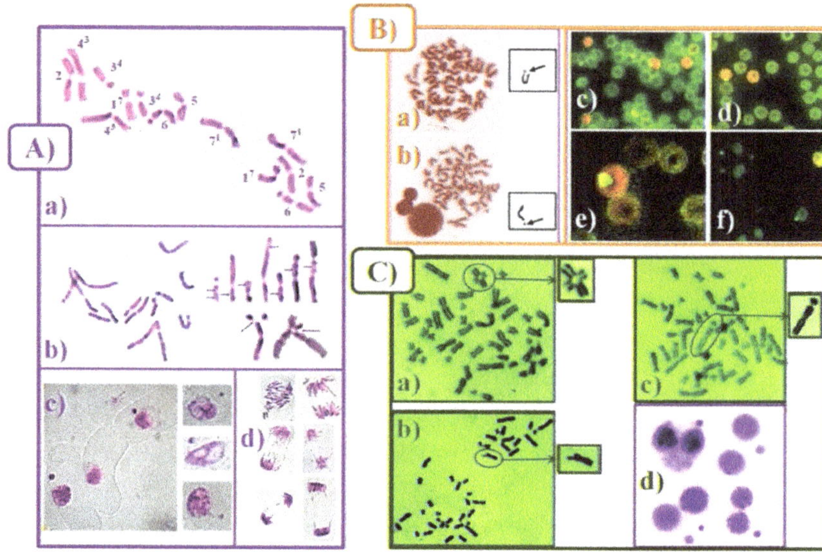

Figure 1. Chromosome aberrations (CA), micronuclei (MN), and mitotic disturbances detected after treatment with hydrosols derived during water–steam distillation of essential oils of Bulgarian *R. damascena* Mill. and *R. alba* L. in: (**A**) plant test system—*H. vulgare* (**a**) reconstructed karyotype, (**b**) different types of chromatid aberrations–isochromatid breaks (above), chromatid translocations (under part), (**c**) MN, and (**d**) spindle defects; (**B**) animal test system (**a**,**b**) chromosome fragments, (**c**–**f**) PC and NC erythrocytes with MN; (**C**) human lymphocyte cells (**a**) translocation, (**b**) chromatid break, (**c**) isochromatid break, and (**d**) MN.

2.5. Statistical Analysis

Each experiment was triplicate. One-way ANOVA with two-tailed Fisher's exact test was used for statistical analysis of the different treatment variants for all three test systems (Microsoft Excel 2010). The statistical differences were assayed as $p > 0.05$ not significant, * $p < 0.05$ significant, ** $p < 0.01$ more significant, and *** $p < 0.001$ extremely significant.

The aberrations hot spots in the reconstructed barley karyotype were calculated followed the protocol of Rieger et al. [49] and Jovtchev et al. [46].

3. Results

3.1. Cytotoxic Effects of Rosa alba L. and Rosa damascena Mill. Hydrosols

3.1.1. Mitotic Index (MI)

No or very low cytotoxic effect was calculated by MI both for hydrosol from *R. alba* L. and *R. damascena* Mill. applied at concentrations of 6–20% (for 1 and 4 h) in *H. vulgare*, compared to the untreated control (Figure 2A). A significant reduction in mitotic activity was obtained after *R. alba* hydrosol application in the bone marrow cells of all experimental animal groups ($p < 0.05$) compared to the control group. Surprisingly, the MI values obtained were close with those calculated in the experimental group treated with MNNG (50 µg/mL) (Figure 2B). In contrast to *R. alba* hydrosol, a lack of antiproliferative action of the *R. damascena* hydrosol solutions was obtained. The human lymphocytes were the most sensitive to both rose hydrosols compared to barley and mice cells. The cytotoxic effect of *R. alba* hydrosol (6%, 11%, 14%, and 20%) assessed by MI value in lymphocytes was not high and decreased in a concentration-dependent manner. The MI values calculated for

R. damascena did not show any dependence on the concentration applied and were slightly increased with the duration of treatment. (Figure 2C). The mitotic activity observed in all three test systems after hydrosols application was much higher compared to the cytostatic action of the MNNG alkylating agent used as a positive control. Data from the *in vivo* and *in vitro* experiments are highly consistent and show that the hydrosols exhibited lower toxicity toward normal cells compared to MNNG ($p < 0.001$) (Figure 2).

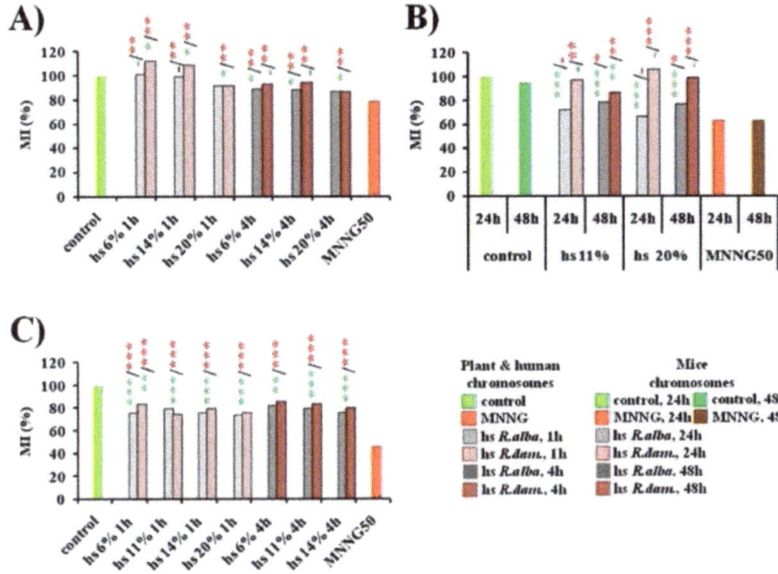

Figure 2. Cytotoxic activity of hydrosols (hs) derived during water–steam distillation of essential oils from *R. alba* L. and *R. damascena* Mill. assessed by the value of mitotic index (MI) in: *H. vulgare* (**A**), mice bone marrow cells (**B**), and human lymphocytes (**C**). Mitotic activity was presented as a percentage of untreated control. * $p < 0.05$, ** $p < 0.01$, *** $p < 0.001$, and non-significantly versus untreated control (before slash) versus positive control MNNG (after slash).

3.1.2. PCE/(PCE+NCE)

The ratio showed significant differences among the *R. alba*-treated ICR mice groups (20% and 11%) and the negative control group ($p < 0.001$) in ICR mice (Figure 3A). In the presence of *R. alba* hydrosol, the percentage of PCE (7.92% ± 1.41 for 11% and 7.64% ± 1.64 for 20%) decreased significantly ($p < 0.05$; $p < 0.01$, respectively) compared with the negative control, which may indicate a suppression of bone marrow proliferation, probably due to mitotic arrest. Comparison between the positive control group with MNNG 50 µg/mL and the two hydrosol concentrations also showed a statistical significance ($p < 0.05$; $p < 0.01$), respectively. Therefore, it can be concluded that the lower, but statistically significant values, calculated after hydrosol treatment, support the assumption of a cell proliferation suppression following *R. alba* hydrosol supplementation. Similar to the mitotic index, the ratio of polychromatic erythrocytes to total number of poly- and normochromatic erythrocytes in the animal test system did not show any dose-dependent effect at the two concentrations of *R. damascena* hydrosol administered ($p > 0.05$). At an exposure of 11% and 20% *R. damascena* hydrosol solution, the corresponding calculated mean values for PCE/(PCE + NCE) (%) were 10.97% ± 2.64 and 11.75% ± 3.30, respectively. These mean values are distinguished significantly from the values in the control group (0.9% NaCl) (14.44% ± 1.27) ($p < 0.01$; $p < 0.05$, respectively). Since there was a well-expressed statistical significance compared to the data of the MNNG group ($p < 0.001$; $p < 0.01$), it can be concluded that the suppression of cell division by *R. damascena* hydrosol assessed by the ratio PCE/(PCE+NCE) is minimal.

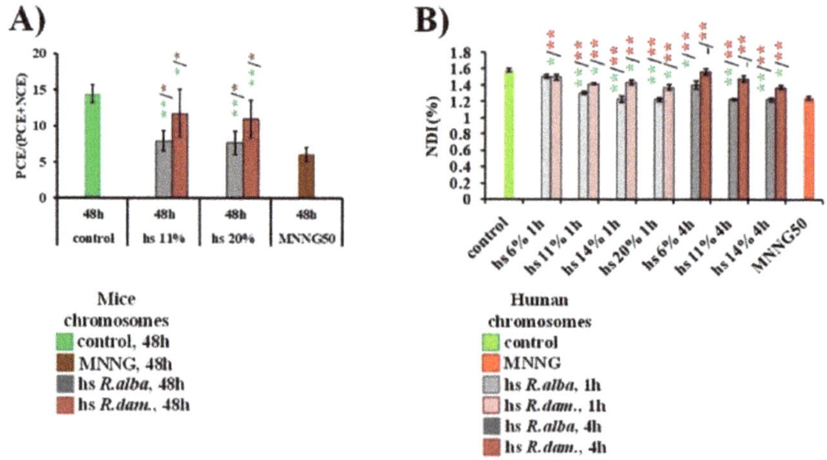

Figure 3. Cytotoxic activity of hydrosols (hs) derived during water–steam distillation of essential oils from *R. alba* L. and *R. damascena* Mill. evaluated by PCE/(PCE + NCE) ratio in ICR mice (**A**) and value of NDI in human lymphocytes (**B**). * $p < 0.05$, ** $p < 0.01$, and non-significantly versus untreated control (before slash) versus positive control MNNG (after slash).

3.1.3. Nuclear Division Index (NDI)

This endpoint used as a second indicator for cytotoxic activity of the hydrosols in lymphocyte cultures was slightly reduced ($p < 0.05$, $p < 0.01$) in rose-hydrosol-treated cells compared to the negative control (Figure 3B). Surprisingly, the higher NDI values were obtained after a 4h hydrosol treatment of *R. damascena*, with values that were similar to those of the untreated control. Data obtained from the NDI showed that *R. damascena* hydrosol has low suppression activity of cell division and slightly higher than of *R. alba*.

3.2. Genotoxic Effects of the Rose Hydrosols

3.2.1. Induction of Chromosome Aberrations

The genotoxic effect of *R. alba* and *R. damascena* hydrosols was investigated by values of metaphases with chromosome aberrations (MwA) as an endpoint.

Treatment with all rose hydrosols' concentrations (6–20%) induced a comparatively low but statistically significant ($p < 0.05$; $p < 0.001$) genotoxic/clastogenic effect compared to the negative controls in the plant test system, animal cells, and human lymphocytes (Figure 4A,E).

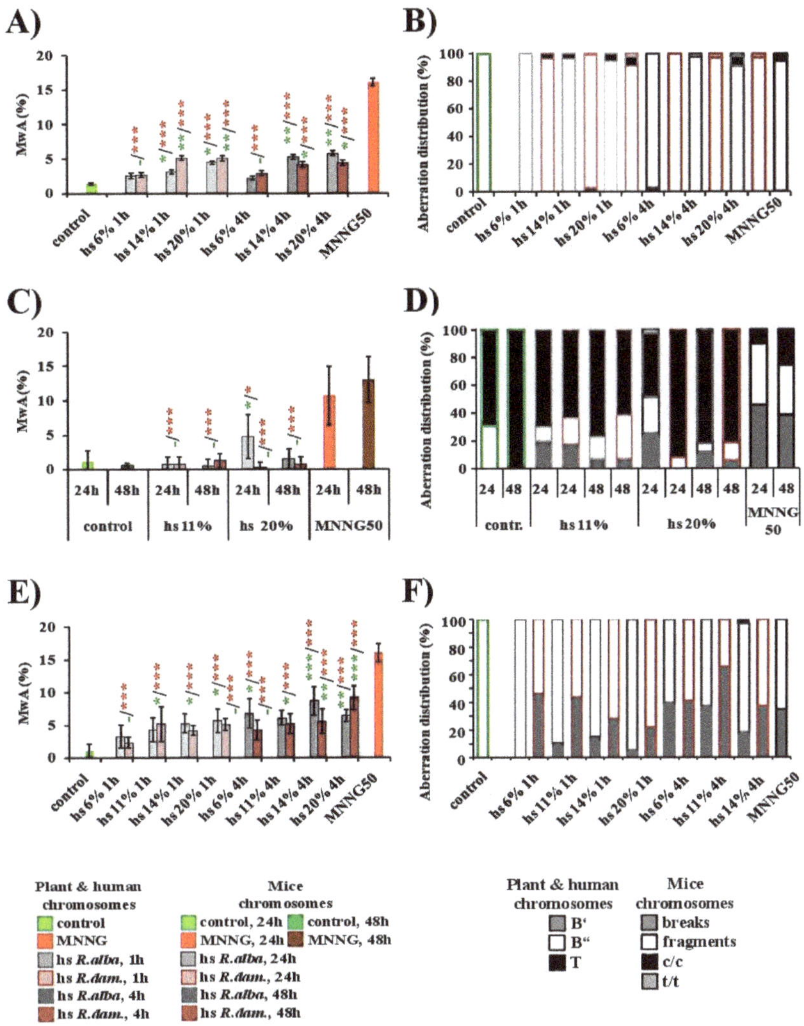

Figure 4. Genotoxic effect of hydrosol (hs) derived after water–steam distillation of essential oils from *R. alba* L., and *R. damascena* Mill., assessed by the frequency of metaphases with aberrations (MwA) in: *H. vulgare* (**A**), mice bone marrow cells (**C**), and human lymphocytes (**E**). Distribution of aberrations observed after treatment with both hydrosols (hs) in *H. vulgare* (**B**), mice bone marrow cells (**D**), and human lymphocytes (**F**). * $p < 0.05$, ** $p < 0.01$, *** $p < 0.001$, and non-significantly versus negative control (before slash) versus positive control MNNG (after slash).

The frequency of aberrations in *H. vulgare* after 1 h treatment with *R. alba* hydrosol ranged from 2.53% ± 0.4 (with 6%) to 4.47% ± 0.47 (with 20%), and treatment for 4 h induced a frequency from 2.2% ± 0.29 (with 6%) to 5.8% ± 0.32 (with 20%), respectively. Similar frequencies were observed after treatment with *R. damascena* hydrosol, namely yields between 2.73% ± 0.31 for 6% (1 h) and 5.07% ± 0.41 for 20% (1 h) and 2.93% ± 0.33 for 6% (4h), 4.40% ± 0.5% for 20% (4h) The effects of two applied concentrations of *R. alba* and *R. damascena* hydrosols (11% and 20%) on the laboratory ICR mice chromosome structures were evaluated using the bone marrow test. The data about the percentage of chromosome aberration frequencies are presented in Figure 4C. The highest percentage of aberrant mitoses (4.75% ± 3.19) was calculated in the experimental mice group, which was treated with 20% *R. alba* hydrosol at the first sampling point (24th from the beginning of the experiment), compared to other experimental variants ($p \leq 0.05$). The clastogenic effect of the 20% hydrosol solution significantly decreases at the 48th hour (1.5% ± 1.41) and does not differ statistically from the data calculated in the 0.9% NaCl control group (1.14% ± 1.57). The results showed a considerable reduction in the percentage of mitoses with aberrations (0.75% ± 1.03 and 0.5% ± 0.92) in the bone marrow mice cells, treated with 11% hydrosol at the 24th and 48th hours compared to the data in the 20%-hydrosol-treated group at the 24th hour ($p < 0.001$). These results as well as the one obtained in the 48th hour with 20% hydrosol, are statistically undistinguishable from the data in the 0.9% NaCl control groups. The results of our experiments with laboratory ICR mice showed that only at the higher applied concentration of *R. alba* hydrosol induced a weak clastogenic effect. A slight dose-dependent effect was observed. This result attested the sensitivity of the assay to detect genotoxicity of various types of genotoxic compounds. A dose-dependent statistically significant difference was not found between the two *R. damascena* hydrosol concentrations tested. The number of aberrations did not increase significantly with the treatment period extension (at 24 and 48 h) ($p > 0.05$), as shown in Figure 4C. All mean values are quite close to the data obtained after the analysis for the untreated control group, and, as expected, there is no statistical reliability ($p > 0.05$). From the results and the performed statistical analysis, it is clear that slightly higher values for the frequency of bone marrow cells with aberrations were obtained in the experimental groups injected with the lower dose of hydrosol (11% solution).

In human lymphocyte cells, both rose hydrosols showed a statistically significant genotoxic effect ($p < 0.05$; $p < 0.01$) with the applied concentrations compared to the untreated control. A dose-dependent effect was observed after treatment with the *R. damascena* and *R. alba* hydrosols, and the frequency of chromosome aberrations was increased with the increasing of the treatment time duration. The frequency of observed chromosomal abnormalities after treatment with *R. alba* hydrosol ranged from 3.30% ± 1.70 (for 6%) to 5.70% ± 1.80 (20%) for 1h. For 4 h treatment, the obtained aberrations varied in a range from 6.8% ± 2.20 (with 6%) to 8.7% ± 2.20 (with 14%), respectively (Figure 4E). Similar effects were observed after *R. damascena* hydrosol treatment, namely yields between 2.40% ± 0.9 for 6% 1h, 5.10% ± 0.9 for 20% 1h, 4.30% ± 1.5 for 6% 4 h, and 9.2% ± 1.8 for 14% 4 h (Figure 4E). Both rose hydrosols had a similar genotoxic effect.

The clastogenic activity of both rose hydrosols at all tested concentrations was far lower ($p < 0.001$) than that of the direct mutagen MNNG (50 µg/mL) in all three test-systems (Figure 4). MNNG produced a statistically significant increase in the number of damaged cells.

In summary, there was no or a slight dependence of treatment duration and the concentrations applied, with respect to the values of induced aberrations in barley as well as in human lymphocytes (Figure 4A,E). Human lymphocytes *in vitro* were more sensitive to both hydrosols than the two *in vivo* test systems (plant and animal cells), using chromosome aberrations as the endpoint for genotoxicity.

The spectrum of the induced chromosome aberrations in *H. vulgare* by different concentrations of *R. alba* and *R. damascena* hydrosol was mainly of isochromatid breaks, a small number of chromatid breaks, translocations, and intercalary deletions (Figure 4B). In mice bone marrow cells, predominantly centromere–centromeric fusions followed by breaks, fragments, and a very small number of telomere-to-telomere fusions were detected (Figure 4D), while, in human lymphocytes, predominantly isochromatid breaks followed by chromatid breaks were observed (Figure 4F).

The aberration hot spots in barley reconstructed karyotype induced by different concentrations of *R. alba* hydrosol showed to be dose-dependent. The lower concentrations have one aberration hot spot, the higher have up to three. The most aberration hot spots were observed in the telomere and centromere regions. The yields of aberration hot spots induced by different concentrations of both hydrosols are significantly lower than the number of hot spots induced by MNNG. As for the positive control, MNNG treatment, 11 aberration hot spots were detected (Figure 5).

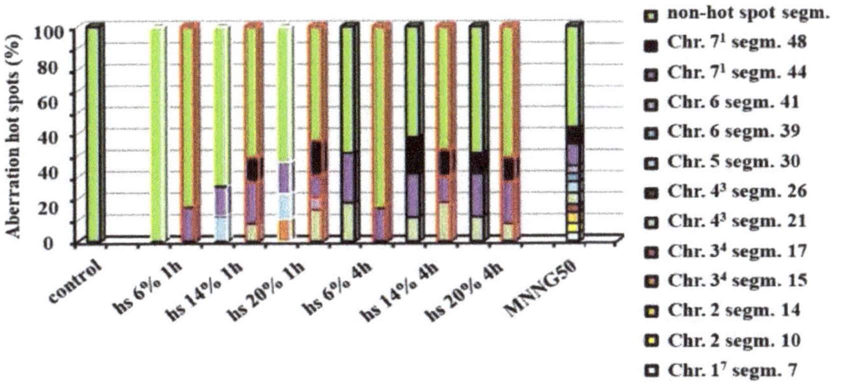

Figure 5. Aberration hot spots observed in the reconstructed barley karyotype MK14/2034* after treatment with different concentrations of *R. alba* and *R. damascena* hydrosols.

* Reconstructed karyotype of *H. vulgare* (MK 14/2034) is a result of the combination of two simple reciprocal translocations between parts of chromosomes 1 and 7 and chromosomes 3 and 4.

3.2.2. Induction of Micronuclei

In the *H. vulgare* test system, the micronuclei induced after treatment with *R. alba* hydrosol ranged from 0.30% ± 0.02 for 6%/1 h to 0.37% ± 0.06 for 20%/1 h. After 4h of treatment, an induction of MN, ranging from 0.08% ± 0.08 (6%) to 0.23% ± 0.08 (20%), was detected (Figure 6A). No clear concentration and treatment-duration dependence were observed. The data for *R. damascena* hydrosol were very close to those for *R. alba* hydrosol, namely 0.17% ± 0.11 for 6% 1h, 0.22% ± 0.16 for 20% 1h, 0.17% ± 0.10 for 6% 4 h, and 0.27% ± 0.14 for 20% 4 h (Figure 6A).

In the *in vivo* animal test system, the number of micronucleated erythrocytes MNPCE observed after treatment with *R. alba* hydrosol was given separately for each treatment group. The results of the micronucleus assay with peripheral blood erythrocytes in laboratory mice are summarized in Figure 6B. The higher frequency of MNPCE was observed in the experimental group, treated with 20% *R. alba* hydrosol (0.09% ± 0.02). The frequencies of MN in PCE were statistically undistinguishable and the results were not dose-dependent between the two hydrosol groups treated with *R. alba* ($p > 0.001$). *R. alba* hydrosol at concentration of 20% significantly increased the frequency of MNPCE from 0.06% ± 0.01 (mean value for the positive control group) to 0.09% ± 0.02 ($p < 0.01$). A significant rise in the frequency of MNPCE compared to the control was also noted following injection of 11% *R. alba* hydrosol (0.08 MNPCE/4000 PCE, $p < 0.01$). It is evident from the results of the rodent erythrocyte MN assay that *R. alba* hydrosol possesses low genotoxic activity, as it produced a slight-but-significant increase in the number of PCE with MN compared to the control group.

The data are far from the values obtained in the experimental animal group, treated with 50 μg/mL MNNG (0.69% ± 0.03). This can be seen as evidence that, despite the statistically positive result in the micronucleus test, *R. alba* hydrosol at the tested concentrations cannot be considered as genotoxic (Figure 6B).

After treatment with *R. damascena* hydrosol, a higher frequency of micronucleated erythrocytes (MNPCE) was observed in the animal group treated with 20% solution (0.07% ± 0.03) compared to that with 11%. The frequency slightly reduced from this value but was not significantly distinct from the control ($p > 0.05$), actually returning to the control level at 11% dose 48 h post-exposure (0.06% ± 0.02). As compared to MNNG (50 μg/mL

(0.01 mg/mL), a significant decrease in the MN frequency ($p < 0.001$) was reported for both hydrosol solution concentrations (Figure 6B).

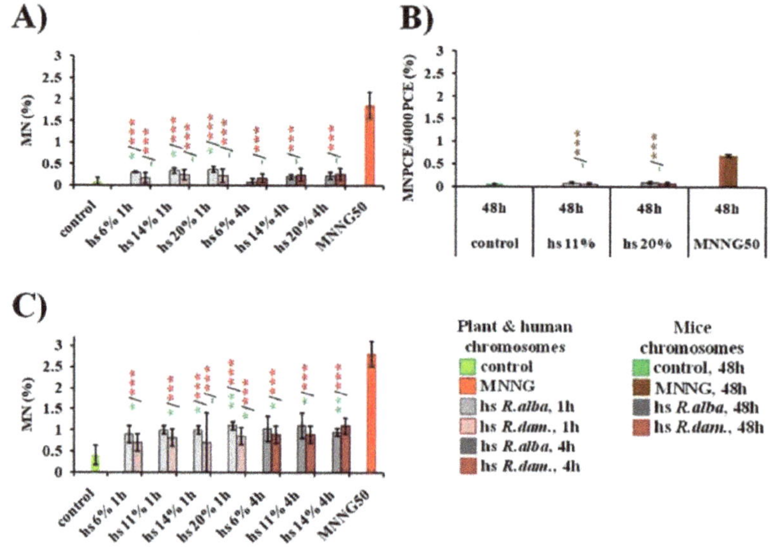

Figure 6. Genotoxic effect of the hydrosols (hs) derived during water–steam distillation of essential oils from *R. alba* L. and *R. damascena* Mill., assessed by the value of induced micronuclei in: *H. vulgare* (**A**), mice bone marrow cells (**B**), and human lymphocytes (**C**). * $p < 0.05$, ** $p < 0.01$, *** $p < 0.001$, and non-significantly versus untreated control (before slash) versus positive control MNNG (after slash).

Our micronuclei analysis study on mice peripheral blood erythrocytes has revealed that *R. damascena* hydrosol solution, based on the concentration and time intervals analyzed, does not cause a variation in chromosome structure but, rather, a slight cytotoxic effect (a weak cytotoxicity) (Figure 6B).

In human lymphocyte cultures, all concentrations (6–20%) of *R. alba* hydrosol applied at both time duration treatments (1 h and 4 h) showed close values of MN (0.09% ± 0.02 for 6% to 1.1% ± 0.01 for 20% 1 h treatment and 1.04% ± 0.03 for 6% to 0.96% ± 0.09 for 14% 4 h treatment). No dose- or duration-dependence were obtained among the frequency of MN after treatment with all concentrations of hydrosol in lymphocyte cells (Figure 6C). A mild statistically significant genotoxic effect was observed for the tested hydrosol concentrations, where the frequency of MN was slightly increased ($p < 0.05$, $p < 0.01$) compared to the control. The same effect was detected also for *R. damascena* hydrosol concentrations. *R. alba* and *R. damascena* hydrosol were not found to have a dose dependence (0.70% ± 0.20 with 6% for 1 h to 0.86% ± 0.20 with 20% for 1 h and 0.90% ± 0.2 with 6% for 4 h to 1.1% ± 0.20 with 14% for 4 h) for the value of MN induced by all tested hydrosol concentrations. The frequencies of MN did not increase significantly with the treatment period (1 and 4 h), as shown in Figure 6C. *R. alba* showed slightly higher genotoxic effect compared with that of *R. damascena* hydrosol assessed by MN induction, but the effect was not significantly proven.

The genotoxic action (expressed by MN) of both rose hydrosols at all tested concentrations was far below than the values obtained for the MNNG (50 μg/mL) ($p < 0.001$) (Figure 6). Our data showed that the genotoxic effect of both hydrosols was most pronounced in human lymphocytes *in vitro*.

4. Discussion

Applying a complex of cytogenetical tests by various endpoints in three different experimental test systems (*in vivo* and *in vitro*), we performed a good assessment of the cytotoxic and genotoxic activities of *R. alba* and *R. damascena* hydrosols. Here, we obtained useful information about the cellular sensitivity and the effect on hereditary material resulting from byproducts application. The use of different test systems with a range of methods makes the evaluation of the effect more representative and useful.

The analysis of the chemical composition of the rose hydrosols is essential for the interpretation of the results obtained. Detailed phytochemical analysis of the *R. alba* L. and *R. damascena* Mill. hydrosols given previously showed variances in the main groups of constituents [8]. There were subtle differences, but they could explain some impact on the effects. The main differences between both hydrosols were in the content of the oxygenated monoterpenes (OM)—76.63% for *R. alba* and 65.87% for *R. damascena*. Geraniol, citronellol, and linalool belong to this group. Monoterpenoid geraniol, presented as trans-geraniol 36.44% and cis-geraniol 6.11%, is the main constituent of *R. alba* hydrosol. *R. damascena* byproduct contained almost 15% less geraniol viz 16.44% trans-geraniol and 10.81% cis-geraniol [8]. β-Citronellol was found to be the main ingredient in *R. damascena* hydrosol, namely 28.7%. This compound was with equal quantity in both rose hydrosols. Another important compound, which belongs to the group of benzenoid compounds (BC), is phenylethyl alcohol. This component is known to vary greatly under certain conditions (time of harvest and degree of freshness of the raw material). Here, the content was very close in both hydrosols, but in *R. alba* hydrosol its amount is slightly higher, 5.95%,compared to 4.95% in *R. damascena*.

Our cytogenetic analysis showed that the hydrosol of *R. alba* slightly reduces the value of MI in *H. vulgare* root tip meristem cells, whereas the tested concentrations of *R. damscena* hydrosol enhanced the cell division. On the other hand, mitotic activity/cell division was significantly reduced in mice bone marrow cells and human lymphocytes *in vitro* after *R. alba* administration. In contrast, no significant alterations of cell division in mice cells and only a weak suppressive affect were observed in human lymphocytes *in vitro* after administration of *R. damascena* hydrosol.

Another endpoint that also provides information about the cytotoxicity, the PCE/(PCE + NCE) ratio in red blood cells of ICR mice, indicates alterations in erythropoiesis in the animal test system after application of hydrosols. *R. damascena* hydrosol reduced this ratio in much weaker extend than that of *R. alba*.

According to our data obtained for all endpoints of cytotoxicity in the three test systems, *R. alba* hydrosol possesses a higher cytotoxic effect than that of *R. damascena*. The scientific literature lacks any information on the cytotoxicity of *R. alba* hydrosol, and there are scarce data on the effect of *R. damascena*. The results obtained by us about *R. damascena* hydrosol are partially consistent with data published by Zamiri et al. [50]. According to these authors, *R. damascena* ethanol extract exhibited cytotoxicity toward the HeLa cell line. Another study also indicated that *R. damascena* and its constituents possess antitumor and anticancerogenic activities [51]. Significantly high cytotoxicity against NB4 and MCF7 cell lines was also demonstrated by Gao et al. [52], applying isolates from the flowers of *Rosa damascena*.

Since rose petals contain some compounds, which exhibit anti-proliferative activity [53], our results, detecting moderate mitotic suppression, provide their explanation. Dose-dependent anti-proliferative activity was reported by Wedler et al. [52] in immortalized human keratinocytes with a half-maximal inhibitory concentration (IC50) of 9.78 μg/mL of polyphenol-enriched fraction of *R. damascena* Mill. Our results concerning the anti-proliferative effects of *R. alba* hydrosol might also be due to the higher quantity of such compounds. Georgieva et al. [8] reported that the content of total phenolic compounds (TPC) of *R. alba* hydrosol is more than two times higher than that of *R. damascena* hydrosol. Carnesecchi et al. [54,55] reported anti-proliferative activity of monoterpenoid geraniol in cancer cells using various assays. More recent studies also identify powerful cytotoxicity of geraniol over tumor cells [56–58]. Our previous study with geraniol also showed its cytotoxic effect on human lymphocytes *in vitro*, which was dependent on the concentration used [31]. It is interesting to note that geraniol in combination with citronellol exhibit some toxic effects on mosquito species *Cx. pipiens* adults, and these compounds are more toxic compared to linalool alone [59]. Another main chemical component of hydrosols, namely phenylethyl alcohol, has demonstrated variation in the developmental toxicity in rats and skin toxicity in humans [60]. These studies showed that the cytotoxic effect of hydrosol clearly depends on the chemical content of the tested by-product, treatment scheme and target organisms. The low toxicity and/or negative results obtained by us in the *in vivo* test systems suggest biotransformation and excretion of the chemical components of the hydrosols by the organism. These factors have been proposed to calculate the realistic safety margin of phenylethyl alcohol for various organisms [60].

Analysis of the genotoxic activity of the two rose hydrosols showed that the value of the induced chromosomal aberrations also depended on the type of rose hydrosols. The yields of structural chromosomal damages increased after treatment with both hydrosols, especially with the *R. alba* hydrosol in laboratory ICR mice and in human lymphocytes *in vitro*. The results from CA showed that *R. damascena* hydrosol failed to elicit significant chromosome anomalies at the tested dose and/or duration of treatment. The present study detected an increase also the micronuclei frequency by both hydrosols in plant cells and cultured human lymphocytes as well as the micronucleated polychromatic erythrocytes in animal cells compared with the non-treated cells. The assessment by this endpoint for genotoxicity also demonstrated lower DNA damage induced by *R. damascena* compared to that by *R. alba* hydrosol. The genotoxic effect observed by us depends on a complex of conditions such as the target cells and the treatment scheme. Human lymphocytes *in vitro* were the most sensitive to both hydrosols, followed by higher plant and animal cells. The lower genotoxic effect of rose hydrosols in the *in vivo* test system for laboratory mice, compared with that in human lymphocytes *in vitro*, is probably due to intact metabolic processes, such as the degradation and excretion of the main hydrosol compounds. In the *in vitro* test system an instantaneous genotoxic effect is registered, without metabolic biotransformation processes. On the other hand, the cell wall in plants makes it more difficult for genotoxins to enter the cell and induce DNA damage.

The combinations of chemical components and their amount in the byproducts are probably of a great importance for the genotoxic activity of the tested hydrosols. The TPC content is reported to depend on the plant raw material (leaves, fruits, flowers, etc.) or the product (oil, hydrosol, wastewater, etc.) [61]. The higher quantity of TPC in *R. alba* probably affects not only the cytotoxicity of the hydrosol but also the observed higher genotoxicity compared to that of *R. damascena*, especially in human lymphocytes *in vitro*. Our study on the effect of monoterpenoid geraniol (10–100 μg/mL) did not show a high genotoxic effect in human lymphocytes *in vitro* and showed a weak one in higher plants [31]. Other authors also reported that geraniol applied in CHO cells at concentrations of 78.1 and 156.3 g/mL, in presence of the S9 fraction, significantly increased the number of cells with structural aberrations. On the other hand, no statistically significant enhancement in the frequency of PCE with micronuclei in bone marrow of mice treated with high concentrations of geraniol was detected [62].

The lower amount of TPC and probably BC suggests a lower or no genotoxic effect of the *R. damascena* hydrosol. Consequently, in the present studies, no or a low clastogenic or aneugenic effect was obtained in the treatment regimen in bone marrow cells, and a low genotoxicity in barley was obtained after application of hydrosol to *R. damascena*. Our findings with the plant test system *in vivo* and animal test system *in vivo* reciprocate earlier investigations reporting a lack of chromosome damage [62–64] in various chemopreventive herbal extracts and phytochemical studies. Since cytogenetic results for geraniol, which together with citronellol are the main constituents of the *R. alba* and *R. damascena* hydrosols, are lacking in the literature, our results are expected and logical.

In our study, we have not observed high positive results for either rose hydrosols in the *in vitro* assays with human lymphocytes, and the consistent negative results or low cytotoxic/genotoxic effects observed in more relevant *in vivo* test systems using chromosomal aberrations and micronucleus analysis suggest that *R. damascena* and *R. alba* hydrosols possess low genotoxic risk to humans when used.

5. Conclusions

This study presents an evaluation of the cytotoxic/genotoxic effects of hydrosols obtained after water–steam distillation of essential oils from Bulgarian *R. alba* L. and *R. damascene* Mill. using tests for genotoxicity in three different *in vivo* and *in vitro* test systems.

The following conclusions can be drawn.

The *R. alba* hydrosol does not show any cytotoxic effect in the plant test system *in vivo*. A comparatively low cytotoxicity was observed in mice bone marrow cells and human lymphocytes *in vitro* at all concentrations compared to the untreated control. It exhibited a low, concentration- and time-dependent, statistically significant genotoxic/clastogenic effect compared to untreated controls, as assessed by chromosomal aberrations and micronuclei in all test systems.

The *R. damascena* hydrosol exhibited weak cytotoxic and clastogenic effects at all concentrations applied in both the *in vivo* and *in vitro* test systems. The observed effect does not depend on the duration of treatment in all three test systems.

The results, obtained here, show that the hydrosols did not possess high cytotoxic and genotoxic activities. The effects depend on the type of rose hydrosols, the concentrations applied, and the sensitivity of the test system used. Both rose hydrosols have the potential to become a part of therapeutic applications. Furthermore, these products can be used for cosmetic purposes and in the food industry. The sensitivity of certain cell types must be considered.

Author Contributions: Conceptualization and writing—G.J., S.G., T.G.; performed cytogenetic experiments, S.G., G.J., T.G., M.T.-A., A.S. and T.A.; provided rose hydrosols, A.D.; wrote—original draft, G.J., S.G. and T.G.; writing—review and editing, G.J., S.G., T.G., M.T.-A. and M.M.; visualization, G.J.; supervision, M.M.; project administration, M.M.; funding acquisition, M.M. All authors have read and agreed to the published version of the manuscript.

Funding: This study was funded by the Bulgarian National Science Fund, project N KP-06-H36/17 "Biological activities of Bulgarian rose oils and approach to valorization of waste obtained from their production".

Institutional Review Board Statement: The experiments were carried out in accordance with the Declaration of Helsinki and were approved by the Commission on Ethics and Academic Unity of Institute of Biodiversity and Ecosystem Research (number: 1, date: 18 February 2022). The animal study protocol was performed in accordance with Bulgaria's Directorate of Health Prevention and Humane Behavior toward Animals. Bulgarian Food Safety Agency (BFSA) published Certificate number 125 and standpoint 45/2015 for a 5-year period, for the use of animals in experiments for the Stephan Angeloff Institute. The Ethical Committee of the Stephan Angeloff Institute approved the experimental design and protocols of the work from 4 October 2020.This certificate was obtained in connection with a project application, and the experiments were carried out from2020–2021.

Informed Consent Statement: All participants involved in the present study signed an informed consent.

Data Availability Statement: All the obtained data of this research are presented in the manuscript.

Acknowledgments: The authors are grateful to all the volunteer donors who participated in the experiments.

Conflicts of Interest: The authors declare no conflict of interest.

References

1. Rajeswara Rao, B.R. Hydrosols and water-soluble essential oils: Medicinal and biological properties. Chapter 6. In *Recent Progress in Medicinal Plants*; Essential Oils, I., Govil, J.N., Bhattacharya, S., Eds.; StudiumPress LLC: Houston, TX, USA, 2013; Volume 36, pp. 119–140.
2. Agarwal, S.G.; Gupta, A.; Kapahi, B.K.; Baleshwar, M.; Thappa, R.K.; Suri, O.P. Chemical Composition of Rose Water Volatiles. *J. Essent. Oil Res.* **2005**, *17*, 265–267. [CrossRef]

3. Verma, R.S.; Padalia, R.C.; Chauhan, A.; Singh, A.; Yadav, A.K. Volatile constituents of essential oil and rose water of damask rose (*Rosa damascena* Mill.) cultivars from North Indian hills. *Nat. Prod. Res. Former. Nat. Prod. Lett.* **2011**, *25*, 1577–1584. [CrossRef]
4. Saxena, M.; Shakya, A.K.; Sharma, N.; Shrivastava, S.; Shukla, S. Therapeutic efficacy of *Rosa damascena* Mill. on acetaminophen-induced oxidative stress in albino rats. *J. Environ. Pathol. Toxicol. Oncol.* **2012**, *31*, 193–201. [CrossRef] [PubMed]
5. Moein, M.; Zarshenas, M.M.; Delnavaz, S. Chemical composition analysis of rose water samples from Iran. *Pharm. Biol.* **2014**, *52*, 1358–1361. [CrossRef]
6. Mahboubi, M. *Rosa damascena* as holy ancient herb with novel applications. *J. Tradit. Complementary Med.* **2016**, *6*, 10–16. [CrossRef]
7. Akram, M.; Riaz, M.; Munir, N.; Akhter, N.; Zafar, S.; Jabeen, F.; Shariati, M.A.; Akhtar, N.; Riaz, Z.; Altaf, S.H.; et al. Chemical constituents, experimental and clinical pharmacology of *Rosa damascena*: A literature review. *J. Pharm. Pharmacol.* **2020**, *72*, 161–174. [CrossRef]
8. Georgieva, A.; Dobreva, A.; Tzvetanova, E.; Alexandrova, A.; Mileva, M. Comparative study of phytochemical profiles and antioxidant properties of hydrosols from Bulgarian *Rosa Alba* L. and *Rosa Damascena* Mill. *JEOP* **2019**, *22*, 1362–1371. [CrossRef]
9. Aćimović, M.G.; Tešević, V.V.; Smiljanić, K.T.; Cvetković, M.T.; Stanković, J.M.; Kiprovski, B.M.; Sikora, V.S. Hydrolates—by-products of essential oil distillation: Chemical composition, biological activity and potential uses. *Adv. Technol.* **2020**, *9*, 54–70. [CrossRef]
10. Xiao, Y.; He, J.; Zeng, J.; Yuan, X.; Zhang, Z.; Wang, B. Application of citronella and rose hydrosols reduced enzymatic browning of fresh-cut taro. *J. Food Biochem.* **2020**, *44*, e13283. [CrossRef]
11. Maruyama, N.; Tansho-Nagakawa, S.; Miyazaki, C.; Shimomura, K.; Ono, Y.; Abe, S. Inhibition of neutrophil adhesion and antimicrobial activity by diluted hydrosol prepared from *Rosa damascena*. *Biol. Pharm. Bull.* **2017**, *40*, 161–168. [CrossRef]
12. Hirulkar, N.B.; Agrawal, M. Antimicrobial activity of rose petals extract against some pathogenic bacteria. *Int. J. Pharmac. Biol. Arch.* **2010**, *1*, 478–484.
13. Bayhan, G.I.; Gumus, T.; Alan, B.; Savas, I.K.; Cam, S.A.; Sahin, E.A.; Arslan, S.O. Influence of *Rosa damascena* hydrosol on skin flora (contact culture) after hand-rubbing. *GMS Hyg. Infect. Control* **2020**, *15*, Doc21. [CrossRef] [PubMed]
14. Georgieva, A.; Ilieva, Y.; Kokanova-Nedialkova, Z.; Zaharieva, M.M.; Nedialkov, P.; Dobreva, A.; Kroumov, A.; Najdenski, H.; Mileva, M. Redox-modulating capacity and antineoplastic activity of wastewater obtained from the distillation of the essential oils of four Bulgarian oil-bearing roses. *Antioxidants* **2021**, *10*, 1615. [CrossRef] [PubMed]
15. Verma, A.; Srivastava, R.; Sonar, P.K.; Yadav, R. Traditional, phytochemical, and biological aspects of *Rosa alba* L.: A systematic review. *Future J. Pharm. Sci.* **2020**, *6*, 114. [CrossRef]
16. Alom, S.; Ali, F.; Bezbaruah, R.; Kakoti, B.B. *Rosa alba* Linn.: A comprehensive review of plant profile, phytochemistry, traditional and pharmacological uses. *WJPR* **2021**, *10*, 798–811. [CrossRef]
17. Boskabady, M.H.; Shafei, M.N.; Saberi, Z.; Amini, S. Pharmacological Effects of *Rosa Damascena*. *Iran J. Basic Med. Sci.* **2011**, *14*, 295–307.
18. Nunes, H.S.; Miguel, M.G. *Rosa damascena* essential oils: A brief review about chemical composition and biological properties. *Trends Phytochem. Res.* **2017**, *1*, 3–111. Available online: https://tpr.shahrood.iau.ir/article_532669_d72d2686983cb5dbbb7e721834197c7c.pdf (accessed on 5 September 2017).
19. Zolotilov, V.; Nevkrytaya, N.; Zolotilova, O.; Seitadzhieva, S.; Myagkikh, E.; Pashtetskiy, V.; Karpukhin, M. The essential-oil-bearing rose collection variability study in terms of biochemical parameters. *Agronomy* **2022**, *12*, 529. [CrossRef]
20. Kovacheva, N.; Rusanov, K.; Atanassov, I. Industrial cultivation of oil bearing rose and rose oil production in Bulgaria during 21st century, directions and challenges. *Biotechnol. Biotechnol. Equip.* **2010**, *24*, 1793–1798. [CrossRef]
21. Jakubczyk, K.; Tuchowska, A.; Janda-Milczarek, K. Plant hydrolates—Antioxidant properties, chemical composition and potential applications. *Biomed. Pharmacother.* **2021**, *142*, 112033. [CrossRef]
22. Başer, K.H.C. Turkish rose oil. *Perfum. Flavor.* **1992**, *17*, 45–52.
23. Ulusoy, S.; Bosgelmez-Tınaz, G.; Secilmis-Canbay, H. Tocopherol, carotene, phenolic contents and antibacterial properties of rose essential oil, hydrosol and absolute. *Curr. Microbiol.* **2009**, *59*, 554–558. [CrossRef] [PubMed]
24. Rusanov, K.; Kovacheva, N.; Rusanova, M.; Atanassov, I. Traditional *Rosa damascena* flower harvesting practices evaluated through GC/MS metabolite profiling of flower volatiles. *Food Chem.* **2011**, *129*, 1851–1859. [CrossRef]
25. Lei, G.; Wang, L.; Liu, X.; Zhang, A. Fast quantification of phenylethyl alcohol in rose water and chemical profiles of rose water and oil of *Rosa damascena* and *Rosa rugosa* from southeast China. *J. Liq. Chromatogr. Relat. Technol.* **2014**, *38*, 823–832. [CrossRef]
26. Popescu, A.; Matei, N.; Roncea, F.; Miresan, H.; Pavalache, G. Determination of caftaric acid in tincture and rose water obtained from Rosaedamascenaeflores. *Ovidius Univ. Ann. Chem.* **2015**, *26*, 12–19. [CrossRef]
27. Moein, M.; Etemadfard, H.; Zarshenas, M.M. Investigation of different Damask rose (*Rosa damascena* Mill.) oil samples from traditional markets in Fars (Iran); Focusing on the extraction method. *Trends PhramaceuticalSci.* **2016**, *2*, 51–58. [CrossRef]
28. Tentative Report for Public Comment. Safety Assessment of Rosa damascena-Derived Ingredients as Used in Cosmetics. 2021. Available online: https://www.cir-safety.org/sites/default/files/Rosa%20damascena_0.pdf (accessed on 8 March 2022).
29. Jovtchev, G.; Stankov, A.; Georgieva, A.; Dobreva, A.; Bakalova, R.; Aoki, I.; Mileva, M. Cytotoxic and genotoxic potential of Bulgarian *Rosa alba* L. essential oil—*In vitro* model study. *Biotechnol. Biotechnol. Equip.* **2018**, *32*, 513–519. [CrossRef]
30. Gateva, S.; Jovtchev, G.; Chanev, C.; Georgieva, A.; Stankov, A.; Dobreva, A.; Mileva, M. Assessment of anti-cytotoxic, anti-genotoxic and antioxidant potential of Bulgarian *Rosa alba* L. essential oil. *Caryologia* **2020**, *73*, 71–88. [CrossRef]

31. Gateva, S.; Jovtchev, G.; Stankov, A.; Georgieva, A.; Dobreva, A.; Mileva, M. The potential of geraniol to reduce cytotoxic and genotoxic effects of MNNG in plant and human lymphocyte test-systems. *S. Afr. J. Bot.* **2019**, *123*, 170–179. [CrossRef]
32. Lei, Y.; Fu, P.; Jun, X.; Cheng, P. Pharmacological properties of geraniol—A review. *Planta Med.* **2019**, *85*, 48–55. [CrossRef]
33. Carrasco, H.A.; Espinoza, L.C.; Cardile, V.; Gallardo, C.; Cardona, W.; Lombardo, L.; Catalán, K.M.; Cuellar, M.F.; Russo, A. Eugenol and its synthetic analogues inhibit cell growth of human cancer cells (Part I). *J. Braz. Chem. Soc.* **2008**, *19*, 543–548. [CrossRef]
34. Manikandan, P.; Vinothini, G.; Priyadarsini, V.R.; Prathiba, D.; Nagini, S. Eugenol inhibits cell proliferation via NF-κB suppression in a rat model of gastric carcinogenesis induced by MNNG. *Investig. New Drugs.* **2011**, *29*, 110–117. [CrossRef] [PubMed]
35. Gateva, S.; Jovtchev, G.; Angelova, T.; Dobreva, A.; Mileva, M. The anti-genotoxic activity of wastewaters produced after water—steam distillation of Bulgarian *Rosa damascena* Mill. And *Rosa alba* L. essential oils. *Life* **2022**, *12*, 455. [CrossRef]
36. Gerasimova, T.; Topashka-Ancheva, M.; Dobreva, A.; Georgieva, A.; Mileva, M. Evaluation of the genotoxic activity of wastewater obtained after steam distillation of essential oil of Bulgarian *Rosa alba* L.—*In vivo* study. *Rom. Biotechnol. Lett.* **2022**, *27*, 3292–3301. [CrossRef]
37. Künzel, G.; Nicoloff, H. Further results on karyotype reconstruction in barley. *Biol. Zentralbl.* **1979**, *98*, 587–592.
38. Jovtchev, G.; Stergios, M.; Schubert, I. A comparison of N-methyl-N-nitrosourea-induced chromatid aberrations and micronuclei in barley meristems using FISH techniques. *Mutat. Res./Genet. Toxicol. Environ. Mutagenesis* **2002**, *517*, 47–51. [CrossRef]
39. Preston, R.; Dean, B.; Galloway, S.; Holden, H.; McFee, A.; Sheldy, M. Mammalian *in vivo* cytogenetic assay analysis of chromosome aberrations in bone marrow cells. *Mutat. Res./Genet. Toxicol.* **1987**, *189*, 157–165. [CrossRef]
40. OECD. *Test No. 474: Mammalian Erythrocyte Micronucleus Test*; OECD Guidelines for the Testing of Chemicals, Section 4; OECD Publishing: Paris, France, 2016. [CrossRef]
41. Rothfuss, A.; Honma, M.; Czich, A.; Aardema, M.J.; Burlinson, B.; Galloway, S.; Hamada, S.; Kirkland, D.; Heflich, R.H.; Howe, J.; et al. Improvement of *in vivo* genotoxicity assessment: Combination of acute tests and integration into standard toxicity testing. *Mutat. Res.* **2011**, *723*, 108–120. [CrossRef]
42. Mitkovska, V.; Chassovnikarova, T.S.; Atanasov, N.; Dimitrov, H. Environmental genotoxicity evaluation using a micronucleus test and frequency of chromosome aberration in free-living small rodents. *J. BioSci. Biotech.* **2012**, *1*, 67–71.
43. Evans, H. *Handbook of Mutagenicity Test Procedure*; Kilbey, B., Legator, M., Nicols, W., Ramel, C., Eds.; Elsevier Science Publishers BV: Amsterdam, The Netherlands, 1984; pp. 405–427. ISBN 9780444600981.
44. Fenech, M. Cytokinesis-block micronucleus cytome assay. *Nat. Protoc.* **2007**, *2*, 1084–1104. [CrossRef]
45. Darzynkiewicz, Z. Cytochemical probes of cycling and quiescent cells applicable to flow cytometry. In *Techniques in Cell Cycle Analysis (Biological Methods)*; Gray, J.J.W., Ed.; Humana Press: Totowa, NJ, USA, 1987; pp. 272–290. [CrossRef]
46. Jovtchev, G.; Gateva, S.; Stergios, M.; Kulekova, S. Cytotoxic and genotoxic effects of paraquat in *Hordeum vulgare* and human lymphocytes *in vitro*. *Environ. Toxicol.* **2010**, *25*, 294–303. [CrossRef] [PubMed]
47. Schmid, W. The micronucleus test. *Mutat. Res.* **1975**, *31*, 9–15. [CrossRef]
48. Hayashi, M.; Sofuni, T.; Ishidate, M. An application of Acridine Orange fluorescent staining to the micronucleus test. *Mutat. Res. Lett.* **1983**, *120*, 241–247. [CrossRef]
49. Rieger, R.; Michaelis, A.; Schubert, I.; Doebel, P.; Jank, H.W. Non-random intrachromosomal distribution of chromatid aberrations induced by X-rays, alkylating agents and ethanol in *Vicia faba*. *Mutat. Res.* **1975**, *27*, 69–79. [CrossRef]
50. Zamiri, A.; Rakhshandeh, H.; Tayarani-Najaran, Z.; Mousavi, S.H. Study of cytotoxic properties of *Rosa damascena* extract in human cervix carcinoma cell line. *Avicenna J. Phytomed.* **2011**, *1*, 74–77. [CrossRef]
51. Zu, Y.; Yu, H.; Liang, L.; Fu, Y.; Efferth, T.; Liu, X.; Wu, N. Activities of ten essential oils towards Propionibacterium acnes and PC-3, A-549 and MCF-7 cancer cells. *Molecules* **2010**, *15*, 3200. [CrossRef]
52. Gao, X.-M.; Yang, L.-Y.; Huang, X.-Z.; Shu, L.-D.; Shen, Y.-Q.; Hu, Q.-F.; Chen, Z.-Y. ChemInform Abstract: Aurones and Isoaurones from the Flowers of *Rosa damascena* and Their Biological Activities. *ChemInform* **2013**, *87*, 583–589. [CrossRef]
53. Wedler, J.; Rusanov, K.; Atanassov, I.; Butterweck, V.A. Polyphenol-enriched fraction of rose oil distillation wastewater inhibits cell proliferation, migration, and TNF-α-induced VEGF secretion in human immortalized keratinocytes. *Planta Med.* **2016**, *82*, 1000–1008. [CrossRef]
54. Carnesecchi, S.; Langley, K.; Exinger, F.; Gosse, F.; Raul, F. Geraniol, a component of plant essential oils, sensitizes human colonic cancer cells to 5-Fluorouracil treatment. *J. Pharmacol. Exp. Ther.* **2002**, *301*, 625–630. [CrossRef]
55. Carnesecchi, S.; Bras-Goncalves, R.; Bradaia, A.; Zeisel, M.; Gosse, F.; Poupon, M.F.; Raul, F. Geraniol, a component of plant essential oils, modulates DNA synthesis and potentiates 5- fluorouracil efficacy on human colon tumor xenografts. *Cancer Lett.* **2004**, *215*, 53–59. [CrossRef]
56. Bakkali, F.; Averbeck, S.; Averbeck, D.; Idaomar, M. Biological effects of essential oils—A review. *Food Chem. Toxicol.* **2008**, *46*, 446–475. [CrossRef] [PubMed]
57. Vieira, A.; Heidor, R.; Cardozo, M.T.; Scolastici, C.; Purgatto, E.; Shiga, T.M.; Barbisan, L.F.; Ong, T.P.; Moreno, F.S. Efficacy of geraniol but not of β-ionone or their combination for the chemoprevention of rat colon carcinogenesis. *Braz. J. Med. Biol. Res.* **2011**, *44*, 538–545. [CrossRef] [PubMed]
58. Lee, S.; Park, Y.R.; Kim, S.H.; Park, E.J.; Kang, M.J.; So, I.; Chun, J.N.; Jeon, J.-H. Geraniol suppresses prostate cancer growth through down-regulation of E2F8. *Cancer Med.* **2016**, *5*, 2899–2908. [CrossRef] [PubMed]

59. Tabari, M.A.; Youssefi, M.R.; Esfandiari, A.; Benelli, G. Toxicity of β-citronellol, geraniol and linalool from *Pelargonium roseum* essential oil against the West Nile and filariasis vector *Culex pipiens* (Diptera: Culicidae). *Res. Vet. Toxicol.* **2017**, *114*, 36–40. [CrossRef]
60. Politano, V.T.; Diener, R.M.; Christian, M.S.; Hoberman, A.M.; Palmer, A.; Ritacco, G.; Adams, T.B.; Anne Marie Api. Oral and dermal developmental toxicity studies of phenylethyl alcohol in rats. *Int. J.Toxicol.* **2013**, *32*, 32–38. [CrossRef]
61. Wong, Y.S.; Sia, C.M.; Khoo, H.E.; Ang, Y.K.; Chang, S.K.; Yim, H.S. Influence of extraction conditions on antioxidant properties of passion fruit (*Passiflora edulis*) peel. *Acta Sci. Pol. Technol. Aliment.* **2014**, *13*, 257–265. [CrossRef]
62. Doppalapudi, R.S.; Riccio, E.S.; Rausch, L.L.; Shimon, J.A.; Lee, P.S.; Mortelmans, K.E.; Kapetanovic, I.M.; Crowell, J.A.; Mirsalis, J.C. Evaluation of chemopreventive agents for genotoxic activity. *Mutat. Res.* **2007**, *629*, 148–160. [CrossRef]
63. Abdel-Hameed, E.; Bazaid, S.; Sabra, A. Total phenolic, *in vitro* antioxidant activity and safety assessment (acute, sub-chronic and chronic toxicity) of industrial taif rose water by-product in mice. *Der Pharm. Lett.* **2015**, *7*, 251–259.
64. Jeong, M.H.; Seong, N.W.; Lee, J.Y.; Kim, Y.J.; Shin, N.R.; Kim, J.C. *In vitro* and *in vivo* evaluation of the genotoxicity of *Eriobotrya japonica* leaf extract. *Regul. Toxicol. Pharmacol.* **2018**, *99*, 238–243. [CrossRef]

life

Article

Snail Mucus Protective Effect on Ethanol-Induced Gastric Ulcers in Mice

Lubomir Petrov [1], Mihail Kachaunov [1], Albena Alexandrova [1,2,*], Elina Tsvetanova [2], Almira Georgieva [2], Aleksander Dolashki [3], Lyudmila Velkova [3] and Pavlina Dolashka [3]

1 National Sports Academy "Vassil Levski", 23, Acad. Stefan Mladenov Str., Studentski Grad, 1700 Sofia, Bulgaria; dr.lubomir.petrov@gmail.com (L.P.); mihailkachaunov@gmail.com (M.K.)
2 Laboratory of Free Radical Processes, Institute of Neurobiology, Bulgarian Academy of Sciences, 23, Acad. Georgi Bonchev Str., 1113 Sofia, Bulgaria; elinanesta@abv.bg (E.T.); almirageorgieva@gmail.com (A.G.)
3 Institute of Organic Chemistry with Centre of Phytochemistry, Bulgarian Academy of Sciences, 9, Acad. Georgi Bonchev Str., 1113 Sofia, Bulgaria; aleksandar.dolashki@orgchm.bas.bg (A.D.); lyudmila_velkova@abv.bg (L.V.); pda54@abv.bg (P.D.)
* Correspondence: a_alexandrova_bas@yahoo.com

Abstract: Nowadays, an increased interest in natural compounds with preventive or therapeutic potential for various diseases has been observed. Given the involvement of oxidative stress in the pathogenesis of gastric ulcer (GU) and the wide range of bioactive compounds isolated from snails, this study aimed to investigate the protective effect of *Cornu aspersum* (Müller, 1774) mucus on ethanol-induced GUs. Male albino mice were divided into Control, Ethanol, Mucus + Ethanol and Mucus + Omeprazole treated groups. The GUs were induced by administration of 96% ethanol (10 mL/kg, per os). One hour before ulcer induction, the mice of Mucus + Ethanol group were pretreated with mucus (20 mg/kg, per os), and the mice of Mucus + Omeprazole group were pretreated with omeprazole (20 mg/kg, per os). Ethanol administration caused grave lesions of gastric mucosa and a significant decrease of glutathione (GSH) and superoxide dismutase (SOD), catalase, and glutathione reductase (GR) activities. In the animals with mucus or omeprazole pre-administration compared to the Ethanol group, the following were observed: only a small number of hemorrhagic fields, significantly reduced GU index with calculated 73% protection by mucus and 78% protection by omeprazole, and significant recovery of mucosal GSH and SOD and GR activities. In addition, the mucus inhibited *Helicobacter pylori* growth. Thus, the protective effect of *C. aspersum* mucus on both gastric mucosa and gastric antioxidant potential in ethanol-induced GU model suggests that it may serve as a good tool for prevention of this disease.

Keywords: gastric ulcers; *Cornu aspersum* mucus; oxidative stress

Citation: Petrov, L.; Kachaunov, M.; Alexandrova, A.; Tsvetanova, E.; Georgieva, A.; Dolashki, A.; Velkova, L.; Dolashka, P. Snail Mucus Protective Effect on Ethanol-Induced Gastric Ulcers in Mice. *Life* **2022**, *12*, 1106. https://doi.org/10.3390/life12081106

Academic Editor: Aglaia Pappa

Received: 11 June 2022
Accepted: 20 July 2022
Published: 22 July 2022

1. Introduction

Digestive system diseases, such as functional gastrointestinal disorders, gastritis, and peptic ulcers, are ubiquitous worldwide. Although not life threatening in general, these pathological conditions can significantly impair the quality of affected people's life and have a broad negative socioeconomic effect [1]. Moreover, it has been found that gastric ulcer patients are at higher risk of developing gastric cancer [2,3].

Gastric ulcer (GU) disease affects about 10% of the population [1,4]. Several classes of drugs are used in GU treatment: proton pump inhibitors, M1-receptor blockers, and H2-receptor antagonists. However, many side effects are associated with these drugs, such as arrhythmia, hematopoietic changes, erectile dysfunction, and gynecomastia [5]. Therefore, new drugs with less side effects than those currently used are needed.

The regenerative capacity of snail mucus against their own shell and skin injuries attracted human attention, and it has been traditionally used therapeutically against gastrointestinal ulcers since ancient Greece [6,7]. Nowadays, studies have demonstrated the protective effect of mucus from the giant African snail in different models of gastric ulcers: induced by indomethacin, histamine, and prolonged exposure of the experimental mice to cold [8,9]. The results show a pronounced beneficial effect, likely due to the cytoprotective and antispasmodic action of the mucus.

Acute gastric ulcers result most frequently from *H. pylori* infection, alcohol consumption, and non-steroidal anti-inflammatory drugs intake, as well as from psychological and physical stress. All of these stressors may induce oxidative stress that is able to initiate and aggravate gastric disorders [10,11]. Oxidative stress (OS) is a condition that is characterized by an imbalance between pro-oxidant processes induced by reactive oxygen species (ROS) and the capacity of the antioxidant protection of organisms to prevent their excess production or overcome the consequences of their action. Much data has shown that increased ROS formation is not only a trigger of inflammation, but the inflammation itself provokes ROS production in gastric disorders [12,13]. In addition, peptic ulcers and other phenotypes of gastrointestinal diseases (e.g., gastroparesis) are known to be related to antioxidant property dysfunction [10].

It has been found that mucus, isolated from the common garden snail *C. aspersum* (synonym *Helix aspersa*), comprises various substances with different bioactive properties with antioxidant potential [14–16]. Therefore, given the involvement of OS in the pathogenesis of GU, as well as the wide range of bioactive compounds that underlies the broad pharmacological activities of *C. aspersum* mucus, this study aimed to investigate the possible protective effect of *C. aspersum* mucus on ethanol-induced GUs in mice and elucidate its efficacy mechanism in terms of OS measures.

2. Materials and Methods

2.1. Snail Mucus Collection

Crude mucus was collected from *C. aspersum* snails grown in Bulgarian eco-farms using patented technology [17,18] without causing suffering in any snail. In brief, the snails were placed in a special device, where electrical stimulation with low voltage caused snails to release mucus without compromising their biological function. After treatment, the snails were returned to the farm. The obtained fresh extract was homogenized and subjected to centrifugation to remove coarse impurities. The purification of supernatant included several steps of filtration, using filters with smaller pore sizes for each subsequent filtration [17–19]. The 2 mucus fractions with molecular weights below 10 and below 20 kDa were obtained by using ultrafiltration membranes, at 4 °C, as follows: discs from Ultracel® Regenerated Cellulose with pore sizes from 10 kDa NMW, (EMD Millipore Corporation, Billerica, MA, USA) and polyethersulfone membrane with pore sizes from 20 kDa (Microdyn Nadir™ from STERLITECH Corporation, Auburn, AL, USA) [18]. Samples were filtered through a syringe filter (Millipore, Millex-HV, 0.22 μm pore size, hydrophilic PVDF membrane) for removing microorganisms. The purified mucus extracts were stored at 4 °C, as previously reported by Trapella et al., 2018 [16]. Thus, to obtain mucus extract and two peptide fractions, non-invasive techniques—a series of centrifugation and filtrations, as well as ultrafiltration, which ensure the production of intact peptides and proteins—were used.

2.2. SDS-PAGE Electrophoresis

The native fresh extract from mucus was analyzed by sodium dodecyl sulfate-polyacrylamide gel electrophoresis (SDS-PAGE) using 5% stacking gel and 12.5% resolving gel according to the Laemmli method [20]. About 20 μg of the sample was loaded on the gel. Protein standard mixture ranging from 10 kDa to 250 kDa (Precision Plus Protein™ Standard All Blue, Bio-Rad Laboratories, Munich, Germany) was used as a molecular marker. Coomassie Brilliant Blue G-250 staining was used for the visualization.

2.3. MALDI-TOF-MS

The isolated mucus extract was lyophilized and analyzed by MALDI-TOF-TOF mass spectrometry on an AutoflexTM III High-Performance MALDI-TOF & TOF/TOF System

(Bruker Daltonics, Bremen, Germany), which uses a 200 Hz frequency-tripled Nd–YAG laser operating at a wavelength of 355 nm. Analysis was carried out after mixing 2.0 μL of the sample with 2.0 μL of matrix solution (7 mg/mL of 3,5-dimethoxy-4-hydroxycinnamic acid (Sinapinic acid)) in 50% CAN containing 0.1% TFA), but only 1.0 μL of the mixture was spotted on a stainless steel 192-well target plate. The samples were allowed to dry at room temperature before being analyzed. A total of 3500 shots were acquired in the MS mode, and collision energy of 4200 was applied. All spectra were obtained using the positive-ion mode. Standard stainless-steel targets obtained from the manufacturer were employed for all analyses. The spectra in this study were calibrated using instrumental calibration based on the parameters determined from the analysis of standard proteins and peptides.

2.4. Antibacterial Activity of Mucus Extracts against H. pylori

The susceptibility of *H. pylori* to mucus extracts was determined by the agar well diffusion method [21]. Three fractions isolated from *C. aspersum* mucus: Fraction 1—native mucus extract (0.5 mg/mL), Fraction 2—mucus containing components with molecular weight (MW) < 20 kDa (0.48 mg/mL), and Fraction 3—mucus containing components MW < 10 kDa (0.45 mg/mL) have been analyzed. The samples were qualitatively tested according to the growth inhibition assay.

Sterile molten Mueller–Hinton agar at 45 °C, supplemented with 5% defibrinated horse blood, was prepared and mixed with three concentrations (0.25, 0.75, and 1.0 mL per 20 mL agar) of native mucus extract. The mixture was aseptically poured into Petri dishes and allowed to set. A sterile cork-borer with a diameter of 3 mm was used to make equidistant wells in the agar. A hundred microliters (100 μL) of a fresh overnight culture of *H. pylori* was poured into the wells. The prepared samples were incubated in a microaerophilic incubator at 37 ± 1 °C for 24–48 h. The plates were observed for inhibition, and the diameter zones of inhibition were measured to the nearest millimeter. All tests were carried out in three repetitions to ensure accuracy

2.5. Animals

A total of 30 male albino mice, weighing 25–35 g (10 weeks of age), were randomly divided into 4 groups of 10 animals each: Control (untreated mice), Ethanol treated (mice with ethanol-induced gastric ulcers), Mucus + Ethanol treated (mice pretreated with snail mucus and ethanol-induced gastric ulcers), and Omeprazole + Ethanol treated (mice pretreated with omeprazole and ethanol-induced gastric ulcers). Each group was kept in a separate cage in a controlled temperature room (24 ± 2 °C) with a 12 h light/dark cycle. The animals had free access to food and water. The mice were maintained and used in accordance with the guidelines of the Care and Use of Laboratory Animals (US National Institute of Health) and the rules of the Ethics Committee of the Institute of Neurobiology, Bulgarian Academy of Sciences (registration FWA 00003059 by the US Department of Health and Human Services).

2.6. Ulcer Damages

Twenty-four hours prior to the experiment, the feeding of mice was withdrawn; however, the animals had free access to water. One hour before ulcer induction, the mice of the Mucus + Ethanol group were pretreated with snail mucus (20 mg/kg, per os), and the mice of the Omeprazole + Ethanol group were pretreated with omeprazole (20 mg/kg, per os). Gastric ulcers were induced by the administration of 96% ethanol (10 mL/kg, per os) in Ethanol, Mucus + Ethanol, and Omeprazole + Ethanol groups. One hour later, mice from all four groups were euthanized by cervical dislocation under pentobarbital (40 mg/kg, intraperitoneal injection) anesthesia. The stomachs were excised, opened over the greater curvature, and rinsed with saline solution (0.9%). Pictures of stoma mucous were taken using a digital camera; the ulcers were quantified, and the gastric damage area (mm^2) was determined using the image analysis program Image J [22].

2.7. Gastric Ulcer Index

The Gastric Ulcer Index (GUI) for each mouse was calculated according to Ganguly [23]: GUI = (TAGU (mm^2) × 100)/(TGA (mm^2)), where TAGU is the total area of gastric ulcers, and TGA is the total gastric area of each mouse. To calculate the protection percentage (PP) [24], the following formula was used: PP = (GUI control − GUI treated)/(GUI control) × 100.

2.8. Tissue Preparation

A small portion of each stomach was cut and homogenized in cold potassium phosphate buffer (0.05 M, pH 7.4). The homogenate was centrifuged at 3000 rpm for 10 min. A part of the obtained supernatant was used to determine the lipid peroxidation (LPO) level and glutathione (GSH) concentrations. The other part was additionally centrifuged at 12,000 rpm for 20 min, and the supernatant was used for measurement of the activities of the antioxidant enzymes superoxide dismutase (SOD), catalase (CAT), and glutathione reductase (GR).

2.9. Biochemical Analyzes

The OS biomarkers were measured spectrophotometrically using commercially available kits: Lipid Peroxidation (MDA) Assay Kit MAK085, Glutathione Assay Kit CS0260, SOD Assay Kit-WST 19160, Catalase Assay Kit CAT100, and Glutathione Reductase Cellular Activity Assay CRSA (Sigma-Aldrich Co., LLC., St. Louis, MO, USA). The manufacturer's working instructions were strictly followed.

2.10. Statistics

Descriptive statistics, the Shapiro–Wilks test of normality and One-Way ANOVA with Tukey post hoc test were applied using the statistical program GraphPad Prism 7.0. In text, all data are presented as the mean ± standard deviation (SD) and in the figures as the mean ± standard error of measurement (SEM).

3. Results

3.1. Composition of *C. aspersum* Mucus

The analysis by 12.5% SDS-PAGE showed that the mucus is a complex mixture of substances with different molecular weights. Several protein bands in the range of 25–35 kDa, 38–40 kDa, 45–50 kDa; 80–90 kDa, and above 250 kDa were detected by SDS-PAGE (Figure 1A). Moreover, by MALDI-MS analysis in the region of 20–80 kDa, the exact molecular masses of mucus proteins were determined (Figure 1B). In the region above 100 kDa, proteins might correspond to glycoproteins and mucins. More details on the *C. aspersum* mucus composition based on our earlier studies [15,18,19,25–27] are presented in Table 1. A number of new peptides have been isolated from the mucus, and their amino acid sequences were determined by de novo sequencing and by MALDI-TOF-MS/MS analyses; some of them are presented in Table 1. Most of the peptides with MW < 3 kDa were characterized by an amphipathic structure, had a positive net charge, pI < 7.0, and displayed generally hydrophobic surfaces (GRAVY > 0) (Table 1).

Table 1. Characterization of peptides in the mucus of the garden snail, *C. aspersum*, determined by de novo MALDI-MS/MS sequencing (GRAVY—grand average of hydropathicity index; pI—Isoelectric point).

No	Amino Acid Sequence of Peptides	Exper. Mass $[M + H]^+$ Da	Calcul. Monois. Mass. Da	pI *	GRAVY *	Net Charge *	Predicted Activity		
							Antibac. (%)	Antiviral. (%)	Antifungal. (%)
1 [a]	γ-ECG (glutathione)	308.058	307.08	3.27	−0.47	−1/0	63.0	54.0	45.0
2 [a]	LGHDVH	677.32	676.33	5.97	−0.383	−1/0	84.0	76	61.0
3 [a,b]	LLMGPEV	758.41	757.40	4.00	+1.171	−1/0	33.0	30.0	12.0
4 [b]	QSGKSPGFGL	977.50	976.50	8.75	−0.520	0/+1	64.0	8.9	26.0

Table 1. *Cont.*

No	Amino Acid Sequence of Peptides	Exper. Mass [M + H]$^+$ Da	Calcul. Monois. Mass. Da	pI *	GRAVY *	Net Charge *	Predicted Activity		
							Antibac. (%)	Antiviral. (%)	Antifungal. (%)
5 [a]	LFSNQLFN	982.50	981.49	5.52	−0.68	0/0	53.0	38.0	43.0
6 [c]	LLFSGGQFNG	1039.52	1038.51	5.52	+0.420	0/0	74.0	28.0	59.0
7 [d]	DLTLNGLSPK	1057.58	1056.58	5.84	−0.300	−1/+1	12.1	17.0	19.0
8 [d]	MPDGALLGGGD	1059.71	1058.47	3.56	+0.058	−2/0	50.0	52.0	32.0
9 [b]	LPDSWEPGGG	1071.56	1070.47	3.67	−0.882	−2/0	7.7	21.0	6.0
10 [e]	LGDLNAEFAAG	1077.67	1076.51	3.67	+0.409	−2/0	27.0	46.0	30.0
11 [a]	LGLGNGGAGGGLVGG	1155.61	1154.60	5.52	+0.687	0/0	86.0	50.8	61.0
12 [b]	YNGFRPGDCY	1191.49	1190.48	5.83	−1.120	−1/+1	43.0	32.0	55.0
13 [e]	AGVGAGGANPSTYVG	1277.91	1276.60	5.57	+0.260	0/0	25.0	7.5	11.0
14 [e]	GAACNLEDGSCLGV	1308.81	1307.55	3.67	+0.564	−2/0	58.0	58.0	53.0
15 [d]	NLVGGSGGGGRGGANPLG	1496.73	1495.75	9.75	- 0.217	0/+1	66.0	33.7	48.2
16 [d]	GLLGGGGGAGGGGLVGGLLNG	1609.94	1608.86	5.52	+0.776	0/+1	90.0	53.6	65.0
17 [d]	MGGLLGGVNGGGKGGGGPGAP	1666.83	1665.83	8.50	+0.005	0/+1	78.6	52.0	61.5
18 [e]	ASKGCGPGSCPPGDTVAGVG	1716.82	1715.76	5.86	+0.005	−1/+1	25.0	12.0	23.0
19 [f]	LFGGHQGGGLVGGLWRK	1738.99	1737.94	11.0	−0.024	0/+2	75.6	41.0	78.5
20 [e]	ACSLLLGGGGVGGGKGGGGHAG	1739.02	1737.86	8.27	+0.409	0/+1	83.0	49.0	67.0
21 [d]	MLLNAKWAPHSTGPPNA	1804.91	1803.91	8.52	−0.400	0/+1	8.5	11.0	6.9
22 [e]	ACLTPVDHFFAGMPCGGGP	1877.14	1875.81	5.08	+0.542	−1/0	32.0	43.0	20.0
23 [e]	NGLFGGLGGGGHGGGGKGPGEGGG	1909.90	1908.88	6.75	−0.487	−1/+1	90.0	67.0	80.0
24 [e]	LLLLMLGGGLVGGLLGGGGKGGG	1966.24	1965.14	8.75	+1.209	0/+1	92.0	57.0	76.0
25 [e]	PFLLGVGGLLGGSVGGGGGGGGAPL	2023.14	2022.09	5.96	+0.912	0/0	69.0	32.0	38.0
26 [e]	GMVVKHCSAPLDSFAEFAGA	2036.93	2035.95	5.32	+0.565	−2/+1	42.0	20.0	26.0
27 [d]	LPFLGLVGGLLGGSVGGGGGGGGPAL	2136.20	2135.17	5.52	+1.023	0/0	69.1	32.0	38.2
28 [d]	DVESLPVGGLGGGGGGAGGGGLVGG NLGGGAG	2479.20	2478.21	3.67	+0.353	−2/0	55.0	42.0	30.0

* Peptides' physicochemical characteristics (isoelectric points (pI), grand average of hydropathicity (GRAVY), and net charge) were determined using the ExPASy ProtParam tool [24]. [a] Vassilev et al., 2020 [19]; [b] Kostadinova et al., 2018 [15]; [c] Beluhova et al., 2022 [25]; [d] Dolashki et al., 2020 [18]; [e] Topalova et al., 2022 [26]; [f] Velkova et al., 2018 [27].

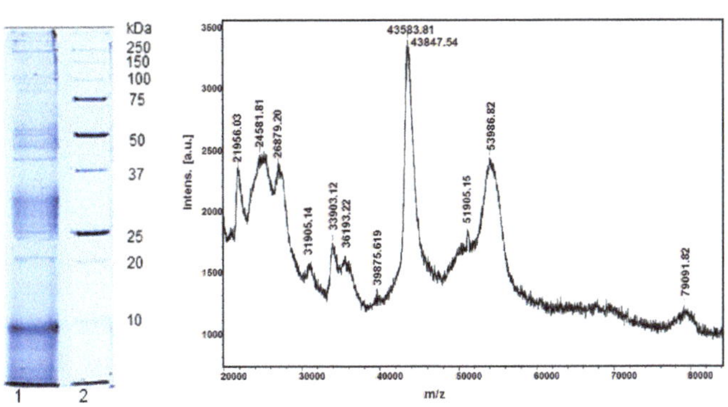

(A) (B)

Figure 1. (**A**) 12.5% SDS–PAGE with Coomassie Brilliant Blue G-250 staining of: 1, crud extract of *C. aspersum* mucus; and 2, standard protein marker (Protein Prestained Standards, Biorad); (**B**) MALDI-TOF-MS spectrum of *C. aspersum* mucus, recorded between 20–80 kDa.

3.2. *In Vitro Susceptibility of H. pylori to Native Mucus Extract*

The antibacterial efficacy of mucus extracts from garden snail *C. aspersum* against bacterial strain *H. pylori* was identified by agar dilution. The growth of the bacterial strain was monitored at 24 and 48 h after treatment with 0.25 mL, 0.75 mL, and 1.0 mL from Fraction 1—native mucus extract, Fraction 2 with MW < 20 kDa, and Fraction 3 with MW < 10 kDa.

The obtained results showed that both fractions, Fraction 2 and Fraction 3, did not affect the growth of *H. pylori*. As shown in Figure 2, only native mucus extract inhibited bacterial growth at 24 and 48 h. *H. pylori* growth was observed in the control sample without extract after 24 h (Figure 2A) and reached its maximum size at 48 h of incubation, with a spot diameter of 10.5 mm (Figure 2B).

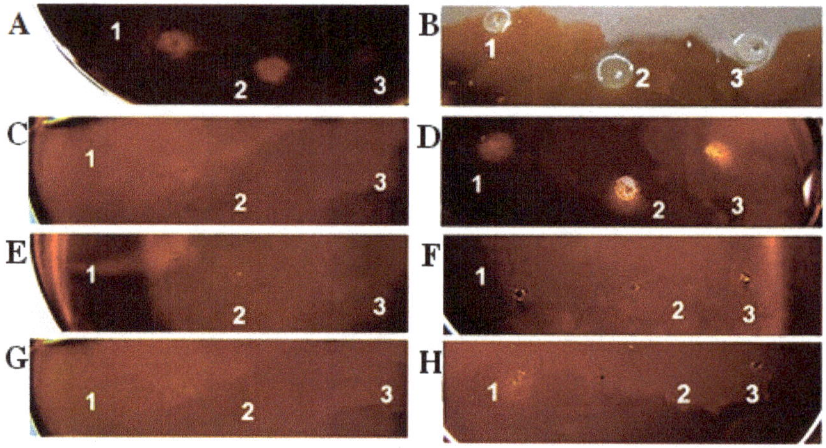

Figure 2. The antibacterial effect on the growth of the bacterial strain *H. pylori* of Fraction 1 in concentrations: (**A**) Control—without extract after 24 h incubation; (**B**) Control—without extract after 48 h incubation; (**C**) 0.25 mL Fraction 1 after 24 h incubation; (**D**) 0.25 mL Fraction 1 after 48 h of incubation; (**E**) 0.75 mL Fraction 1 after 24 h incubation; (**F**) 0.75 mL Fraction 1 after 48 h of incubation; (**G**) 1.0 mL Fraction 1 after 24 h incubation; (**H**) 1.0 mL Fraction 1 after 48 h of incubation.

Testing of Fraction 1 with the lowest concentration (0.25 mL per 20 mL agar) showed significant inhibition of bacterial growth at 24 h (Figure 2C) as opposed to 48 h (Figure 2D). Despite the observed bacterial growth at 48 h, the growth zone decreased from 10.5 mm in diameter to 8.0 mm compared to the control. Antibacterial tests performed with higher concentrations, 0.75 mL and 1.0 mL per 20 mL agar of the natural mucus extract, showed inhibition of bacterial growth at both 24 (Figure 2E,G, respectively) and 48 (Figure 2F,H, respectively) h. Application of Fraction 1 with the highest concentration resulted in the most significant anti-*H. pylori* effect.

Since the application of native mucus extract resulted in the most significant anti-H. pylori effect, the in vivo studies were performed only with it.

3.3. *Effect of C. aspersum Mucus on Ethanol-Induced Gastric Ulcers*

As shown in Figure 3, the animals that had received ethanol (10 mL/kg; 96%) had grave lesions with large hemorrhagic necrosis of the gastric mucosa. In the animals with pre-administration of snail mucus (20 mg/kg) or omeprazole (20 mg/kg), only a small number of hemorrhagic fields were observed.

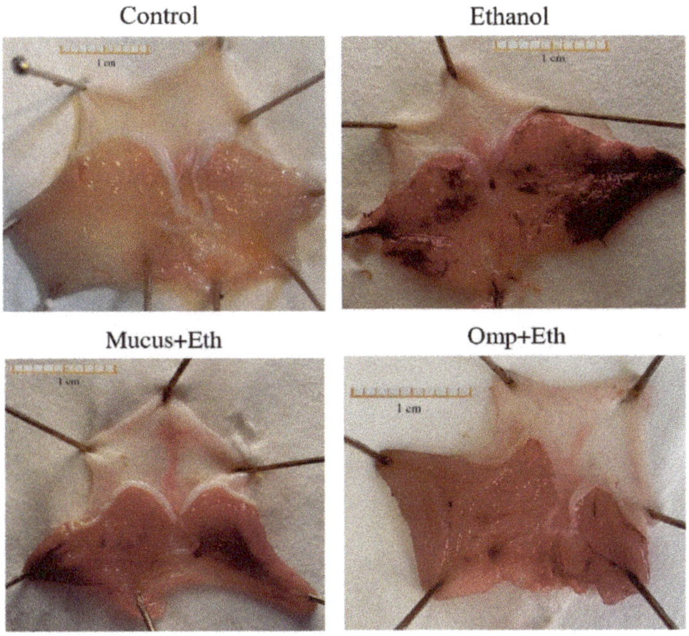

Figure 3. Effect of snail mucus and omeprazole on the macroscopic morphology in mice with ethanol-induced gastric ulcers.

The attenuation of the hemorrhagic mucosal lesions in the snail-mucus-pretreated mice was notable in comparison with the Ethanol group. Figure 4 shows that GUI was significantly reduced in the mice pretreated with snail mucus compared to the Ethanol group (4.7% vs. 17.3%, respectively) ($p < 0.001$). The protection percentage (PP) was calculated to be 73%. The effect of snail mucus was similar to those of omeprazole in which PP was calculated to be 78%.

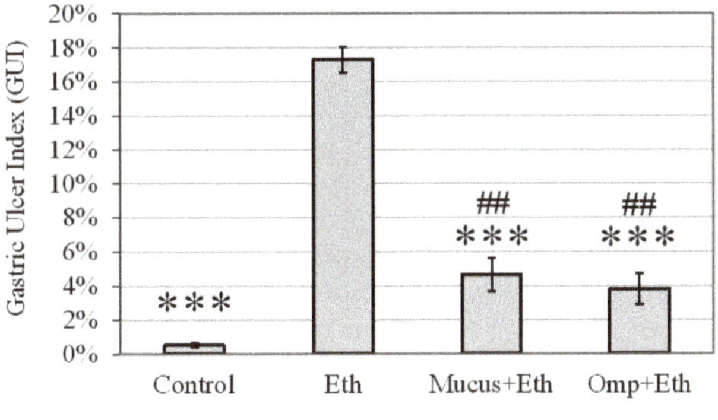

Figure 4. Average Gastric Ulcer Index (GUI) in the four experimental groups: Control, Ethanol (Eth), Mucus + Ethanol (Mucus + Eth), and Omeprazole + Ethanol (Omp + Eth); ***—significance $p < 0.001$ vs. Ethanol group, ##—significance $p < 0.01$ vs. Control group.

3.4. Effect of *C. aspersum* Mucus on Oxidative Stress Parameters

The level of MDA (as an LPO index), the concentration of the GSH, and the activity of antioxidant enzymes (SOD, CAT, and GR) in gastric tissue were evaluated (Figure 5). Gastric MDA was significantly higher in the Ethanol group versus the Control, and the pretreatment with snail mucus led to a decreased MDA concentration. Significant depletion of the GSH concentration in the Ethanol group was observed, and the administration of snail mucus led to the prevention of a decrease in its amount. The activities of the antioxidant enzymes SOD, CAT, and GR in the Ethanol group significantly decreased compared to the Control group. The pretreatment of mice with snail mucus significantly protected SOD and GR activity in relation to the Ethanol group, and CAT activity was slightly (insignificantly) raised. The effect of snail mucus pretreatment was similar to those of omeprazole in terms of all tested parameters of the oxidative status of stomach homogenate of mice with ethanol-induced ulcers.

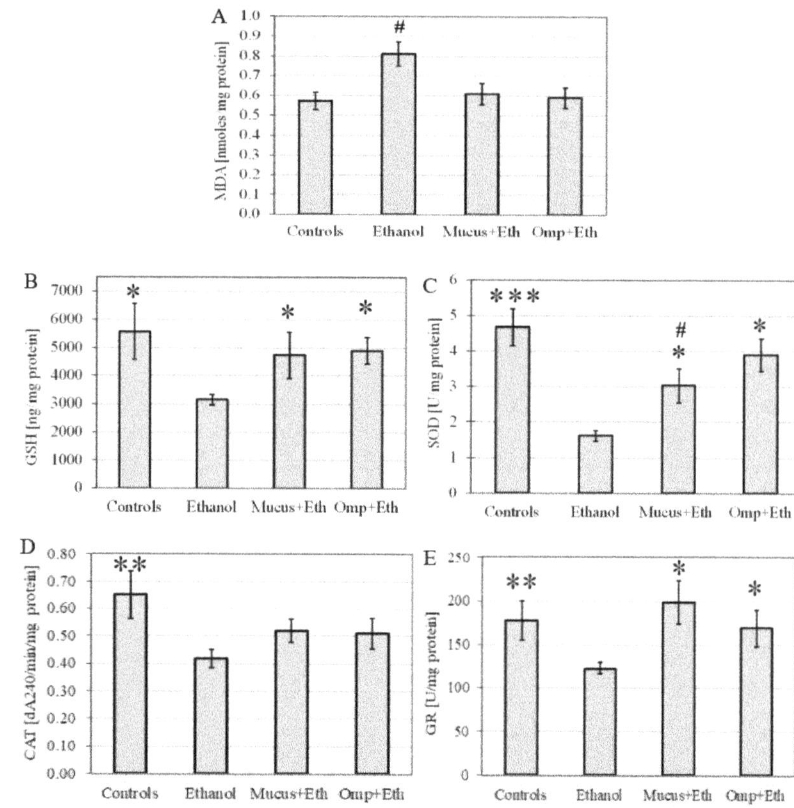

Figure 5. Lipid peroxidation level (**A**); GSH concentration (**B**); and antioxidant enzymes activity of SOD (**C**); CAT (**D**); and GR (**E**); in stomach homogenate from the four experimental groups: Control, Ethanol, Mucus + Ethanol (Mucus + Eth) and Omeprazole + Ethanol (Omp + Eth); *—significance $p < 0.05$ vs. Ethanol group; ** significance—$p < 0.01$ vs. Ethanol group; ***—significance $p < 0.001$ vs. Ethanol group, #—$p < 0.05$ vs. Control group.

4. Discussion

The ethanol-induced GU animal model is a classical method used for screening compounds that possess anti-ulcer activity. The effect of ethanol on gastric mucosa is multifactorial and complicated. Ethanol exfoliates gastric cells and eventually provokes bleeding. The macroscopic observations in this study showed clearly visible bloody lesions after ethanol treatment. These cause disruption of the mucosal barrier integrity and an increase in the mucosal permeability. This model proves that a strong inflammatory response is associated with augmented neutrophil infiltration and myeloperoxidase activation [28], modulation of the NO signaling pathways [29], and perturbation of inflammatory/anti-inflammatory cytokine balance [30]. All these events trigger excess ROS production and cause oxidative damage with cellular deleterious effects [31,32]. The results obtained in this study are in agreement with the published data for increased levels of LPO, reduced GSH concentration, and decreased antioxidant enzymes' activities after ethanol administration [33,34]. In this study, the pretreatment of mice with snail mucus led to a notable gastric lesion reduction along with significant preservation of antioxidant potential, probably due to the suppression of OS development after ulcer induction. The effects of snail mucus were comparable to pretreatment with omeprazole, a proton pump inhibitor, used as a positive control in this study. Omeprazole is a widely used drug for GU treatment and has been applied in numerous studies to provide a gastro-protective effect [35,36]. It has been demonstrated that omeprazole has a dual action in gastrointestinal protection by providing potent antioxidant properties in addition to its major role as an acid-suppression agent [37].

Snail mucus is a complex, evolutionary determined multi-component mixture that includes substances that determine its antioxidant properties. The antioxidant activity could be due to amino acid composition and sequence, the steric structure of the peptides and proteins, glycation of the proteins, and the presence of enzymatic and non-enzymatic antioxidants in snail mucus. The peptide and protein bands detected by SDS-PAGE in this research are in accordance with previously obtained results [18,38,39]. In the MW < 10 kD fraction isolated from the mucus of *C. aspersum*, the amino acid sequences of peptides (Table 1) clearly demonstrate the predominant presence of amino acid residues, such as glycine, proline, leucine, valine, tryptophan, aspartic acid, phenylalanine, and arginine, refs. [40,41] some of which are associated with antioxidant activity [15].

A key factor in the ability of the peptide to trap radicals is considered to be the high proportion of hydrophobic amino acids compared to hydrophilic amino acids [41]. Research has shown that some peptide features, namely, low molecular mass, the presence of antioxidant amino acids such as tyrosine, phenylalanine, proline, alanine, histidine, and leucine; hydrophobicity amino acids (histidine, tryptophan, phenylalanine, proline, lysine, leucine, and valine); indole/imidazole/pyrrolidine ring; along with the steric structure at the C- and N-termini, play an important role for peptide antioxidant properties [41]. Thus, the presence of peptide structures 3, 7, 9, 12, 13, 17, 18, 21, and 22 in the applied snail mucus can be related to its antioxidant activity. Some of these peptides contain three Pro residues (Table 1). Several publications have shown that the higher proline content in substances of plant [41] and animal [42] origin is associated with higher antioxidant activity. Peptide antioxidant activity is determined not only by amino acid composition, but also by amino acid sequences and interactions between amino acids in polypeptide chains, the steric structure at the C- and N-termini, and molecular mass. The higher content of hydrophobic amino acids in peptides determines their good radical scavenging effect in comparison to those with a higher content of hydrophilic amino acids. Low peptide molecular mass also contributes to a significant antioxidant effect [43]. It was suggested that the amphiphilic nature of peptides enhances the radical-scavenging activities of peptides by increasing their solubility and in this way facilitating the interaction and proton exchanges with radical species. A large portion of the amino acids in snail mucus proteins are glycated. In the region above 100 kDa, proteins might correspond to mucins and glycoproteins [44]. The healthy stomach lining produces protective mucins, and it can be assumed that the mucins contained in the snail mucus complement their effect. Furthermore, it is possible that snail mucus coats the gastric mucosa, building an extra layer that protects against ethanol exposure. Concerning glycoproteins, it has been reported that they have antioxidant properties [45–47]. Several studies [45,46] have shown that the addition of purified glycoproteins prevents DNA damage provoked by hydroxyl radicals. However, it is not clarified which part of the glycoproteins—glycan or protein—is responsible for this activity. It has been suggested that there are differences in the radical scavenging ability of glycoproteins that might be attributed to the glycan part or protein part [47].

Antioxidant enzymes such as SOD and CAT, as well as the non-enzymatic antioxidant GSH, detected in the *C. aspersum* mucus [15,18] probably contribute most to its antioxidant potential. SOD and CAT constitute the first line of defense against ROS [48]. The GSH is an essential part of the antioxidant system, able to directly neutralize ROS such as superoxide radicals, lipid peroxyl and hydroxyl radicals, peroxynitrite, and hydrogen peroxide [49]. GSH is involved in the reduction of other antioxidants in the cell such as vitamin C and indirectly vitamin E [50]. In addition, GSH is a mandatory cofactor of glutathione peroxidases (which reduce H_2O_2 and organic peroxides) and glutathione transferases (which reduce organic peroxides) [51]. Oxidized peroxiredoxin that directly reduces peroxides and peroxynitrites is also reduced by GSH [49]. Precise kinetic measurements have shown that peroxiredoxin reduces more than 90% of peroxides in cells [52], surpassing catalase and glutathione peroxidase in this function.

The antioxidant properties of snail mucus are only part of its overall beneficial effect on GUs. The identified in snail mucus hyaluronic acid, chondroitin sulfate, hyaluronan, dermatan sulfate, heparin, and heparin sulfate [53] promote dermal regeneration and epidermal proliferation. The presence of components such as collagen, elastin, glycolic acid, and allantoin as well as the mucopolysaccharide components are associated with the protective effect of snail mucus on the gastric mucosa [53]. Allantoin, collagen, elastin, and glycolic acid that promote collagen synthesis are also found in snail mucus [53]. The collagen and elastin contained in the mucus have a structure close to those in human derma and are in a similar ratio. Therefore, these compounds probably also contribute to the good gastro-protective effect of the snail mucus established in this research.

In addition, in this study, we found that the native mucus extract applied in the highest concentration tested manifested a significant anti-*H. pylori* effect that effectively inhibited bacterial growth at 24 and 48 h, whereas both fractions with MW 10 and MW 20 kDa were not effective. Therefore, the presence of low molecular weight metabolites, peptides, and glycopeptides with antimicrobial activity and antioxidant properties do not sufficiently inhibit the growth of *H. pylori*. This fact leads us to the hypothesis that anti-*H. pyloric* activity is due to the synergy between different components found in the mucus such as secondary metabolites, peptides, proteins, and glycoproteins with antimicrobial activity and the specific ratio between them. Trapella et al. [16] also suggested that snail mucus's potential as a therapeutic agent in wound repair was attributable to the synergistic activity of several molecules. In mucus, such as we have used in the present paper, the primary structures of novel antimicrobial peptides with molecular weights mainly between 1–3 kDa were identified by tandem mass spectrometry (Table 1), and their antibacterial activity was demonstrated [18,26]. The *C. aspersum* mucus contains both cationic and anionic and neutral peptides, but cationic are dominant. That is considered a prerequisite for observed mucus antimicrobial activity—destruction of biological membranes and/or direct cell lysis [18,26]. The antimicrobial activity was predicted based on the identified primary structures using iAMPpred software (online prediction server available at http://cabgrid.res.in:8080/amppred, accessed on 10 June 2022). The presented results (Table 1) showed that some peptides (№ 2, 11, 16, 17, 19, 20, 23–25, and 27 from Table 1) could be predicted to have antimicrobial activity. Most of these peptides have high levels of glycine (Gly) and leucine (Leu) residues as well as one or two proline (Pro) residues and belong to a new class of Gly/Leu-rich antimicrobial peptides [18,27,40]. It is known that cationic antimicrobial peptides kill microbes via mechanisms that predominantly involve interactions between the peptide's positively charged residues and anionic components of the target cell's membranes. Moreover, some of the positively charged peptides may penetrate into the cell to bind intracellular molecules that are crucial to the life of the cell [26]. Recently, different peptide and protein fractions from the *C. aspersum* mucus demonstrated antimicrobial and antioxidant activity as well as regenerative properties [18,19,36]. Dolashki et al. [18] found some bioactive components with MW 10–30 kDa exhibited predominant antibacterial activity against *B. laterosporus* and *E. coli*, and the fraction with MW > 20 kDa manifested a promising antibacterial effect against *C. perfringens*. *H. aspersa* mucus was also effective against three laboratory strains of *P. aeruginosa* [39].

5. Conclusions

In conclusion, the positive effect of *C. aspersum* mucus we found in this model of gastric injury is probably the result of the complex action of many factors, including the protection by molecules with antioxidant properties, substances that ensure tissue regeneration, and compounds with antimicrobial effects which may contribute to the beneficial effect of snail mucus in gastric ulcers and make it an appropriate substance for the prevention of stomach diseases. Obviously, further studies using different injury-inducing models are needed to reveal the particular mechanisms and compounds responsible for the snail mucus's healthy effects.

Author Contributions: Conceptualization, L.P. and A.A.; methodology, L.V., P.D. and L.P.; formal analysis, L.P. and M.K.; investigation, L.P., M.K., A.A., E.T., A.G., A.D. and L.V.; resources, P.D. and A.D.; data curation, L.P. and L.V.; writing—original draft preparation, A.A. and L.P.; writing—review and editing, A.A. and P.D.; visualization, L.P.; supervision, A.A.; funding acquisition, P.D. and A.A. All authors have read and agreed to the published version of the manuscript.

Funding: This research was funded by the Bulgarian Ministry of Education and Science (Grant D01-217/30.11.2018) under the National Research Programme "Innovative Low-Toxic Bioactive Systems for Precision Medicine (BioActiveMed)" approved by DCM#658/14.09.2018.

Institutional Review Board Statement: The experiments were performed according to the "Principles of Laboratory Animal Care" (NIH publication No. 85-23, revised 1985) and the rules of the Ethics Committee of the Institute of Neurobiology, Bulgarian Academy of Sciences (registration FWA 00003059 by the US Department of Health and Human Services.

Informed Consent Statement: Not applicable.

Data Availability Statement: The authors confirm that the data supporting the findings of this study are available within the article. Row data that support the findings of this study are available from the corresponding author upon reasonable request.

Conflicts of Interest: The authors declare no conflict of interest. The funders had no role in the design of the study; in the collection, analyses, or interpretation of data; in the writing of the manuscript, or in the decision to publish the results.

References

1. Wen, Z.; Li, X.; Lu, Q.; Brunson, J.; Zhao, M.; Tan, J.; Wan, C.; Lei, P. Health related quality of life in patients with chronic gastritis and peptic ulcer and factors with impact: A longitudinal study. *BMC Gastroenterol.* **2014**, *14*, 149. [CrossRef] [PubMed]
2. Hansson, L.E.; Nyren, O.; Hsing, A.W.; Bergstrom, R.; Josefsson, S.; Chow, W.H.; Fraumeni, J.F., Jr.; Adami, H.O. The risk of stomach cancer in patients with gastric or duodenal ulcer disease. *N. Engl. J. Med.* **1996**, *335*, 242–249. [CrossRef] [PubMed]
3. Thomopoulos, K.C.; Melachrinou, M.P.; Mimidis, K.P.; Katsakoulis, E.C.; Margaritis, V.G.; Vagianos, C.E.; Nikolopoulou, V.N. Gastric ulcers and risk for cancer. Is follow-up necessary for all gastric ulcers? *Int. J. Clin. Pract.* **2004**, *58*, 675–677. [CrossRef]
4. Xie, X.; Ren, K.; Zhou, Z.; Dang, C.; Zhang, H. The global, regional and national burden of peptic ulcer disease from 1990 to 2019: A population-based study. *BMC Gastroenterol.* **2022**, *22*, 58. [CrossRef]
5. Schubert, M.L. Physiologic, pathophysiologic, and pharmacologic regulation of gastric acid secretion. *Curr. Opin. Gastroenterol.* **2017**, *33*, 430–438. [CrossRef]
6. Bonnemain, B. Helix and drugs: Snails for Western health care from antiquity to the present. *Evid Based Complement Alternat Med.* **2005**, *2*, 25–28. [CrossRef]
7. Quave, C.L.; Pieroni, A.; Bennett, B.C. Dermatological remedies in the traditional pharmacopoeia of Vulture-Alto Bradano, inland southern Italy. *J. Ethnobiol. Ethnomed.* **2008**, *4*, 5. [CrossRef] [PubMed]
8. Adikwu, M.U.; Okolie, C.O.; Agbok, A.A. The effect of snail mucin on the ulcer healing rate of clarithromycin. *J. Pharm. Res.* **2009**, *8*, 6–8. [CrossRef]
9. Nwodo, N.J.; Okonta, J.M.; Attama, A.A. Antiulcer potencial of phylum mollusca (tropical snail) slime. *Asian Pac. J. Trop. Med.* **2009**, *2*, 23–28.
10. Suzuki, H.; Nishizawa, T.; Tsugawa, H.; Mogami, S.; Hibi, T. Roles of oxidative stress in stomach disorders. *J. Clin. Biochem. Nutr.* **2012**, *50*, 35–39. [CrossRef]
11. Tandon, R.; Khanna, H.D.; Dorababu, M.; Goel, R.K. Oxidative stress and antioxidants status in peptic ulcer and gastric carcinoma. *Indian J. Physiol. Pharm.* **2004**, *48*, 115–118.
12. Jasna, D.; Dražen, S. Oxidative stress pathway driven by inflammation in gastric mucosa. In *Gastritis and Gastric Cancer—New Insights in Gastroprotection, Diagnosis and Treatments*; Tonino, P., Ed.; IntechOpen: London, UK, 2011; Available online: https://www.intechopen.com/chapters/19868 (accessed on 10 June 2022).

13. Gloire, G.; Legrand-Poels, S.; Piette, J. NF-kappaB activation by reactive oxygen species: Fifteen years later. *Biochem. Pharm.* **2006**, *72*, 1493–1505. [CrossRef] [PubMed]
14. Dolashka, P.; Dolashki, A.; Velkova, L.; Stevanovic, S.; Molin, L.; Traldi, P.; Velikova, R.; Voelter, W. Bioactive compounds isolated from garden snails. *J. BioSci. Biotechnol.* **2015**, *SE/Online*, 147–155.
15. Kostadinova, N.; Voynikov, Y.; Dolashki, A.; Krumova, E.; Abrashev, R.; Kowalewski, D.; Stevanovic, S.; Velkova, L.; Velikova, R.; Dolashka, P. Antioxidative screening of fractions from the mucus of garden snail Cornu aspersum. *Bulg. Chem. Commun.* **2018**, *50*, 176–183.
16. Trapella, C.; Rizzo, R.; Gallo, S.; Alogna, A.; Bortolotti, D.; Casciano, F.; Zauli, G.; Secchiero, P.; Voltan, R. HelixComplex snail mucus exhibits pro-survival, proliferative and pro-migration effects on mammalian fibroblasts. *Sci. Rep.* **2018**, *8*, 17665. [CrossRef]
17. Dolashka, P. Device for Collecting Garden Snail Extract. BG Utility Model 2097. 2015. Available online: https://portal.bpo.bg/bpo_online/-/bpo/utility-model-detail (accessed on 10 June 2022).
18. Dolashki, A.; Velkova, L.; Daskalova, E.; Zheleva, N.; Topalova, Y.; Atanasov, V.; Voelter, W.; Dolashka, P. Antimicrobial activities of different fractions from mucus of the garden snail Cornu aspersum. *Biomedicines* **2020**, *8*, 315. [CrossRef] [PubMed]
19. Vassilev, N.G.; Simova, S.D.; Dangalov, M.; Velkova, L.; Atanasov, V.; Dolashki, A.; Dolashka, P. An (1)H NMR- and MS-based study of metabolites profiling of garden snail Helix aspersa mucus. *Metabolites* **2020**, *10*, 360. [CrossRef] [PubMed]
20. Laemmli, U.K. Cleavage of structural proteins during the assembly of the head of bacteriophage T4. *Nature* **1970**, *227*, 680–685. [CrossRef]
21. Lawal, T.O.; Igbokwe, C.O.; Adeniyi, B.A. Antimicrobial activities andthe bactericidal kinetics of *Allium ascalonicum* Linn. (wholeplant) against standard and clinical strains of *Helicobacter pylori*: Support for ethnomedical use. *J. Nat. Sci. Res.* **2014**, *4*, 48–56.
22. Khan, H.A. Computer-assisted visualization and quantitation of experimental gastric lesions in rats. *J. Pharm. Toxicol. Methods* **2004**, *49*, 89–95. [CrossRef]
23. Ganguly, A.K. A method for quantitative assessment of experimentally produced ulcers in the stomach of albino rats. *Experientia* **1969**, *25*, 1224. [CrossRef] [PubMed]
24. Adinortey, M.B.; Ansah, C.; Galyuon, I.; Nyarko, A. In vivo models used for evaluation of potential antigastroduodenal ulcer agents. *Ulcers* **2013**, *2013*, 796405. [CrossRef]
25. Belouhova, M.; Daskalova, E.; Yotinov, I.; Topalova, Y.; Velkova, L.; Dolashki, A.; Dolashka, P. Microbial diversity of garden snail mucus. *MicrobiologyOpen* **2022**, *11*, e1263. [CrossRef]
26. Topalova, Y.; Belouhova, M.; Velkova, L.; Dolashki, A.; Zheleva, N.; Daskalova, E.; Kaynarov, D.; Voelter, W.; Dolashka, P. Effect and Mechanisms of Antibacterial Peptide Fraction from Mucus of *C. aspersum* against *Escherichia coli* NBIMCC 8785. *Biomedicines* **2022**, *10*, 672. [CrossRef] [PubMed]
27. Velkova, L.; Nissimova, A.; Dolashki, A.; Daskalova, E.; Dolashka, P.; Topalova, Y. Glycine-rich peptides from *Cornu aspersum* snail with antibacterial activity. *Bulg. Chem. Commun.* **2018**, *50*, 169–175.
28. Tepperman, B.L.; Besco, J.M.; Kiernan, J.A.; Soper, B.D. Relationship between myeloperoxidase activity and the ontogenic response of rat gastric mucosa to ethanol. *Can. J. Physiol. Pharm.* **1991**, *69*, 1882–1888. [CrossRef] [PubMed]
29. Whittle, B.J.R.; Lopez-Belmonte, J. *Gastric Mucosal Damage and Protection: Involvement of Novel Endothelium-Derived Mediators*; Springer: Berlin/Heidelberg, Germany, 1993. [CrossRef]
30. Du, Y.; Zhao, W.; Lu, L.; Zheng, J.; Hu, X.; Yu, Z.; Zhu, L. Study on the antiulcer effects of *Veronicastrum axillare* on gastric ulcer in rats induced by ethanol based on tumor necrosis factor-α (TNF-α) and endothelin-1 (ET-1). *Asian Pac. J. Trop. Biomed.* **2013**, *3*, 925–930. [CrossRef]
31. Arda-Pirincci, P.; Bolkent, S.; Yanardag, R. The role of zinc sulfate and metallothionein in protection against ethanol-induced gastric damage in rats. *Dig. Dis. Sci.* **2006**, *51*, 2353–2360. [CrossRef]
32. Pan, J.S.; He, S.Z.; Xu, H.Z.; Zhan, X.J.; Yang, X.N.; Xiao, H.M.; Shi, H.X.; Ren, J.L. Oxidative stress disturbs energy metabolism of mitochondria in ethanol-induced gastric mucosa injury. *World J. Gastroenterol.* **2008**, *14*, 5857–5867. [CrossRef]
33. Almasaudi, S.B.; El-Shitany, N.A.; Abbas, A.T.; Abdel-dayem, U.A.; Ali, S.S.; Al Jaouni, S.K.; Harakeh, S. Antioxidant, anti-inflammatory, and antiulcer potential of manuka honey against gastric ulcer in rats. *Oxid. Med. Cell. Longev.* **2016**, *2016*, 3643824. [CrossRef]
34. Cuevas, V.M.; Calzado, Y.R.; Guerra, Y.P.; Yera, A.O.; Despaigne, S.J.; Ferreiro, R.M.; Quintana, D.C. Effects of grape seed extract, vitamin C, and vitamin E on ethanol- and aspirin-induced ulcers. *Adv. Pharm. Sci.* **2011**, *2011*, 740687. [CrossRef] [PubMed]
35. Strand, D.S.; Kim, D.; Peura, D.A. 25 Years of Proton Pump Inhibitors: A Comprehensive Review. *Gut Liver* **2017**, *11*, 27–37. [CrossRef] [PubMed]
36. Li, W.S.; Lin, S.C.; Chu, C.H.; Chang, Y.K.; Zhang, X.; Lin, C.C.; Tung, Y.T. The Gastroprotective effect of naringenin against ethanol-induced gastric ulcers in mice through inhibiting oxidative and inflammatory responses. *Int. J. Mol. Sci.* **2021**, *22*, 11985. [CrossRef] [PubMed]
37. Abed, M.N.; Alassaf, F.A.; Jasim, M.H.M.; Alfahad, M.; Qazzaz, M.E. Comparison of antioxidant effects of the proton pump-inhibiting drugs Omeprazole, Esomeprazole, Lansoprazole, Pantoprazole, and Rabeprazole. *Pharmacology* **2020**, *105*, 645–651. [CrossRef] [PubMed]
38. Pitt, S.J.; Graham, M.A.; Dedi, C.G.; Taylor-Harris, P.M.; Gunn, A. Antimicrobial properties of mucus from the brown garden snail Helix aspersa. *Br. J. Biomed. Sci.* **2015**, *72*, 174–181. [CrossRef] [PubMed]

39. Pitt, S.J.; Hawthorne, J.A.; Garcia-Maya, M.; Alexandrovich, A.; Symonds, R.C.; Gunn, A. Identification and characterisation of anti—*Pseudomonas aeruginosa* proteins in mucus of the brown garden snail, *Cornu aspersum*. *Br. J. Biomed. Sci.* **2019**, *76*, 129–136. [CrossRef]
40. Dolashki, A.; Nissimova, A.; Daskalova, E.; Velkova, L.; Topalova, Y.; Hristova, P.; Traldi, P.; Voelter, W.; Dolashka, P. Structure and antibacterial activity of isolated peptides from the mucus of garden snail *Cornu aspersum*. *Bulg. Chem. Commun.* **2018**, *50*, 195–200.
41. Karagozler, A.A.; Erdag, B.; Emek, Y.C.; Uygun, D.A. Antioxidant activity and proline content of leaf extracts from *Dorystoechas hastata*. *Food Chem.* **2008**, *111*, 400–407. [CrossRef]
42. Saidi, S.; Deratani, A.; Belleville, M.P.; Amar, R.B. Antioxidant properties of peptide fractions from tuna dark muscle protein by-product hydrolysate produced by membrane fractionation process. *Food Res. Int.* **2014**, *65*, 329–336. [CrossRef]
43. Zou, T.B.; He, T.P.; Li, H.B.; Tang, H.W.; Xia, E.Q. The Structure-Activity Relationship of the Antioxidant Peptides from Natural Proteins. *Molecules* **2016**, *21*, 72. [CrossRef]
44. Gabriel, U.I.; Mirela, S.; Ionel, J. Quantification of mucoproteins (glycoproteins) from snails mucus, *Helix aspersa* and *Helix pomatia*. *J. Agroaliment. Process. Technol.* **2011**, *17*, 410–413.
45. Kim, E.Y.; Kim, Y.R.; Nam, T.J.; Kong, I.S. Antioxidant and DNA protection activities of a glycoprotein isolated from a seaweed, *Saccharina japonica*. *Int. J. Food Sci. Technol.* **2012**, *47*, 1020–1027. [CrossRef]
46. Rafiquzzaman, S.M.; Kim, E.Y.; Kim, Y.R.; Nam, T.J.; Kong, I.S. Antioxidant activity of glycoprotein purified from *Undaria pinnatifida* measured by an in vitro digestion model. *Int. J. Biol. Macromol.* **2013**, *62*, 265–272. [CrossRef] [PubMed]
47. Senthilkumar, D.; Jayanthi, S. Antioxidant activities of purified glycoprotein extracted from *Codium decorticatum*. *J. Appl. Pharm. Sci.* **2015**, *5*, 101–104. [CrossRef]
48. Ighodaro, O.M.; Akinloye, O.A. First line defence antioxidants-superoxide dismutase (SOD), catalase (CAT) and glutathione peroxidase (GPX): Their fundamental role in the entire antioxidant defence grid. *Alex. J. Med.* **2018**, *54*, 287–293. [CrossRef]
49. Forman, H.J.; Zhang, H.; Rinna, A. Glutathione: Overview of its protective roles, measurement, and biosynthesis. *Mol. Asp. Med.* **2009**, *30*, 1–12. [CrossRef]
50. Meister, A. Glutathione, ascorbate, and cellular protection. *Cancer Res.* **1994**, *54*, 1969s–1975s.
51. Brigelius-Flohe, R.; Maiorino, M. Glutathione peroxidases. *Biochim. Biophys. Acta BBA Gen. Subj.* **2013**, *1830*, 3289–3303. [CrossRef]
52. Adimora, N.J.; Jones, D.P.; Kemp, M.L. A model of redox kinetics implicates the thiol proteome in cellular hydrogen peroxide responses. *Antioxid. Redox Signal.* **2010**, *13*, 731–743. [CrossRef]
53. Cilia, G.; Fratini, F. Antimicrobial properties of terrestrial snail and slug mucus. *J. Complement. Integr. Med.* **2018**, *15*. [CrossRef]

life

Article

Anti-Coronavirus Efficiency and Redox-Modulating Capacity of Polyphenol-Rich Extracts from Traditional Bulgarian Medicinal Plants

Neli Vilhelmova-Ilieva [1], Zdravka Petrova [1,2], Almira Georgieva [1,3], Elina Tzvetanova [1,3], Madlena Trepechova [1] and Milka Mileva [1,*]

[1] The Stephan Angeloff Institute of Microbiology, Bulgarian Academy of Sciences, 26 Georgi Bonchev, 1113 Sofia, Bulgaria; nelivili@gmail.com (N.V.-I.); zdr.z1971@abv.bg (Z.P.); al.georgieva@inb.bas.bg (A.G.); elinaroum@yahoo.com (E.T.); madi_trepechova@yahoo.com (M.T.)
[2] Institute of Morphology, Pathology and Anthropology with Museum, Bulgarian Academy of Sciences, 25 Georgi Bonchev, 1113 Sofia, Bulgaria
[3] Institute of Neurobiology, Bulgarian Academy of Sciences, 23 Acad. G. Bontchev St., 1113 Sofia, Bulgaria
* Correspondence: milkamileva@gmail.com; Tel.: +359-899-151-169

Abstract: Background: The use of various herbal therapists as part of traditional medicine in different parts of the world, including Bulgaria, is due to the knowledge accumulated over the centuries by people about their valuable biological activities. In this study, we investigate extracts from widely used Bulgarian medicinal plants for their ability to prevent the coronavirus infection of cells by testing different mechanisms of antiviral protection, their polyphenol content, and redox-modulating capacity. **Methods:** The influence on the stage of viral adsorption, the inhibition of extracellular virions, and the protective effect on uninfected cells of the plant's extracts were reported by the end-point dilution method, and virus titer (in Δ lgs) was determined as compared to the untreated controls. The total content of polyphenols and flavonoids was also determined. We tested the antioxidant power of the extracts by their ability to inhibit the generation of superoxide anionic radicals and to scavenge DPPH radicals. We determined their iron-reducing, copper-reducing, and metal-chelating antioxidant powers. **Results:** Most of the extracts tested suppress the extracellular virions of HCov. They also inhibit the stage of viral adsorption to the host cell to varying degrees and have a protective effect on healthy cells before being subjected to viral invasion. The examined extracts contained significant levels of polyphenols and quercetin-like flavonoids and showed remarkable antioxidant, radical, and redox-modulating effects. **Conclusions:** All of these 13 extracts from Bulgarian medicinal plants tested can act as antioxidants and antiviral and symptomatic drugs for the management of coronavirus infection.

Keywords: natural extracts; coronavirus infection; virucidal activity; viral adsorption; antiradical and metal-chelating capacity

Citation: Vilhelmova-Ilieva, N.; Petrova, Z.; Georgieva, A.; Tzvetanova, E.; Trepechova, M.; Mileva, M. Anti-Coronavirus Efficiency and Redox-Modulating Capacity of Polyphenol-Rich Extracts from Traditional Bulgarian Medicinal Plants. *Life* **2022**, *12*, 1088. https://doi.org/10.3390/life12071088

Academic Editors: Kousuke Hanada and Francois Lefort

Received: 13 June 2022
Accepted: 18 July 2022
Published: 20 July 2022

1. Introduction

Natural products have been used for millennia for the treatment of human diseases. The diversity of plant species and the knowledge accumulated by people over the centuries for their proper use as herbal medicines makes possible the existence of traditional medicine in different parts of the world. Of the approximately 250,000 flowering plant species described, between 50,000 and 70,000 are known to be used in traditional and modern medicine worldwide [1]. Recently, we have witnessed a resurgence of global interest in herbal medicines. More and more researchers are turning to the study of herbal products in healthcare.

Bulgaria is a country with a small territory and with great potential in terms of herbal drugs. The Bulgarian flora is rich in medicinal plants— it contains more than 4300 plant species, and over 500 are rare or endemic to the countries of the Balkan area. Only 750–770 of all the plant species are medicinal plants, and most of them are wild-growing and occur freely in nature. People collect and use these species that are known for their medicinal properties, mostly in the treatment of viral and bacterial infections.

It is widely known that viral and bacterial infections are the most common causes of disease in human society. A wide range of a specific and nonspecific chemotherapeutics have been developed to treat sundry viral infections, but their use is followed by side effects, which are followed by the development of different degrees of resistant strains that are not susceptible to the therapy used [2].

In retrospect, herbal medicinal plants have been used for thousands of years as a medicinal approach for the treatment of various viral illnesses. Products derived from medicinal plants are more easily absorbed by the body because of their natural origin and show fewer side effects because they are structurally close to normal cellular components [3–5]. The potential for the development of resistant strains to natural antiviral agents is significantly reduced compared to chemotherapeutics due to their complicated chemical structure and too often because of their multistage mode of action.

Many plant sources are also used in cooking as spices and food additives, and so they are involved indirectly in human metabolism through the daily food intake, manifesting their healing effect.

For example, medicinal plants such as *O. basilici*, *A. sativumi*, *T. vulgaris*, *R. canina*, *A. officinalis* and others act in a similar way [6–8]. There is evidence that *A. officinalis* is used purposefully when coughing, *A. annua* is used in "fever" conditions, and in various forms of inflammation *A. hippocastani*, *A. officinalis*, *H. perforati*, *M. chamomillae*, *O. basilicum*, *P. lanceolata*, *P. reptans*, *R. caninae*, *T. vulgaris* and others are recommended [9–16]. *H. perforata*, *R. canina*, and *T. vulgaris* have a neuroprotective and calming effect, and extracts of *A. annua*, *G. glabra*, *T. vulgaris*, *P. reptans*, *R. canina*, and *P. lanceolata* have an antispasmodic effect [11,15–18].

Many extracts of medicinal plants have been studied for their antitumor activity against various forms of cancer (*A. annuae*, *T. vulgaris*, *P. reptans*, *R. caninae*, *P. lanceolata*, etc.) [11,15–18]. Very good antifungal, antibacterial, and antiviral activity were shown in extracts of *P. reptans*, *O. basilicum*, *G. glabra*, and *A. Annuae* [6,16,17,19].

The combination of honey and *O. basilici* against Gram-positive and Gram-negative bacteria shows a significant antibacterial effect [20]. *P. lanceolata* has been shown to have anti-inflammatory activity in *E. coli* infection, probably due to the high content of polyphenols in the extract. [21]. *R. caninae* has an inhibitory effect on multidrug-resistant bacterial strains, including *S. aureus* [22]. Extracts of *H. perforati*, *A. annua*, *P. reptans*, *T. vulgaris*, *G. glabra* and others also show antibacterial activity [9,12,13,16,19].

The activity of some extracts on the replication of various viruses, as well as on the viability of extracellular virions, was also studied. *A. hippocastani* extract showed virucidal and antiviral activity against enveloped HSV-1, VSV, and dengue viruses in vitro, as well as virucidal activity against RSV and antiviral activity in vivo [23]. *A. sativum* is widely used for the prevention and treatment of many viral diseases in humans, animals, and plants by inhibiting viral RNA-pol, reverse transcriptase, and DNA synthesis [24]. *S. nigra* shows activity against influenza virus by blocking viral glycoproteins, as well as against *Infectious bronchitis virus* (IBV)—pathogenic chicken coronavirus [25]. Antiviral activity against RNA viruses, including CoVs—feline infectious peritonitis (FIP)—feline coronavirus, is also exhibited in *T. vulgaris* extract [26]. Extracts of *T. vulgaris* and *O. basilici* show anti-HIV-1 activity, and the same extracts also show anti-HSV-1 activity [6,7]. *G. glabra* extract also demonstrates antiherpes activity [19].

Some of the antibacterial and antiviral activities are due to the immunomodulatory action of the extracts in the multicellular organism [7,10,24,25].

Extracts, essential oils, or tinctures of different parts of plants can suppress the symptoms of some illnesses due to the scavenging of free radicals in a different way or to inhibit their generation. From a chemical point of view, a free radical is a relatively stable structure that contains one or more unpaired electrons. It has the ability with other molecules, either by taking it away from another molecule to increase stability or by donating its unpaired electron to another molecule starting, and a chain process is initiated; one radical gives rise to another radical.

In a biological context, free radicals relate to reactive oxygen species, which are formed at mitochondrial respiration [27]; they are the results of a variety of enzymatic reactions such as NADPH oxidases, xanthine oxidase, nitric oxide synthase, lipoxygenases, and cyclooxygenases; they may be products of ionizing and UV radiation, or by the metabolism of a wide range of drugs and xenobiotics [28]. Stationary (physiological) levels of ROS are intrinsic to the normal functioning of cells, fulfilling the functions of cell signaling and homeostasis [29]. On the other hand, in pathological conditions when they are produced in excess or when cellular defenses are not able to metabolize them, oxidative stress damage occurs. At this moment, antioxidants take their place—(i) endogenous, which are enzymatic or nonenzymatic produced by the body; or (ii) exogenous, which are taken through food, as supplements, or in combination with conventional drugs. As antioxidants can reduce the risk of inflammatory disease development by counterbalancing an excess of reactive species, they are vital for human health [30].

It has been reported that extracts of *A. annua, G. glabra, A. hippocastani, A. officinalis, H. perforati, M. chamomillae, O. basilici, P. lanceolata, P. reptans, R. canina, T. vulgaris*, etc. are rich in polyphenols and especially quercetin-like flavonoids, so they are able to control the symptoms of the so-called "oxidative stress diseases", i.e., inflammations and respiratory virus infections [7,8,11,15–17,19,31–33].

The last three years showed humanity a significant example of the dangers and seriousness of the prevention, treatment, and control of viral diseases via the COVID-19 pandemic situation. The search for new approaches to control the pathogenesis of this virus infection, with extremely diverse symptoms, led us to study the antiviral activities of a panel of Bulgarian medicinal plants that are widely used to relieve symptoms often found in patients with COVID-19 (Table 1).

Table 1. Clearly expressed healing antiviral and symptomatic effects in the context of anti-coronavirus symptoms of the Bulgarian medicinal plants tested.

Plant Species/Drug	Control of the Respiratory Infections					
	Cold	Bronchitis	Cough	Pain	Fever	Viral Diseases
S. nigra	[34]	[35]	[36]	[37,38]	[34]	[34,36,39,40]
A. sativum	[41]	[42]	[43]	[44]	[45]	[24]
P. reptans	[46]	-	-	-	[47]	[47]
R. canina	-	[48]	[48]	[49]	[50]	[51]
M. chamomilla	[52]	[53]	[54]	-	[54]	[52,55]
A. hippocastani	[56]	-	[57]	[58]	[59]	[57]
G. glabra	[60]	[61]	[62]	[63]	[64]	[65,66]
P. lanceolata	[67]	[68]	[69]	[70]	[71]	[67]
H. perforatum	-	[72]	[73]	[74]	[73]	[73,75]
T. vulgaris	[76]	[7]	[76]	[77]	[76]	[78,79]
A. annua	[80]	[81]	[80]	[82]	[83]	[84,85]
A. officinalis	[86]	[86]	[86]	[87]	[88]	[55]
O. basilicum	[89]	[90]	[90]	[91]	[92]	[89,93]

As it turns out, the therapeutic effects of medicinal plants depend first on their ingredients and their solubility in various solvents (water, alcohol, fat, etc.). In turn, their solubility determines the drug forms for their preparation and administration—infusions, decocts, tinctures, syrups, oils, tablets, ointments, and balms or directly on the skin. These herbal products are known among doctors and pharmacists as phytopharmaceuticals that are prepared from different plant parts (roots, stems, leaves, flowers, fruits) [3].

The aim of the present study was to investigate extracts from some widely used Bulgarian medicinal plants: (i) for their ability to prevent the coronavirus infection of cells by testing different mechanisms of antiviral protection; (ii) to be analyzed their polyphenol content, and (iii) to be studied their redox-modulating capacity.

2. Materials and Methods

2.1. Antiviral Investigations

This work was supported by the National Science Fund at the Republic Bulgaria approved by Research Grant No. КП-06-ДК1/3 of the Ministry of Education and Science.

Antiviral research conducted at the Stephan Angeloff Institute of Microbiology, Bulgarian Academy of Sciences, which is a member of the World Organization of Institutes of the Louis Pasteur Network, meets all the necessary requirements for good practices in virology and microbiology, with safety level 2. The researchers certify their responsibilities.

2.1.1. Host Cell Culture

Diploid cell line MCR-5 derived from normal lung tissue was purchased from the American Type Culture Collection (ATCC). Cells were incubated at 37 °C in the presence of 5% CO_2 using Eagle's Minimum Essential Medium (Capricorn Scientific GmbH, Ebsdorfergrund, Germany), supplemented with 10% Fetal bovine serum and (Gibco BRL, St. Louis, MO, USA) and 100 IU penicillin and 0.1 mg streptomycin/mL (Sigma-Aldrich, St. Louis, MO, USA).

2.1.2. Virus

Human coronavirus (HCoV)—229E strain (ATCC: VR-740) strain was replicated in monolayer MRC-5 cells in Eagle's Minimum Essential Medium supplemented with 2% Fetal bovine serum, 100 U/mL penicillin, and 100 μg/mL streptomycin. The cells were incubated with the virus for 5 days at 35 °C and 5% CO_2 and then lysed by double freezing and thawing. The virus was titrated by the method of Reed and Muench. Viral aliquots were stored at −80 °C.

2.1.3. Researched Plant Material

A total of 13 dry water-ethanol (15 °C) extracts of Bulgarian medicinal plants were produced and provided by Extractpharma Ltd.,Sofia, Bulgaria, one of the leading companies in Bulgaria for production and trade in food supplements, herbal products, and liquid and dry extracts of medicinal plants, Sofia, Bulgaria (Table 2).

Table 2. Plant species and cytotoxicity of the extracts.

Plant Species		Area of the Collected Material	Cytotoxicity (μg/mL)	
			CC_{50}	MTC
Sambucus nigra (elderberry)		Fruit	1750 ± 35.2	1000

Table 2. *Cont.*

Plant Species		Area of the Collected Material	Cytotoxicity (μg/mL)	
			CC_{50}	MTC
Allium sativum (garlic)		Root	1700 ± 33.2	1000
Potentilla reptans (Creeping cinquefoil)		Steam	1600 ± 13.6	200
Rosa canina L. (rosehip)		Fruit	1480 ± 32.4	1000
Matricaria chamomilla L. (chamomile)		Flower	820 ± 8.5	1000
Aesculus hippocastanum (horse chestnut)		Seed	1220 ± 23.6	800

Table 2. *Cont.*

Plant Species		Area of the Collected Material	Cytotoxicity (μg/mL)	
			CC_{50}	MTC
Glycyrrhiza glabra L. (licorice)		Root	1700 ± 42.7	1000
Plantago lanceolata (Ribwort plantain)		Stem	1560 ± 41.3	1000
Hypericum perforatum (St. John's Wort)		Steam	830 ± 12.4	500
Thymus vulgaris (thyme)		Steam	880 ± 18.4	320
Artemisia annua (sweet wormwood)		Steam	1150 ± 36.1	320

Table 2. *Cont.*

Plant Species		Area of the Collected Material	Cytotoxicity (µg/mL)	
			CC_{50}	MTC
Althaea officinalis (Marsh Mallow)		Steam	2100 ± 42.1	1500
Ocimum basilicum (basil)		Stem	1250 ± 19.3	1000

2.1.4. Production of Dry Extract

The standardized herbal mixture (according to the recipe) was extracted in an extraction installation on the principle of countercurrent, by performing a classic solid–liquid extraction. The extractant was water-ethanol 15 °C. The extraction was carried out at atmospheric pressure and temperature of 40 °C for 18 h. The resulting liquid herbal extract was concentrated to a certain percentage of dry matter, depending on the type of herb, and dried in dryers: (1) vacuum-drying plant and (2) powder-drying plant to obtain a dry extract. The extracted herb was not reused.

2.1.5. Cytotoxicity Assay

Confluent monolayer cell culture in 96-well plates (Costar®, Corning Inc., Kennebunk, ME, USA) was treated with 0.1 mL/well support medium containing decreasing concentrations of tested extracts. The cells were incubated at 37 °C and 5% CO_2 for 120 h. After microscopic evaluation, the medium containing the test extracts was removed, and the cells were washed and incubated with neutral red at 37 °C for 3 h. After incubation, the neutral-red dye was removed, and the cells were washed with PBS and 0.15 mL/well desorbing solution (1% glacial acetic acid and 49% ethanol in distilled water) was added. The optical density (OD) of each well was read at 540 nm in a microplate reader (Biotek Organon, West Chester, PA, USA). A total of 50% cytotoxic concentration (CC_{50}) was defined as the concentration of the material that reduces cell viability by 50% compared to untreated controls. Each sample was tested in triplicate with four wells for cell culture on a test sample.

The maximum tolerable concentration (MTC) of the extracts was also determined, which is the concentration at which they do not affect the cell monolayer, and in the samples, it looks like the cells in the control sample (untreated with extract).

2.1.6. Virucidal Assay

Samples of 1 mL containing HCoV (with 10^5 cell-culture infectious dose 50 ($CCID_{50}$)) and tested extract in its maximal tolerable concentration (MTC) were contacted in a 1:1 ratio and subsequently stored at room temperature for different time intervals (15, 30, 60, 90, and

120 min). Then, the residual infectious virus content in each sample was determined by the end-point dilution method, and Δlgs as compared to the untreated controls was evaluated.

2.1.7. Effect on Viral Adsorption

Some 24-well plates containing monolayer cell culture from MRC-5 cells were pre-cooled to 4 °C and inoculated with 10^4 $CCID_{50}$ of human coronavirus. In parallel, they are treated with tested extracts at their maximum tolerable concentration (MTC) and incubated at 4 °C for the time of virus adsorption. At various time intervals (15, 30, 60, 90 and 120 min), the cells were washed with PBS to remove both the compound and the unattached virus, then coated with support medium and incubated at 35 °C for 120 h. After freezing and thawing three times, the infectious viral titer of each sample was determined by the final dilution method. Δlg was determined compared to the viral control (untreated with the compounds). Each sample was prepared in four replicates.

2.1.8. Pretreatment of Healthy Cells

Cell monolayers grown in 24-well cell-culture plates (CELLSTAR, Greiner Bio-One) were treated for different time intervals—15, 30, 60, 90, and 120 min at maximum tolerable concentration (MTC) of extracts in maintenance medium (1 mL/well). After the above time intervals, the compounds were removed, and the cells were washed with phosphate buffered saline (PBS) and inoculated with human coronavirus (1000 $CCID_{50}$ in 1 mL/well). After 120 min of adsorption, the nonadsorbed virus was removed, and the cells were coated with a maintenance medium. Samples were incubated at 35 °C for 120 h and, after freezing and thawing three times, infectious virus titers were determined by the final dilution method. Δlg was determined compared to the viral control (untreated with the compounds). Each sample was prepared in four replicates.

2.2. Redox-Modulating Capacity

All reagents used for the preparation of reaction mixtures, namely potassium salts, thiobarbituric acid, riboflavin, methionine, nitro- blue tetrazolium, and 1,1-diphenyl-2-picryl-hydrazyl (DPPH), 2,4,6-tripyrydil-s-triazine, hydrochloric acid, and ferric chloride were obtained from Sigma-Aldrich (Darmstadt, Germany).

2.2.1. Determination of the Total Flavonoid Content

The total flavonoid content (TFC) was determined using the aluminum chloride colorimetric method reported by Gouveia and Castilho (2011) [94], with slight modifications: 0.30 mL of methanol, 0.02 mL of 10% $AlCl_3$, and 0.56 mL of distilled water were combined with 0.5 mL of appropriately diluted samples. For 30 min, the vials were incubated at room temperature. TFC was estimated as micrograms of quercetin equivalents (QE) per gram of plant extract using a 415 nm absorbance.

2.2.2. Determination of the Total Polyphenol Content

The Folin–Ciocalteu assay was used to assess the total phenol content (TPC) of the extracts [95]. To 0.150 mL samples in disposable test tubes were added 0.750 mL Folin–reagent Ciocalteu's (diluted 1:10 with deionized water) and 0.600 mL sodium carbonate (7.5 percent). The tubes were incubated for 10 min at 50 °C, and the absorbance was measured at 760 nanometers. TPC was calculated as microgram gallic acid equivalents (GAE) per gram of plant extract.

2.2.3. Superoxide Anion Radical Generating System ($^{\bullet}O_2^-$)

We used the method of Beauchamp and Fridovich (1971) [96] for the photochemically generation of superoxide anion radicals ($^{\bullet}O_2^-$), in a medium containing 50 mM potassium phosphate buffer, pH 7.8; 1.17×10^{-6} M riboflavin; 0.2 mM methionine; 2×10^{-5} M KCN and 5.6×10^{-5} M nitro-blue tetrazolium (NBT). The NBT reduction by $^{\bullet}O_2^-$ to a blue formazan product, in the absence (control) and the presence of different concentrations

(5; 2.5; 1.25; 0.625; 0.3125 mg/mL) of the tested substances, was measured at 560 nm. The antioxidant capacity of the plant extracts was expressed as a half-maximal inhibitory concentration IC_{50}.

2.2.4. DPPH Radical-Scavenging Assay

The measurement of the DPPH radical-scavenging activity was performed according to Brand-Williams (1995) [97]. The tested substances in different concentrations as follows (5; 2.5; 1.25; 0.625; 0.3125; 0.156; 0.0781; 0.04 mg/mL) reacted with the stable DPPH in methanol solution. The reduction of $DPPH^{\bullet}$ to DPPHH led to changes in color (from deep violet to light yellow), which was read at 517 nm after 30 min incubation at room temperature. A mixture of methanol and sample served as blank, and a mixture of methanol and DPPH radical solution served as control. The scavenging activity percentage (AA%) was determined as follows:

$$AA\% = 100 - [(\text{Abs (sample)} - \text{Abs (blank)} \times 100)/\text{Abs (control)}]$$

2.2.5. Ferric-Reducing Antioxidant Power (FRAP)

The FRAP assay was performed according to Benzie and Strain (1996) [98] with some modifications using the following solutions: (1) 30 mM acetate buffer at pH 3.6; (2) 1 mM TPTZ (2,4,6-Tri(2-pyridyl)-s-triazine) in 40 mM HCl; (3) 1.5 mM $FeCl_3$. Thus, prepared solutions were mixed in the following ratio: 10 parts (1): 1 part (2): 20 parts (3). To a 50 mL sample put in a disposable test tube was added 1.5 mL of reaction mixture: blank–reaction mixture + 50 mL of deionized H_2O. The method is based on the reduction in Fe (III) ion if the sample contains a reductant (antioxidant) to Fe (II) at low pH. The colorless Fe (III)-TPTZ complex is transformed into the blue Fe (II)-TPZ complex after incubation for 4′ at 37 °C. Absorption was measured at 593 nm. The results were expressed as μmol Trolox equivalents per g of extract.

2.2.6. Cupric-Reducing Antioxidant Capacity (CUPRAC)

The cupric-reducing antioxidant capacity (CUPRAC) was performed according to Apak et al. (2004) [99] with some adaptation. In the ratio of 1:1:1, the following solutions were mixed: 10 mM $CuCl_2$ in ddH_2O, 1.0 M ammonium acetate buffer pH 7.0, and 7.5 mM neocuproin (NC) in 96% ethanol. In 96-well plates, 80 μL was pipetted from different concentrations (5; 2.5; 1.25; 0.625; 0.3125 mg/mL) of the tested substances, then 220 μL of the reaction mixture was added. After incubation at 50 °C for 20 min, absorbance was measured at 450 nm against a blank (220 μL reaction mixture and 80 μL H_2O). The standard curve was prepared with Trolox at various concentrations ranging from 0.1 to 1.0 mM, and the results were expressed as μM Trolox equivalent/1 g extract.

2.2.7. Iron-Chelating Power

The reaction of iron (II) ions with ferrozine produces a pink complex with a maximum absorption wavelength of 562 nm. The addition of a sample containing a chelating agent lowers the observed absorbance. Procedure: 0.2 mL of sample solution was mixed with 0.74 mL of 0.1 M sodium/acetate buffer (pH 5.23) and 0.02 mL of 2 mM $FeSO_4$ in 0.2 M HCl. After 10–15 s, 0.04 mL of 5 mM ferrozine was added. After 10 min of staying in the dark, the absorbance was measured. The formula for determining the Fe (II) chelating capability of the tested material is:

$$\text{Activity (\%)} = 100\ (Ac - As)/(Ac),$$

where Ac is the absorbance of the blank probe, instead containing all of the sample −200 μL sodium/acetate buffer. As is the absorbance of the sample solution [100]. The Fe (II)-chelating capability is expressed as mM EDTA equivalent per 1 g extract.

2.3. Statistical Analysis

Data on cytotoxicity were analyzed statistically. The values of CC_{50} were presented as means ± SD. All measurements of redox-modulating activities were made in triplicate and the data in graphics were presented as a means ± standard deviation.

3. Results

3.1. Anticoronavirus Acitvities

3.1.1. Cytotoxicity of the Thirteen Extracts against the MRC-5 Cell Line

The cytotoxicity of the thirteen extracts against the MRC-5 cell line was determined. Cytotoxic concentrations of 50% (CC_{50}) and maximum tolerable concentrations (MTC) of the extracts were determined. When comparing the cytotoxicity of the studied extracts, it was noticed that the lowest cytotoxicity was shown by *A. officinalis* extract with values for CC_{50} = 2100 µg/mL. The cytotoxicities of extracts of *S. nigra* (CC_{50} = 1750 µg/mL), *A. sativum* and *G. glabra* (CC_{50} = 1700 µg/mL), *P. reptans* (CC_{50} = 1600 µg/mL), and *P. lanceolata* were also relatively low (CC_{50} = 1560 µg/mL). For all extracts, the highest cytotoxicity was shown in *M. chamomilla* extract with CC_{50} = 820 µg/mL and *H. perforatum* (CC_{50} = 830 µg/mL). The maximum tolerable concentrations of the extracts required for many of the following experiments were also determined (Table 2).

3.1.2. Virucidal Activity of Extracts against Human Coronavirus Virions

The next step of our investigation was to monitor whether the extracts we studied could protect healthy cells from viral invasion. The subjects of the study were MRC-5 cells and human coronavirus. Three model systems were used. In the first, the direct effect of the extracts on extracellular virions was determined. Four of the extracts showed virucidal activity in the first study time interval (15 min). The most pronounced effect was observed in the extract of *T. vulgaris* (Δlg = 2.5), followed by the extracts of *M. chamomillae* (Δlg = 2.25), *A. sativum* (Δlg = 2.0), and *P. reptans* (Δlg = 1.75), and this effect remains significant at all studied time intervals. In *A. sativum* and *M. chamonillae*, after 90 min of exposure the inhibitory effect is slightly enhanced. At a 30 min exposure time, *G. glabra* extract (Δlg = 1.75) also showed significant activity, which increased with increasing exposure time. At a contact time of 90 min, the extract of *O. basilicum* significantly reduced the viral titer by Δlg = 2.25, and the extract of *A. hippocastani* caused a significant decrease in viral yield after exposure of 120 min.

Extracts of *P. lanceolata*, *H. perforatum*, *A. officinalis* (Δlg = 1.5), and *A. annua* (Δlg = 1.25) showed minimal effects. *S. nigra* extract showed no effect on extracellular viral particles (Table 3).

Table 3. Virucidal activity of extracts against human coronavirus virions.

Extract	Δlg				
	15 min	30 min	60 min	90 min	120 min
Sambucus nigra	0	0	0.5	0.5	0.5
Allium sativum	2.0	2.0	2.0	2.25	2.25
Potentilla reptans	1.75	1.75	1.75	1.75	1.75
Rosa canina L.	1.0	1.0	1.25	1.25	1.25
Matricaria chamomilla L.	2.25	2.25	2.25	2.5	2.5
Aesculus hippocastanum	1.0	1.0	1.25	1.5	1.75
Glycyrrhiza glabra L.	1.5	1.75	1.75	2.0	2.0
Plantago lanceolata	1.5	1.5	1.5	1.5	1.5

Table 3. *Cont.*

Extract	Δlg				
	15 min	30 min	60 min	90 min	120 min
Hypericum perforatum	1.5	1.5	1.5	1.5	1.5
Thymus vulgaris	2.5	2.5	2.5	2.5	2.5
Artemisia annua	1.25	1.25	1.25	1.25	1.25
Althaea officinalis	1.25	1.25	1.5	1.5	1.5
Ocimum basilicum	1.25	1.25	1.5	2.25	2.25

3.1.3. Effect of the Extracts on the Adsorption Step of HCoV Virions to Host Cells

In the second experimental setup, we traced the effect of the extracts on the adsorption step of HCoV virions to host cells. The effect of the extracts was reported at various time intervals. In the first-time interval (15 min) studied, only four of the extracts significantly inhibited viral adsorption, the most notable being the effect of *T. vulgaris* extract, which reduced viral yield by Δlg = 2.5. The activity of *M. chamomilla* extract (Δlg = 2.25), *A. sativum* (Δlg = 2.0) and *P. reptans* (Δlg = 1.75) was also strong. The *S. nigra* extract showed no activity at an exposure time of 15 min, while the other extracts inhibited the adsorption step of HCoV on sensitive cells, but to a lesser extent. As the exposure time increased, the initially reported effect of 15 min was either maintained or slightly enhanced. The extract from *T. vulgaris* at 120 min retained the same activity together with that from chamomile extract (Δlg = 2.5). *A. sativum* slightly enhanced its activity by lowering the viral titer by Δlg = 2.25. The activity of extracts from *G. glabra*, *A. annua* and *A. officinalis* (Δlg = 2.0), as well as *P. reptans* and *O. basilicum* (Δlg = 1.75) is also significant. Extracts of *R. canina*, *A. hippocastani*, *P. lanceolata* and *H. perforatum* weakly inhibited the adsorption step of HCoV, and *S. nigra* extract did not affect this step at any of the observed time intervals (Table 4).

Table 4. Influence of the extracts on the stage of adsorption of HCoV to sensitive MRC-5 cells.

Extract	Δlg				
	15 min	30 min	60 min	90 min	120 min
Sambucus nigra	0	0	0.5	0.5	0.5
Allium sativum	2.0	2.0	2.0	2.25	2.25
Potentilla reptans	1.75	1.75	1.75	1.75	1.75
Rosa canina L.	1.0	1.0	1.25	1.25	1.25
Matricaria chamomilla L.	2.25	2.25	2.25	2.5	2.5
Aesculus hippocastanum	1.0	1.0	1.25	1.25	1.25
Glycyrrhiza glabra L.	1.5	1.75	1.75	2.0	2.0
Plantago lanceolata	1.5	1.5	1.5	1.5	1.5
Hypericum perforatum	1.5	1.5	1.5	1.5	1.5
Thymus vulgaris	2.5	2.5	2.5	2.5	2.5
Artemisia annua	1.25	1.25	1.5	1.5	2.0
Althaea officinalis	1.25	1.25	1.5	1.5	2.0
Ocimum basilicum	1.25	1.25	1.5	1.5	1.75

3.1.4. Protective Effect of the Studied Extracts on Pretreated Healthy Cells

In the next experimental model that we used to monitor the protective effect of the studied extracts, we pretreated healthy cells with the extracts at their maximum tolerable concentration followed by HCoV infection. Almost all tested extracts had varying degrees of protection on uninfected cells, reducing their susceptibility to attachment, entry, and production of new viral progeny in the host cell. The strongest protection for 15 min of treatment was demonstrated by garlic and rosehip extracts (Δlg = 2.5). With close activity were extracts from *P. reptans* (Δlg = 2.25), *A. officinalis* and *O. basilicum* (Δlg = 2.0). The effect of the extract from *A. hippocastani*, *G. glabra*, and *H. perforati* was also significant (Δlg = 1.75). When treated for longer intervals, the protective result of the extracts increased. In the last reported time interval—120 min—the most pronounced was the protective effect of the extract from *A. hippocastani*, which almost completely prevented subsequent coronavirus infection by lowering the viral titer by Δlg = 4.5. Also remarkable was the protection of extracts from *S. nigra* and *M. chamomilla* (Δlg = 3.5), as well as extracts from *A. sativum* and *P. lanceolata* (Δlg = 3.25). The protective activity of *P. reptans* (Δlg = 2.75), *G. glabra*, *H. perforatum*, *A. officinalis* and *O. basilicum* (Δlg = 2.0) is also significant. At this stage of protecting the cell from viral infection, extracts of *T. vulgaris* and *A. annua* did not show significant activity (Table 5).

Table 5. Protective effect of pretreatment of extracts on healthy MRC-5 cells and subsequent HCoV infection.

Extract	Δlg				
	15 min	30 min	60 min	90 min	120 min
Sambucus nigra	1.5	2.0	3.25	3.25	3.5
Allium sativum	2.5	2.5	3.0	3.25	3.25
Potentilla reptans	2.25	2.25	2.5	2.75	2.75
Rosa canina L.	2.5	2.5	2.5	2.5	2.5
Matricaria chamomilla L.	1.0	2.5	3.25	3.25	3.5
Aesculus hippocastanum	1.75	1.75	4.25	4.5	4.5
Glycyrrhiza glabra L.	1.75	1.75	2.0	2.0	2.0
Plantago lanceolata	1.5	1.5	3.25	3.25	3.25
Hypericum perforatum	1.75	1.75	2.0	2.0	2.0
Thymus vulgaris	0.5	0.5	0.5	1.0	1.0
Artemisia annua	0.5	0.5	0.5	1.0	1.0
Althaea officinalis	2.0	2.0	2.0	2.0	2.0
Ocimum basilicum	2.0	2.0	2.0	2.0	2.0

3.2. Content of Polyphenols, Total Polyphenols, and Redox-Modulating Properties

All tested extracts contained significant amounts of polyphenols comparable to gallic acid and quercetin—similar flavonoids—and showed significant antioxidant, radical scavenging, and redox-modulating effects (Table 6). The extracts of *Ocimum basilicum* and *Hypericum perforatum* proved to be the richest in polyphenols in quantities corresponding to 7.2–7.3 mg gallic acid/g extract, followed by *Sambucus nigra* and *Rosa canina* L. They also showed the best iron-reducing effect, together with *Thymus vulgaris*.

Ocimum basilicum extract has proven to be an excellent radical cleansing, chelating metal, iron, and copper-reducing agent. It showed the best properties as an acceptor of DPPH radicals and the lowest IC_{50} as an inhibitor of superoxide anion radical generation. Our results revealed another also highly active redox modulator, the extract of *Rosa canina* L., followed by *Sambucus nigra*, *Potentilla reptans*, *Thymus vulgaris*, and *Artemisia annua*. As can be seen in Table 6, there is a strong relationship between the content of polyphenols, flavonoids, and the manifested redox-modulating capacity of the studied extracts. It has been shown that the healing properties of plant extracts are mainly due to the presence of these compounds, which have the ideal molecular structure of excellent antioxidants, redox modulators, and medicinal substances [101,102].

Table 6. Total polyphenols, total flavonoids content, radical-scavenging, and metal-chelating activities of extracts.

Plant Species	Total Polyphenols * [mg Gallic Acid/g Extract]	Total Flavonoids ** [mg Quercetin/g Extract]	FRAP #	CUPRAC ##	Fe (II) Chelating ++	DPPH Scavenging Activity, %	Inhibition of Superoxide Generation, IC_{50} [mg/mL]
Sambucus nigra	5.92 ± 0.23	0.98 ± 0.05	8.20 ± 0.13	126.80 ± 5.87	1.50 ± 0.03	53.93	2.85
Allium sativum	0.36 ± 0.05	0.12 ± 0.00	-	2.23 ± 0.85	1.07 ± 0.05	5.53	3.02
Potentilla reptans	4.14 ± 0.05	0.84 ± 0.11	7.32 ± 0.77	116.79 ± 13.99	1.56 ± 0.4	50.19	1.72
Rosa canina L.	5.69 ± 1.64	0.44 ± 0.07	14.44 ± 1.46	217.72 ± 9.35	1.83 ± 0.14	48.48	1.12
Matricaria chamomilla L.	2.07 ± 0.04	0.58 ± 0.10	2.07 ± 0.30	30.59 ± 0.00	1.22 ± 0.07	25.14	1.64
Aesculus hippocastanum	2.05 ± 0.10	0.79 ± 0.11	1.41 ± 0.35	110.31 ± 2.00	2.43 ± 0.61	16.26	27.35
Glycyrrhiza glabra L.	3.26 ± 0.06	1.81 ± 0.09	1.65 ± 0.16	135.09 ± 0.00	2.43 ± 0.61	16.26	10.76
Plantago lanceolata	1.77 ± 0.02	0.40 ± 0.10	3.74 ± 0.20	108.22 ± 0.00	0.63 ± 0.12	25.76	0.66
Hypericum perforatum	7.24 ± 0.23	1.60 ± 0.09	12.00 ± 1.43	172.93 ± 16.11	1.89 ± 0.11	64.36	1.05
Thymus vulgaris	4.87 ± 0.08	1.04 ± 0.14	9.70 ± 0.77	141.88 ± 16.95	1.09 ± 0.13	52.53	7.33
Artemisia annua	3.76 ± 0.20	1.05 ± 0.12	6.22 ± 0.54	46.62 ± 11.55	1.26 ± 0.18	48.64	1.1
Althaea officinalis	1.45 ± 0.29	0.36 ± 0.10	1.46 ± 0.70	14.75 ± 2.05	1.17 ± 0.28	8.79	9.49
Ocimum basilicum	7.32 ± 0.25	1.01 ± 0.15	14.24 ± 1.59	235.49 ± 22.50	0.98 ± 0.10	64.36	0.19

Legend: * The total polyphenolic content is calculated from the gallic acid calibration curve and is expressed as µg gallic acid/mg extract; ** The total flavonoid content was calculated from a quercetin calibration curve and was expressed as µg quercetin/mg extract. # Fe (III)-reducing activity is calculated from a Trolox calibration curve and is expressed as µM Trolox equivalent/1 g extract. ## Cu (II) reducing activity is calculated from a Trolox calibration curve and is expressed as µM Trolox equivalent/1 g extract. ++ Fe-chelating activity is calculated from the EDTA calibration curve and is expressed as mM EDTA equivalent/1 g extract; DPPH-capture activity is calculated at sample concentration 0.3125 mg/mL extract.

4. Discussion

To the best of our knowledge, this is the first systematic study on polyphenol analysis, antiviral activity, and redox-modulating capacity activity of these Bulgarian representatives of plants. All products derived from natural sources are a mix of secondary metabolites (alkaloids, flavonoids, organic acids, tannins, polysaccharides, lipids, saponins, glycosides, anthraquinones, terpenes, etc.) [6,11,14,15]. The content of these products is due to the therapeutic effect of herbalists: anticancer, antioxidant, antidiabetic, immunosuppressive, antifungal, anti-inflammatory, antimalarial, antibacterial, antipyretic, antidiabetic, insecticidal, antiviral and others [17,31,33,101–106]. Plant bioactive components can be used as therapeutics with new targets other than those targeted by existing synthetic therapists. In Bulgaria, as well as worldwide, there is an abundance of medicinal plants, but their use as potential antivirals is poorly studied. For example, *A. annua* affects some viral diseases and the artemisinin contained in it shows SARS-CoV-1 and 2 activities [13]. Electron microscopy has shown that the polyphenol contained in *S. nigra* compromises the viral envelopes of infectious bronchitis virus (IBV), pathogenic chicken coronavirus. It is assumed that this is the probable mechanism of action of the extract of this plant.

It is possible that the mechanism of action of the extracts we studied is similar, which explains their virucidal activity and inhibition of stage of viral adsorption. Most likely, the components contained in them affect the structure of the viral envelope, thereby impairing the ability of the virus to attach and enter the host cell. On the other hand, natural metabolites apparently could bind to components of the cell membrane—receptors and other structures necessary for the recognition and penetration of the virus—without adversely affecting the cell. In this way, the cell becomes unsusceptible to the virus. Such a mechanism of action can explain the protective effect that we observed in pretreated cells, as well as the inhibition of the viral adsorption step.

New mutating viruses require new pharmacological approaches, especially during the ongoing COVID-19 pandemic and due to the lack of efficient treatment. Recently, special attention has been paid to studying plants rich in polyphenols that can be efficient against coronavirus infections [6,11,14,15,101–106].

Finding new products to inhibit the penetration and replication of pathogens from the coronavirus family, especially the spread of SARS-CoV-2 (severe acute respiratory syndrome) is of great importance for human health just now. In this case, studied plant extracts from Bulgarian flora that are rich in quercetin-like compounds, gallic acid, and other polyphenols and inhibit the growth cycle of viruses are welcome in therapeutic schemes for treatment and prevention.

It has been proven that polyphenols protect from viral infections, and in case of infection, support the healing process by various mechanisms: (i) they block the entry into the host cells; (i) inhibit the multiplication of the virus; (iii) support anti-inflammatory functions and the human body's defense by modulating immune regulation and inhibiting cytokine storms; (iv) they also inhibit pro-inflammatory cytokines and model cellular immunity, so they act as immunomodulators; (v) support anti-inflammatory functions and the human body's defense by modulating immune regulation and inhibiting cytokine storms; (vi) act as free radical scavengers [101–106].

Both COVID-19 and all viral infections have been reported to cause oxidative stress in infected tissues, following a possible cytokine storm, blood clotting, and exacerbation of hypoxia—a life-threatening systemic inflammatory syndrome involving elevated levels of circulating cytokines—as well as iron overload [107]. Therefore, antioxidant therapy and the removal of excess iron from the body with special drugs can be useful in the prevention of oxidative stress symptoms.

As a rule, antioxidants can have different functions depending on the type of radical species produced and of the target molecule protected. They can be divided into three subgroups according to their defense mechanisms: (i) preventive, (ii) radical scavenger, and (iii) enzyme-repairing.

Plant polyphenols are secondary metabolites with strong antioxidant capacities. Plants synthesize them for their own defense against oxidative stress, but these compounds retain the ability to act as antioxidants ex planta and thus largely contribute to the biological properties of plant-derived drugs and supplements [108]. In this context, the characterization of polyphenols is an important point for characterizing the antioxidant properties of medicinal drugs. Flavonoids, as well as many other plant polyphenols, possess a chemical structure ideal for free-radical scavenging [109]. Their antioxidant properties include reactivity to a variety of reactive oxygen species, as well as metal chelating.

One of the most dangerous effects of overproduction of ROS is ferroptosis. It is a type of iron-dependent, oxidative cell death that can be caused by a variety of factors. Ferroptosis is different from apoptosis but is also the result of dysfunction of antioxidant defense, leading to loss of cellular redox homeostasis [110,111].

An increase in ROS in the presence of iron ions has been shown to lead to ferroptosis. There are data showing that the overgeneration of ROS and ferroptosis can be controlled by treatment with deferoxamine as an iron chelator [112].

The mechanisms of FRAP and CUPRAC methods are based on a single electron transfer. The reducing power of FRAP cannot detect antioxidants that act by radical quenching (H transfer). The FRAP method is based on the reduction of Fe (III) to Fe (II), and the CUPRAC method is based on the reduction of Cu (II) to Cu (I) [113]. When iron (III) and copper (II) are in a "free" form they can catalyze the production of highly toxic hydroxyl radicals by dint of one electron donating, so the revocation of one electron is critical importance for the normal cell's functions [114].

Antioxidants that could chelate and reduce iron (III) ions are potential candidates for controlling ferroptosis and its destructive effects on healthy cells. The high metal-reduction capacity in our study suggests that extracts from Bulgarian medicinal plants can be the first line of defense against copper toxicity and serve as a copper chelator by sequestering the metal in a non-redox-active form.

The analysis of oxidative-stress biomarkers and components of the antioxidative defense system in various forms of disease induced by COVID-19 infection, as well as in different aspects of pathogenicity during the disease, could indicate the role of oxidative stress and can suggest timely intervention using various antioxidants. Correlations of oxidative stress with the immune response could provide insight into the interaction between the immune system and oxidative damage.

This mechanism of antioxidant activity is beneficial to live organisms due to the prevention of oxidative damage to the cellular membranes and is essential for cell survival. According to our investigations, the studied 13 plant extracts might perform an essential detoxification function against ions of copper and iron. This function would be beneficial for maintaining the metal homeostasis and protecting the function of cellular structures against the damaging effects of reactive oxygen species.

5. Conclusions

In this investigation, we screened herbal medicinal plants from Bulgarian flora as potential anti-coronavirus medications. Based on the literature databases, more of them are described by modes of action as symptomatic drugs on the viral infections. We added some data to these findings by the testing different mechanisms of antiviral protection against coronavirus infection by (i) cell cytotoxicity, based on maximum tolerable concentration of extracts; (ii) inhibition of viral attachment and penetration in the host cells on the data of their residual infectious virus content; and (iii) virucidal assay, based on decreasing of virus titer of the samples.

The study of the mechanism of antioxidant activity of drugs and substances supplemented in condition of the diseases associated with viral pathogenesis is beneficial to live organisms due to the prevention of oxidative damage to the cellular membranes, and because it is essential for cell survival. According to our results, the 13 studied plant extracts, along with the typical healing benefits, might perform an essential detoxification function against ions of copper and iron. This function would be beneficial for maintaining metal homeostasis and protecting the function of cellular structures against the damaging effects of reactive oxygen species. Plant extracts from this investigation can act as antioxidants and antiviral and symptomatic drugs for the management of coronavirus infection. We hope this finding will help researchers and clinicians to identify the source of appropriate antiviral drugs from plants in combating coronaviruses, and ultimately, save millions of affected human lives.

Author Contributions: Conceptualization, N.V.-I. and M.M.; methodology, N.V.-I., Z.P., A.G., E.T. and M.T.; software, N.V.-I., M.M. and A.G.; validation, N.V.-I., A.G. and E.T.; formal analysis N.V.-I. and M.M.; investigation, N.V.-I., Z.P., A.G., E.T. and M.T.; resources, N.V.-I. and M.M.; data curation, N.V.-I. and M.M.; writing—original draft preparation, N.V.-I. and M.M.; writing—review and editing, N.V.-I. and M.M.; visualization, N.V.-I., M.M. and A.G.; supervision, N.V.-I.; project administration, N.V.-I. and M.M.; funding acquisition, N.V.-I. All authors have read and agreed to the published version of the manuscript.

Funding: This work was supported by the National Science Fund at the Ministry of Education and Science, Bulgaria approved by Research Grant No. КП-06-ДК1/3 "Biopolymer-based functional platforms for advanced in vitro target and co-delivery of therapeutic payloads for the treatment of coronavirus infection".

Institutional Review Board Statement: Not applicable.

Informed Consent Statement: Not applicable.

Data Availability Statement: Not applicable.

Acknowledgments: The authors thank Extractpharma Ltd. and its manager, Eng. Marin Penkov, for the provided extracts, which are part of a wide range of products of Extractpharma Ltd. and Mirta-Medicus Ltd., offered commercially.

Conflicts of Interest: The authors declare no conflict of interest.

References

1. Schippmann, U.; Leaman, D.; Cunningham, A.B. *Medicinal and Aromatic Plants*; Springer: Dordrecht, The Netherlands, 2006.
2. Piret, J.; Boivin, G. Resistance of herpes simplex viruses to nucleoside analogues: Mechanisms, prevalence, and management. *Antimicrob. Agents Chemother.* **2011**, *55*, 459–472. [CrossRef] [PubMed]
3. Jarić, S.; Kostić, O.; Mataruga, Z.; Pavlović, D.; Pavlović, M.; Mitrović, M.; Pavlović, P. Traditional wound-healing plants used in the Balkan region (Southeast Europe). *J. Ethnopharmacol.* **2018**, *211*, 311–328. [CrossRef] [PubMed]
4. Todorov, D.; Hinkov, A.; Shishkova, K.; Shishkov, S. Antiviral potential of Bulgarian medicinal plants. *Phytochem. Rev.* **2014**, *13*, 525–538. [CrossRef]
5. Mukhtar, M.; Arshad, M.; Ahmad, M.; Pomerantz, R.J.; Wigdahl, B.; Parveen, Z. Antiviral potentials of medicinal plants. *Virus Res.* **2008**, *131*, 111–120. [CrossRef]
6. Chiang, L.C.; Ng, L.T.; Cheng, P.W.; Chiang, W.; Lin, C.C. Antiviral activities of extracts and selected pure constituents of *Ocimum basilicum*. *Clin. Exp. Pharm. Physiol.* **2005**, *32*, 811–816. [CrossRef]
7. Patil, S.M.; Ramu, R.; Shirahatti, P.S.; Shivamallu, C.; Amachawadi, R.G. A systematic review on ethnopharmacology, phytochemistry and pharmacological aspects of *Thymus vulgaris* Linn. *Heliyon* **2021**, *7*, e07054. [CrossRef]
8. Tahmouzi, S.; Salek Nejat, M.R. New infertility therapy effects of polysaccharides from Althaea officinalis leaf with emphasis on characterization, antioxidant and anti-pathogenic activity. *Int. J. Biol. Macromol.* **2020**, *145*, 777–787. [CrossRef]
9. Aygül, A.; Şerbetçi, T. The antibacterial and antivirulent potential of *Hypericum lydium* against *Staphylococcus aureus*: Inhibition of growth, biofilm formation, and hemolytic activity. *Eur. J. Integr. Med.* **2020**, *35*, 101061. [CrossRef]
10. Bonaterra, G.A.; Bronischewski, K.; Hunold, P.; Schwarzbach, H.; Heinrich, E.U.; Fink, C.; Aziz-Kalbhenn, H.; Muller, J.; Kinscherf, R. Anti-inflammatory and Anti-oxidative Effects of Phytohustil((R)) and Root Extract of *Althaea officinalis* L. on Macrophages in vitro. *Front. Pharm.* **2020**, *11*, 290. [CrossRef]
11. Brunakova, K.; Balintova, M.; Henzelyova, J.; Kolarcik, V.; Kimakova, A.; Petijova, L.; Cellarova, E. Phytochemical profiling of several Hypericum species identified using genetic markers. *Phytochemistry* **2021**, *187*, 112742. [CrossRef]
12. de Oliveira, J.R.; de Jesus Viegas, D.; Martins, A.P.R.; Carvalho, C.A.T.; Soares, C.P.; Camargo, S.E.A.; de Oliveira, L.D. *Thymus vulgaris* L. extract has antimicrobial and anti-inflammatory effects in the absence of cytotoxicity and genotoxicity. *Arch. Oral Biol.* **2017**, *82*, 271–279. [CrossRef]
13. Nair, M.S.; Huang, Y.; Fidock, D.A.; Polyak, S.J.; Wagoner, J.; Towler, M.J.; Weathers, P.J. *Artemisia annua* L. extracts inhibit the in vitro replication of SARS-CoV-2 and two of its variants. *J. Ethnopharmacol.* **2021**, *274*, 114016. [CrossRef]

14. Osei Akoto, C.; Acheampong, A.; Boakye, Y.D.; Naazo, A.A.; Adomah, D.H. Anti-inflammatory, antioxidant, and anthelmintic activities of *Ocimum basilicum* (Sweet Basil) fruits. *J. Chem.* **2020**, *2020*, 2153534. [CrossRef]
15. Samuelsen, A.B. The traditional uses, chemical constituents and biological activities of *Plantago major* L. A review. *J. Ethnopharmacol.* **2000**, *71*, 1–21. [CrossRef]
16. Tomczyk, M.; Latte, K.P. Potentilla—A review of its phytochemical and pharmacological profile. *J. Ethnopharmacol.* **2009**, *122*, 184–204. [CrossRef]
17. Jakovljević, M.R.; Grujičić, D.; Vukajlović, J.T.; Marković, A.; Milutinović, M.; Stanković, M.; Vuković, N.; Vukić, M.; Milošević-Djordjević, O. In vitro study of genotoxic and cytotoxic activities of methanol extracts of *Artemisia vulgaris* L. and *Artemisia alba* Turra. *S. Afr. J. Bot.* **2020**, *132*, 117–126. [CrossRef]
18. Niksic, H.; Becic, F.; Koric, E.; Gusic, I.; Omeragic, E.; Muratovic, S.; Miladinovic, B.; Duric, K. Cytotoxicity screening of *Thymus vulgaris* L. essential oil in brine shrimp nauplii and cancer cell lines. *Sci. Rep.* **2021**, *11*, 13178. [CrossRef]
19. Kaur, R.; Kaur, H.; Dhindsa, A.S. Glycyrrhiza glabra: A phytopharmacological review. *Int. J. Pharm. Sci. Res.* **2013**, *4*, 2470–2477.
20. Khalil, A.T.; Khan, I.; Ahmad, K.; Khan, Y.A.; Khan, M.; Khan, M.J. Synergistic antibacterial effect of honey and Herba Ocimi Basilici against some bacterial pathogens. *J. Tradit. Chin. Med.* **2013**, *33*, 810–814. [CrossRef]
21. Zubair, M.; Widen, C.; Renvert, S.; Rumpunen, K. Water and ethanol extracts of Plantago major leaves show anti-inflammatory activity on oral epithelial cells. *J. Tradit. Complement. Med.* **2019**, *9*, 169–171. [CrossRef]
22. Oyedemi, S.O.; Oyedemi, B.O.; Prieto, J.M.; Coopoosamy, R.M.; Stapleton, P.; Gibbons, S. In vitro assessment of antibiotic-resistance reversal of a methanol extract from *Rosa canina* L. *S. Afr. J. Bot.* **2016**, *105*, 337–342. [CrossRef]
23. Salinas, F.M.; Vazquez, L.; Gentilini, M.V.; Donohoe, A.O.; Regueira, E.; Nabaes Jodar, M.S.; Viegas, M.; Michelini, F.M.; Hermida, G.; Alche, L.E.; et al. *Aesculus hippocastanum* L. seed extract shows virucidal and antiviral activities against respiratory syncytial virus (RSV) and reduces lung inflammation in vivo. *Antivir. Res.* **2019**, *164*, 1–11. [CrossRef]
24. Rouf, R.; Uddin, S.J.; Sarker, D.K.; Islam, M.T.; Ali, E.S.; Shilpi, J.A.; Nahar, L.; Tiralongo, E.; Sarker, S.D. Antiviral potential of garlic (*Allium sativum*) and its organosulfur compounds: A systematic update of pre-clinical and clinical data. *Trends Food Sci. Technol.* **2020**, *104*, 219–234. [CrossRef]
25. Torabian, G.; Valtchev, P.; Adil, Q.; Dehghani, F. Anti-influenza activity of elderberry (*Sambucus nigra*). *J. Funct. Foods* **2019**, *54*, 353–360. [CrossRef]
26. Catella, C.; Camero, M.; Lucente, M.S.; Fracchiolla, G.; Sblano, S.; Tempesta, M.; Martella, V.; Buonavoglia, C.; Lanave, G. Virucidal and antiviral effects of *Thymus vulgaris* essential oil on feline coronavirus. *Res. Vet. Sci.* **2021**, *137*, 44–47. [CrossRef]
27. Juan, C.A.; Perez de la Lastra, J.M.; Plou, F.J.; Perez-Lebena, E. The Chemistry of Reactive Oxygen Species (ROS) Revisited: Outlining Their Role in Biological Macromolecules (DNA, Lipids and Proteins) and Induced Pathologies. *Int. J. Mol. Sci.* **2021**, *22*, 4642. [CrossRef]
28. Collin, F. Chemical Basis of Reactive Oxygen Species Reactivity and Involvement in Neurodegenerative Diseases. *Int. J. Mol. Sci.* **2019**, *20*, 2407. [CrossRef]
29. Sies, H.; Jones, D.P. Reactive oxygen species (ROS) as pleiotropic physiological signalling agents. *Nat. Rev. Mol. Cell Biol.* **2020**, *21*, 363–383. [CrossRef] [PubMed]
30. Halliwell, B. Antioxidant characterization. Methodology and mechanism. *Biochem. Pharm.* **1995**, *49*, 1341–1348. [CrossRef]
31. Al-Dabbagh, B.; Elhaty, I.A.; Elhaw, M.; Murali, C.; Al Mansoori, A.; Awad, B.; Amin, A. Antioxidant and anticancer activities of chamomile (*Matricaria recutita* L.). *BMC Res. Notes* **2019**, *12*, 3. [CrossRef] [PubMed]
32. Oliveira, A.I.; Pinho, C.; Sarmento, B.; Dias, A.C. Neuroprotective Activity of Hypericum perforatum and Its Major Components. *Front. Plant Sci.* **2016**, *7*, 1004. [CrossRef] [PubMed]
33. Owczarek, A.; Kolodziejczyk-Czepas, J.; Wozniak-Serwata, J.; Magiera, A.; Kobiela, N.; Wasowicz, K.; Olszewska, M.A. Potential Activity Mechanisms of Aesculus hippocastanum Bark: Antioxidant Effects in Chemical and Biological In Vitro Models. *Antioxidant* **2021**, *10*, 995. [CrossRef] [PubMed]
34. Mahboubi, M. Sambucus nigra (black elder) as alternative treatment for cold and flu. *Adv. Tradit. Med.* **2021**, *21*, 405–414. [CrossRef]
35. Chen, C.; Zuckerman, D.M.; Brantley, S.; Sharpe, M.; Childress, K.; Hoiczyk, E.; Pendleton, A.R. Sambucus nigra extracts inhibit infectious bronchitis virus at an early point during replication. *BMC Vet. Res.* **2014**, *10*, 24. [CrossRef]
36. Wieland, L.S.; Piechotta, V.; Feinberg, T.; Ludeman, E.; Hutton, B.; Kanji, S.; Seely, D.; Garritty, C. Elderberry for prevention and treatment of viral respiratory illnesses: A systematic review. *BMC Complement. Med. Ther.* **2021**, *21*, 112. [CrossRef]
37. Asaadi, K.; Moradi, P.; Amini, K.; Habibi Lahiji, S. Investigating the most effective compounds in medicinal plant of Sambucus nigra in Azarbayjan region. *Iran. J. Plant Physiol.* **2012**, *2*, 485–488.
38. Jabbari, M.; Daneshfard, B.; Emtiazy, M.; Khiveh, A.; Hashempur, M.H. Biological effects and clinical applications of dwarf elder (*Sambucus ebulus* L): A review. *J. Evid.-Based Complement. Altern. Med.* **2017**, *22*, 996–1001. [CrossRef]
39. Zakay-Rones, Z.; Thom, E.; Wollan, T.; Wadstein, J. Randomized study of the efficacy and safety of oral elderberry extract in the treatment of influenza A and B virus infections. *J. Int. Med. Res.* **2004**, *32*, 132–140. [CrossRef]
40. Zakay-Rones, Z.; Varsano, N.; Zlotnik, M.; Manor, O.; Regev, L.; Schlesinger, M.; Mumcuoglu, M. Inhibition of several strains of influenza virus in vitro and reduction of symptoms by an elderberry extract (*Sambucus nigra* L.) during an outbreak of influenza B Panama. *J. Altern. Complement. Med.* **1995**, *1*, 361–369. [CrossRef]

41. Teshome, A.; Adane, A.; Girma, B.; Mekonnen, Z.A. The impact of vitamin D level on COVID-19 infection: Systematic review and meta-analysis. *Front. Public Health* **2021**, *9*, 169. [CrossRef]
42. Redondo, G.L.M.; Loria, A.; Barrantes, J.B.; Navarro, M.P.; Monge, M.C.O.; Vargas, M.J.C. General aspects about *Allium sativum*—A review. *Ars Pharm.* **2021**, *62*, 471–481. [CrossRef]
43. El-Saber Batiha, G.; Magdy Beshbishy, A.; Wasef, L.G.; Elewa, Y.H.A.; Al-Sagan, A.A.; Abd El-Hack, M.E.; Taha, A.E.; Abd-Elhakim, Y.M.; Prasad Devkota, H. Chemical Constituents and Pharmacological Activities of Garlic (*Allium sativum* L.): A Review. *Nutrients* **2020**, *12*, 872. [CrossRef]
44. Kazempor, S.F. The analgesic effects of different extracts of aerial parts of Coriandrum Sativum in mice. *Int. J. Biomed. Sci. IJBS* **2015**, *11*, 23–28.
45. Daka, D. Antibacterial effect of garlic (*Allium sativum*) on Staphyloccus aureus: An in vitro study. *Afr. J. Biotechnol.* **2011**, *10*, 666–669.
46. Ahangar, N.; Bakhshi Jouybari, H.; Davoodi, A.; Shahani, S. Phytochemical Screening and Antinociceptive Activity of the Hydroalcoholic Extract of *Potentilla reptans* L. *Pharm. Biomed. Res.* **2021**, *7*, 277–284. [CrossRef]
47. Uysal, S.; Zengin, G.; Mahomoodally, M.F.; Yilmaz, M.A.; Aktumsek, A. Chemical profile, antioxidant properties and enzyme inhibitory effects of the root extracts of selected Potentilla species. *S. Afr. J. Bot.* **2019**, *120*, 124–128. [CrossRef]
48. Ayati, Z.; Amiri, M.S.; Ramezani, M.; Delshad, E.; Sahebkar, A.; Emami, S.A. Phytochemistry, traditional uses and pharmacological profile of rose hip: A review. *Curr. Pharm. Des.* **2018**, *24*, 4101–4124. [CrossRef]
49. Christensen, R.; Bartels, E.M.; Altman, R.D.; Astrup, A.; Bliddal, H. Does the hip powder of *Rosa canina* (rosehip) reduce pain in osteoarthritis patients?—A meta-analysis of randomized controlled trials. *Osteoarthr. Cartil.* **2008**, *16*, 965–972. [CrossRef]
50. Parandin, R.; Mohammadi, L. Anti-inflammatory and Anti-nociceptive Activities of Hydroalcoholic Extract of *Rosa canina* L. *Fruit Male Mice J. Ardabil. Univ. Med. Sci.* **2019**, *19*, 161–171.
51. Sargin, S.A. Potential anti-influenza effective plants used in Turkish folk medicine: A review. *J. Ethnopharmacol.* **2021**, *265*, 113319. [CrossRef]
52. Raal, A.; Volmer, D.; Soukand, R.; Hratkevits, S.; Kalle, R. Complementary treatment of the common cold and flu with medicinal plants—Results from two samples of pharmacy customers in Estonia. *PLoS ONE* **2013**, *8*, e58642. [CrossRef]
53. Segal, R.; Pilote, L. Warfarin interaction with *Matricaria chamomilla*. *CMAJ* **2006**, *174*, 1281–1282. [CrossRef]
54. Jarrahi, M. An experimental study of the effects of Matricaria chamomilla extract on cutaneous burn wound healing in albino rats. *Nat. Prod. Res.* **2008**, *22*, 422–427. [CrossRef]
55. Azimi, M.; Mojahedi, M.; Mokaberinejad, R.; Hasheminasab, F.S. Ethnomedicine Knowledge of Iranian Traditional Healers and the Novel Coronavirus Disease 2019 (COVID-19). *J. Adv. Med. Biomed. Res.* **2021**, *29*, 238–245. [CrossRef]
56. Ruther, J.; Hilker, M. Attraction of forest cockchafer Melolontha hippocastani to (Z)-3-hexen-1-ol and 1, 4-benzoquinone: Application aspects. *Entomol. Exp. Appl.* **2003**, *107*, 141–147. [CrossRef]
57. Rabinovich, A. Medicinal plants and their application in the USSR. In Proceedings of the International Symposium on Medicinal and Aromatic Plants, XXIII IHC 306, Hungary, Italy, 4 September 1990; pp. 161–168.
58. Dudek-Makuch, M.; Studzińska-Sroka, E. Horse chestnut–efficacy and safety in chronic venous insufficiency: An overview. *Rev. Bras. Farmacogn.* **2015**, *25*, 533–541. [CrossRef]
59. Ianovici, N.; Latis, A.A.; Radac, A.I. Foliar traits of *Juglans regia*, *Aesculus hippocastanum* and *Tilia platyphyllos* in urban habitat. *Rom. Biotechnol. Lett.* **2017**, *22*, 12400–12408.
60. Soltani Delroba, N.; Karamian, R.; Ranjbar, M. Interactive effects of salicylic acid and cold stress on activities of antioxidant enzymes in *Glycyrrhiza glabra* L. *J. Med. Herbs* **2011**, *2*, 7–13.
61. Wahab, S.; Ahmad, I.; Irfan, S.; Siddiqua, A.; Usmani, S.; Ahmad, M.P. Pharmacological Efficacy and Safety of *Glycyrrhiza glabra* in the treatment of respiratory tract infections. *Mini Rev. Med. Chem.* **2021**, 1476—1494. [CrossRef]
62. Thakur, A.K.; Raj, P. Pharmacological perspective of *Glycyrrhiza glabra* Linn: A mini-review. *J. Anal. Pharm. Res.* **2017**, *5*, 00156. [CrossRef]
63. Rahimian, G.; Babaeian, M.; Kheiri, S.; Moradi, M.T.; Rafieian-Kopaei, M. Effect of *Glycyrrhiza glabra* (D-reglis tablet) on pain and defecation of patients with irritable bowel syndrome. *J. Birjand Univ. Med. Sci.* **2010**, *17*, 240–248.
64. Parvaiz, M.; Hussain, K.; Khalid, S.; Hussnain, N.; Iram, N.; Hussain, Z.; Ali, M.A. A review: Medicinal importance of *Glycyrrhiza glabra* L. (Fabaceae family). *Glob. J. Pharm.* **2014**, *8*, 8–13.
65. Alexyuk, P.G.; Bogoyavlenskiy, A.P.; Alexyuk, M.S.; Turmagambetova, A.S.; Zaitseva, I.A.; Omirtaeva, E.S.; Berezin, V.E. Adjuvant activity of multimolecular complexes based on *Glycyrrhiza glabra* saponins, lipids, and influenza virus glycoproteins. *Arch. Virol.* **2019**, *164*, 1793–1803. [CrossRef] [PubMed]
66. Pandey, D.N.; Rastogi, S.; Agarwal, G.G.; Lakhotia, S.C. Influenza like illness related clinical trial on AYUSH-64 requires cautious interpretation. *J. Ayurveda Integr. Med.* **2022**, *13*, 100346. [CrossRef]
67. Adebayo, S.A.; Ondua, M.; Shai, L.J.; Lebelo, S.L. Inhibition of nitric oxide production and free radical scavenging activities of four South African medicinal plants. *J. Inflamm. Res.* **2019**, *12*, 195–203. [CrossRef]
68. Rapior, S.; Weber, H.R.; Françoise, F. Evaluation of Efficacy of Plantago Species for Human Affections of Digestive System. In Proceedings of the iCEPS e-Conference 2021, Toulouse, France, 1–2 April 2021.
69. Ozkol, H.U.; Akdeniz, N.; Ozkol, H.; Bilgili, S.G.; Calka, O. Development of phytophotodermatitis in two cases related to *Plantago lanceolata*. *Cutan. Ocul. Toxicol.* **2012**, *31*, 58–60. [CrossRef]

70. Ondua, M.; Adebayo, S.A.; Shai, L.J.; Lebelo, S.L. The Anti-Inflammatory and Anti-Nociceptive Activities of Some Medicinal Plant Species Used to Treat Inflammatory Pain Conditions in Southern Africa. *UPSpace* **2016**, *8*, 1571–1575.
71. Fayera, S.; Babu, G.N.; Dekebo, A.; Bogale, Y. Phytochemical investigation and antimicrobial study of leaf extract of *Plantago lanceolata*. *Nat. Prod. Chem. Res.* **2018**, *6*, 1–8. [CrossRef]
72. Chen, H.; Muhammad, I.; Zhang, Y.; Ren, Y.; Zhang, R.; Huang, X.; Diao, L.; Liu, H.; Li, X.; Sun, X.; et al. Antiviral activity against infectious bronchitis virus and bioactive components of *Hypericum perforatum* L. *Front. Pharmacol.* **2019**, *10*, 1272. [CrossRef]
73. Dastagir, G.; Ahmed, R.; Shereen, S. Elemental, nutritional, phytochemical and biological evaluation of *Hypericum perforatum* Linn. *Pak. J. Pharm. Sci.* **2016**, *29*, 547–555.
74. Raak, C.; Scharbrodt, W.; Berger, B.; Bussing, A.; Geissen, R.; Ostermann, T. Hypericum perforatum to improve post-operative Pain Outcome after monosegmental Spinal microdiscectomy (HYPOS): A study protocol for a randomised, double-blind, placebo-controlled trial. *Trials* **2018**, *19*, 253. [CrossRef]
75. Mehrbod, P.; Safari, H.; Mollai, Z.; Fotouhi, F.; Mirfakhraei, Y.; Entezari, H.; Goodarzi, S.; Tofighi, Z. Potential antiviral effects of some native Iranian medicinal plants extracts and fractions against influenza A virus. *BMC Complement. Med. Ther.* **2021**, *21*, 246. [CrossRef]
76. Sardari, S.; Mobaiend, A.; Ghassemifard, L.; Kamali, K.; Khavasi, N. Therapeutic effect of thyme (*Thymus vulgaris*) essential oil on patients with COVID-19: A randomized clinical trial. *J. Adv. Med. Biomed. Res.* **2021**, *29*, 83–91.
77. Taherian, A.A.; Babaei, M.; Vafaei, A.A.; Jarrahi, M.; Jadidi, M.; Sadeghi, H. Antinociceptive effects of hydroalcoholic extract of *Thymus vulgaris*. *Pak. J. Pharm. Sci.* **2009**, *22*, 83–89.
78. Hosseinzadeh, S.; Jafarikukhdan, A.; Hosseini, A.; Armand, R. The application of medicinal plants in traditional and modern medicine: A review of *Thymus vulgaris*. *Int. J. Clin. Med.* **2015**, *6*, 635. [CrossRef]
79. Wani, A.R.; Yadav, K.; Khursheed, A.; Rather, M.A. An updated and comprehensive review of the antiviral potential of essential oils and their chemical constituents with special focus on their mechanism of action against various influenza and coronaviruses. *Microb. Pathog.* **2021**, *152*, 104620. [CrossRef]
80. Iqbal, S.; Younas, U.; Chan, K.W.; Zia-Ul-Haq, M.; Ismail, M. Chemical composition of *Artemisia annua* L. leaves and antioxidant potential of extracts as a function of extraction solvents. *Molecules* **2012**, *17*, 6020–6032. [CrossRef]
81. Bussmann, R.W.; Batsatsashvili, K.; Kikvidze, Z.; Khajoei Nasab, F.; Ghorbani, A.; Paniagua-Zambrana, N.Y.; Tchelidze, D. Artemisia absinthium L. *Artemisia annua* L. *Artemisia dracunculus* L. *Artemisia leucodes* Schrenk Artemisia scoparia Waldst. & Kit. *Artemisia vulgaris* L. *Eclipta prostrata* (L.) L. Asteraceae. In *Ethnobotany of the Mountain Regions of Far Eastern Europe: Ural, Northern Caucasus, Turkey, and Iran*; Batsatsashvili, K., Kikvidze, Z., Bussmann, R.W., Eds.; Springer: Cham, Switzerland, 2020; pp. 1–16.
82. Stebbings, S.; Beattie, E.; McNamara, D.; Hunt, S. A pilot randomized, placebo-controlled clinical trial to investigate the efficacy and safety of an extract of Artemisia annua administered over 12 weeks, for managing pain, stiffness, and functional limitation associated with osteoarthritis of the hip and knee. *Clin. Rheumatol.* **2016**, *35*, 1829–1836. [CrossRef]
83. Feng, X.; Cao, S.; Qiu, F.; Zhang, B. Traditional application and modern pharmacological research of *Artemisia annua* L. *Pharmacol. Ther.* **2020**, *216*, 107650. [CrossRef]
84. Haq, F.U.; Roman, M.; Ahmad, K.; Rahman, S.U.; Shah, S.M.A.; Suleman, N.; Ullah, S.; Ahmad, I.; Ullah, W. Artemisia annua: Trials are needed for COVID-19. *Phytother. Res.* **2020**, *34*, 2423–2424. [CrossRef]
85. Lawan, K.M.; Bharti, J.; Kargo, M.A.; Bello, U.R. Impact of medicinal plants on treatment of SARS-CoV, SARS-CoV-2 and influenza virus in India. *Asian J. Pharm. Pharmacol.* **2020**, *6*, 306–311. [CrossRef]
86. Mahboubi, M. Marsh Mallow (*Althaea officinalis* L.) and Its Potency in the Treatment of Cough. *Complement. Med. Res.* **2020**, *27*, 174–183. [CrossRef]
87. Valiei, M.; Shafaghat, A.; Salimi, F. Chemical composition and antimicrobial activity of the flower and root hexane extracts of *Althaea officinalis* in Northwest Iran. *J. Med. Plants Res.* **2011**, *5*, 6972–6976.
88. Kianitalaei, A.; Feyzabadi, Z.; Hamedi, S.; Qaraaty, M. *Althaea officinalis* in Traditional Medicine and modern phytotherapy. *J. Adv. Pharm. Educ. Res.* **2019**, *9*, 155.
89. Bilal, A.; Jahan, N.; Ahmed, A.; Bilal, S.N.; Habib, S.; Hajra, S. Phytochemical and pharmacological studies on *Ocimum basilicum* Linn—A review. *Int. J. Curr. Res. Rev.* **2012**, *4*, 73–83.
90. Eftekhar, N.; Moghimi, A.; Mohammadian Roshan, N.; Saadat, S.; Boskabady, M.H. Immunomodulatory and anti-inflammatory effects of hydro-ethanolic extract of Ocimum basilicum leaves and its effect on lung pathological changes in an ovalbumin-induced rat model of asthma. *BMC Complement. Altern. Med.* **2019**, *19*, 349. [CrossRef]
91. Venancio, A.M.; Onofre, A.S.; Lira, A.F.; Alves, P.B.; Blank, A.F.; Antoniolli, A.R.; Marchioro, M.; Estevam Cdos, S.; de Araujo, B.S. Chemical composition, acute toxicity, and antinociceptive activity of the essential oil of a plant breeding cultivar of basil (*Ocimum basilicum* L.). *Planta Med.* **2011**, *77*, 825–829. [CrossRef] [PubMed]
92. Shahrajabian, M.H.; Sun, W.; Cheng, Q. Chemical components and pharmacological benefits of Basil (*Ocimum basilicum*): A review. *Int. J. Food Prop.* **2020**, *23*, 1961–1970. [CrossRef]
93. Choi, H.J. Chemical Constituents of Essential Oils Possessing Anti-Influenza A/WS/33 Virus Activity. *Osong Public Health Res. Perspect* **2018**, *9*, 348–353. [CrossRef]
94. Gouveia, S.; Castilho, P.C. Characterisation of phenolic acid derivatives and flavonoids from different morphological parts of *Helichrysum obconicum* by a RP-HPLC–DAD-(−)–ESI-MSn method. *Food Chem.* **2011**, *129*, 333–344. [CrossRef]

95. McDonald, S.; Prenzler, P.D.; Antolovich, M.; Robards, K. Phenolic content and antioxidant activity of olive extracts. *Food Chem.* **2001**, *73*, 73–84. [CrossRef]
96. Beauchamp, C.; Fridovich, I. Superoxide dismutase: Improved assays and an assay applicable to acrylamide gels. *Anal. Biochem.* **1971**, *44*, 276–287. [CrossRef]
97. Brand-Williams, W.; Cuvelier, M.E.; Berset, C.L.W.T. Use of a free radical method to evaluate antioxidant activity. *LWT-Food Sci. Technol.* **1995**, *28*, 25–30. [CrossRef]
98. Benzie, I.F.; Strain, J.J. The ferric reducing ability of plasma (FRAP) as a measure of "antioxidant power": The FRAP assay. *Anal. Biochem.* **1996**, *239*, 70–76. [CrossRef] [PubMed]
99. Apak, R.; Güçlü, K.; Özyürek, M.; Karademir, S.E. Novel total antioxidant capacity index for dietary polyphenols and vitamins C and E, using their cupric ion reducing capability in the presence of neocuproine: CUPRAC method. *J. Agric. Food Chem.* **2004**, *52*, 7970–7981. [CrossRef]
100. Venditti, E.; Bacchetti, T.; Tiano, L.; Carloni, P.; Greci, L.; Damiani, E. Hot vs. cold water steeping of different teas: Do they affect antioxidant activity? *Food Chem.* **2010**, *119*, 1597–1604. [CrossRef]
101. Aliyu, A.B.; Koorbanally, N.A.; Moodley, B.; Singh, P.; Chenia, H.Y. Quorum sensing inhibitory potential and molecular docking studies of sesquiterpene lactones from *Vernonia blumeoides*. *Phytochemistry* **2016**, *126*, 23–33. [CrossRef]
102. Kazemian, H.; Ghafourian, S.; Heidari, H.; Amiri, P.; Yamchi, J.K.; Shavalipour, A.; Houri, H.; Maleki, A.; Sadeghifard, N. Antibacterial, anti-swarming and anti-biofilm formation activities of *Chamaemelum nobile* against *Pseudomonas aeruginosa*. *Rev. Soc. Bras. Med. Trop.* **2015**, *48*, 432–436. [CrossRef]
103. Bacha, K.; Tariku, Y.; Gebreyesus, F.; Zerihun, S.; Mohammed, A.; Weiland-Brauer, N.; Schmitz, R.A.; Mulat, M. Antimicrobial and anti-Quorum Sensing activities of selected medicinal plants of Ethiopia: Implication for development of potent antimicrobial agents. *BMC Microbiol.* **2016**, *16*, 139. [CrossRef]
104. Ma, L.; Yao, L. Antiviral Effects of Plant-Derived Essential Oils and Their Components: An Updated Review. *Molecules* **2020**, *25*, 2627. [CrossRef]
105. Reyes-Jurado, F.; Navarro-Cruz, A.R.; Ochoa-Velasco, C.E.; Palou, E.; Lopez-Malo, A.; Avila-Sosa, R. Essential oils in vapor phase as alternative antimicrobials: A review. *Crit. Rev. Food Sci. Nutr.* **2020**, *60*, 1641–1650. [CrossRef]
106. Vilhelmova-Ilieva, N.; Jacquet, R.; Deffieux, D.; Pouysegu, L.; Sylla, T.; Chassaing, S.; Nikolova, I.; Quideau, S.; Galabov, A.S. Anti-Herpes Simplex Virus Type 1 Activity of Specially Selected Groups of Tannins. *Drug Res. (Stuttg)* **2019**, *69*, 374–381. [CrossRef]
107. Kumar, D.S.; Hanumanram, G.; Suthakaran, P.K.; Mohanan, J.; Nair, L.D.V.; Rajendran, K. Extracellular Oxidative Stress Markers in COVID-19 Patients with Diabetes as Co-Morbidity. *Clin. Pract.* **2022**, *12*, 168–176. [CrossRef]
108. Csepregi, K.; Neugart, S.; Schreiner, M.; Hideg, E. Comparative Evaluation of Total Antioxidant Capacities of Plant Polyphenols. *Molecules* **2016**, *21*, 208. [CrossRef]
109. Winterbourn, C.C. Toxicity of iron and hydrogen peroxide: The Fenton reaction. *Toxicol. Lett.* **1995**, *82–83*, 969–974. [CrossRef]
110. Yang, W.S.; SriRamaratnam, R.; Welsch, M.E.; Shimada, K.; Skouta, R.; Viswanathan, V.S.; Cheah, J.H.; Clemons, P.A.; Shamji, A.F.; Clish, C.B.; et al. Regulation of ferroptotic cancer cell death by GPX4. *Cell* **2014**, *156*, 317–331. [CrossRef]
111. Yang, W.S.; Stockwell, B.R. Synthetic lethal screening identifies compounds activating iron-dependent, nonapoptotic cell death in oncogenic-RAS-harboring cancer cells. *Chem. Biol.* **2008**, *15*, 234–245. [CrossRef]
112. Dixon, S.J.; Lemberg, K.M.; Lamprecht, M.R.; Skouta, R.; Zaitsev, E.M.; Gleason, C.E.; Patel, D.N.; Bauer, A.J.; Cantley, A.M.; Yang, W.S.; et al. Ferroptosis: An iron-dependent form of nonapoptotic cell death. *Cell* **2012**, *149*, 1060–1072. [CrossRef]
113. Prior, R.L.; Wu, X.; Schaich, K. Standardized methods for the determination of antioxidant capacity and phenolics in foods and dietary supplements. *J. Agric. Food Chem.* **2005**, *53*, 4290–4302. [CrossRef]
114. Gutteridge, J.M.C.; Halliwell, B. Mini-Review: Oxidative stress, redox stress or redox success? *Biochem. Biophys. Res. Commun.* **2018**, *502*, 183–186. [CrossRef]

life

MDPI

Article

Chromatographic Profile and Redox-Modulating Capacity of Methanol Extract from Seeds of *Ginkgo biloba* L. Originating from Plovdiv Region in Bulgaria

Lubomir Petrov [1], Albena Alexandrova [1,*], Mariana Argirova [2], Teodora Tomova [2], Almira Georgieva [3], Elina Tsvetanova [3] and Milka Mileva [4,*]

[1] National Sports Academy "Vassil Levski", 21, Acad. Stefan Mladenov, Studentski Urad, 1700 Sofia, Bulgaria; dr.lubomir.petrov@gmail.com
[2] Department of Chemical Sciences, Faculty of Pharmacy, Medical University of Plovdiv, 4000 Plovdiv, Bulgaria; mariyana.argirova@mu-plovdiv.bg (M.A.); teodora.tomova@mu-plovdiv.bg (T.T.)
[3] Institute of Neurobiology, Bulgaria Academy of Sciences, 23 Acad. G. Bonchev Str., 1113 Sofia, Bulgaria; al.georgieva@inb.bas.bg (A.G.); elinaroum@yahoo.com (E.T.)
[4] The Stephan Angeloff Institute of Microbiology, Bulgarian Academy of Sciences, 26 Acad. G. Bontchev Str., 1113 Sofia, Bulgaria
* Correspondence: a_alexandrova_bas@yahoo.com (A.A.); milkamileva@gmail.com (M.M.); Tel.: +359-893396570 (A.A.); +359-899151169 (M.M.)

Abstract: Oxidative stress underlies the pathogenesis of many diseases, which determines the interest in natural substances with antioxidant properties. *Ginkgo biloba* L. leaves are well known and widely used in the pharmaceutical industry, but the therapeutic properties of the seeds are less studied. This study aimed to identify the chromatographic profile and to evaluate the antioxidant properties of methanol extract from seeds of *G. biloba* (GBSE). In the GBSE, flavonoids and terpenes were found as terpenes predominated. The GBSE antioxidant capacity determined by 2,2 azino-bis (3-ethylbenzothiazoline-6-sulfonic acid) (ABTS) and 1-diphenyl-2-picrylhydrazyl (DPPH) methods were equal to 1.34% and 0.58% of the activity of reference substance Trolox, respectively. The results of the ferric reducing antioxidant power method showed that the effect of concentration 1 mg/mL (*w*/*v*) GBSE was equal to 7.418 mM $FeSO_4$ used as a standard. The cupric reducing antioxidant capacity activity of the GBSE was found to be 215.39 μmol Trolox/g GBSE and is presented as Trolox equivalent. The metal chelation effect of 1 mg/mL (*w*/*v*) GBSE was equal to that obtained for 0.018 mM EDTA. In conclusion, GBSE showed a good ability to neutralize ABTS and DPPH radicals and could have a beneficial effect in pathological conditions with oxidative stress etiology.

Keywords: *Ginkgo biloba* seeds extract; phytochemical profile; antiradical activity; metal-reducing and metal chelating effect

Citation: Petrov, L.; Alexandrova, A.; Argirova, M.; Tomova, T.; Georgieva, A.; Tsvetanova, E.; Mileva, M. Chromatographic Profile and Redox-Modulating Capacity of Methanol Extract from Seeds of *Ginkgo biloba* L. Originating from Plovdiv Region in Bulgaria. *Life* **2022**, *12*, 878. https://doi.org/10.3390/life12060878

Academic Editor: Stefania Lamponi

Received: 15 May 2022
Accepted: 2 June 2022
Published: 12 June 2022

Publisher's Note: MDPI stays neutral with regard to jurisdictional claims in published maps and institutional affiliations.

1. Introduction

Ginkgo biloba L. is one of the oldest tree species on Earth, having existed for more than 200 million years. The beneficial properties of extracts from the leaves and seeds of the plant have been known and used to treat and maintain the well-being of people since ancient times. *G. biloba* tree is widely cultivated around the world. In Bulgaria, it has been cultivated since the end of the 19th century, and Plovdiv is the city with the most specimens [1,2].

Although the leaves, seeds, and nuts are often used in traditional Chinese medicine, modern research primarily focuses on ginkgo leaf extracts [3]. One of the reasons for avoiding seeds is the unpleasant odor emitted by ripened fleshy seed coats. Nevertheless, the seeds are rich in secondary metabolites, which are considered responsible for the beneficial properties of ginkgo leaves [4]. The study of the pharmacological properties of seed extracts may offer new perspectives for the use of these neglected sources of biologically active substances.

Ginkgo biloba leaf extracts have been found to improve mainly attention/concentration and memory, and they are part of several medications and supplements with neuroprotective effects [5,6]. Along with the possibility of using extracts for the treatment of neurodegenerative disorders such as dementia [7], Alzheimer's disease [7,8], and Parkinson's disease [9], pronounced positive effects on anxiety, schizophrenia, and depression have been found [10]. Extracts from the leaves of the plant improve cardiovascular and peripheral vascular disorders by improving blood circulation, strengthening capillary walls, and reducing clot formation [11,12]. Modern pharmacological studies have revealed that *G. biloba* extracts and individual substances extracted from the plant have an antitumor effect [13]. In addition, they are used against asthma and bronchitis [14] and have an anti-inflammatory effect [15].

The broad spectrum of compounds is responsible for the therapeutic effects of *G. biloba* extracts [16]. Among them, flavonol glycosides and terpene trilactones are present in high concentrations [17]. The significant levels of flavonoids and terpenoids determine the manifestation of antioxidant and radical-scavenger properties of extracts. These activities are based on their ability to be excellent donors of hydrogen, which is accepted by reactive radicals to yield much less active radical and non-radical species [18]. The interest in various substances with antioxidant properties stems from the participation of oxidative stress (OS) in many pathologies. Oxidative stress is a condition characterized by an increase in pro-oxidant processes and/or a decrease in antioxidant protection, leading to a pro/antioxidant imbalance in cells with consequences at higher organism levels. In living cells under normal physiological conditions, there is a balance between the rate of generation of reactive oxygen species (ROS) and their neutralization by the cell's antioxidants [19,20]. The redox state of a cell determines its cellular functioning and is usually kept within a narrow range under normal conditions [19]. The fine balance between the useful and detrimental effects of ROS is due to the metabolic reactions that use oxygen and constitute a very important part of protecting living organisms and maintaining "redox homeostasis" by controlling the redox regulation in vivo [21]. In principle, the antioxidant effects of complex plant extracts can include various mechanisms of antioxidant defense such as free radical scavenging, the termination of oxidative chain reactions, reducing capacity, and binding of pro-oxidant metal ions [22]. Numerous studies have shown that a wide range of diseases such as cardiovascular, neurodegenerative, metabolic, pulmonary, renal, and neoplasms have OS etiology [23] and the application of antioxidants from plant or animal origin has beneficial effects; therefore, the search for new sources of antioxidants with therapeutic potential is important for human health.

The review of the existing data shows that most studies have focused on the effects of *G. biloba* leaf extracts, while the pharmacological effects of seed extracts have been poorly studied. In this context, the aim of this study was to investigate the phytochemical profile and the redox modulating potential—radical-scavenging, metal-reducing, and metal-chelating properties of methanol extract from seeds of *G. biloba* growing in Bulgaria's Plovdiv region.

2. Materials and Methods

2.1. Preparation of *Ginkgo biloba* Seed Extract (GBSE)

Mature seeds of *Ginkgo biloba*, from trees grown on the campus of the Medical University in Plovdiv, Bulgaria. The seeds were peeled, and the endosperm was homogenized in a high-speed tissue homogenizer. The resulting slurry was mixed with anhydrous methanol (1:10 *w/v*) and stirred for 12 h at room temperature in a light-protected flask. The extract was centrifuged at 6000× *g* for 10 min, and the supernatant was separated. The extraction was carried out two more times, using 70% methanol as an extractant [4]. Thereafter, the combined extracts were evaporated at 30 °C under a vacuum [4]. The amount of dry matter obtained was determined gravimetrically. For the purposes of the present study, the dry extract was dissolved in distilled water. From the obtained *Ginkgo biloba* seed extract (GBSE) serial dilutions: 2.0, 1.0, 0.5, 0.25, 0.125, and 0.0625 mg/mL were prepared.

2.2. *Chromatographic Determination of GBSE Composition*

The GBSE composition was determined as 12 batches with an average dry matter content of 6–8% and was quantified by liquid chromatography-mass spectrometry (LC-MS) as described by Tomova et al. [4]. The chromatographic system Thermo Dionex Ultimate 3000 with a triple quadrupole mass detector, Thermo TSQ Quantum Access MAX and HESI-ionizator (Heated Electrospray Ionization) was used. The analytical conditions are given in Table 1.

2.3. *ABTS$^{\bullet+}$ Assay*

ABTS (2,2′-azino-bis (3-ethylbenzthiazoline-6-sulphonic acid) (7.0 mM) was mixed with potassium persulfate (2.45 mM) to obtain the cationic radical (ABTS$^{\bullet+}$) [24]. The solution was diluted in methanol (2 mL ABTS$^{\bullet+}$ + 58 mL methanol). The absorption at 743 nm of the resulting working solution was 1.1 $\pm$ 0.02 AU. Then, 1.425 mL of ABTS$^{\bullet+}$ working solution was added to 75 µL of the test extract, and the absorption at 743 nm was measured after 15 min of incubation at 37 °C against methanol. A blank containing 75 µL of water instead of the test extract also was measured against methanol. A calibration curve was obtained with different concentrations (10; 20; 30; 40; 50 mM) of Trolox dissolved in methanol. The antioxidant activity of the test substance relative to that of Trolox was calculated as the ratio between the concentrations of the test substance and of the Trolox, which gave 50% inhibition of the cationic radical ABTS$^{\bullet+}$.

2.4. *DPPH Assay*

DPPH analysis was conducted by the method described by Brand-Williams et al. [25]. First, 500 µL of the test solutions were added to 500 µL of a freshly prepared solution of 0.1 mM DPPH in methanol and incubated in the dark for 30 min. Then, the absorbance at 517 nm (A_{517}) was measured against a mixture of DPPH solution and methanol (1:1) as a control. Antioxidant activity determined by the DPPH was calculated as follows:

$$\text{Antioxidant activity (\%)} = [(A_{517}\ \text{control} - A_{517}\ \text{probe})/A_{517}\ \text{control}] \times 100$$

A calibration curve with different concentrations (10; 20; 30; 40; 50 mM) of Trolox dissolved in methanol was also obtained. The antioxidant activity of the test substance relative to that of Trolox was calculated as the ratio between the concentrations of the test substance and Trolox that gave 50% inhibition of the DPPH radical.

2.5. *FRAP Assay*

FRAP analysis was performed according to Benzie et al. [26] with some modifications. The following solutions: 0.03 M acetate buffer, 1.0 mM TPTZ, and 1.5 mM $FeCl_3$ were mixed in the ratio 10:1:20, respectively. To 1.5 mL of the reaction mixture, 50 µL of the sample was added. After incubation for 4 min at 37 °C, absorbance was measured at 593 nm against the blank sample. A calibration curve was prepared with $FeSO_4$, instead of a sample in concentrations: 0.1, 0.2, 0.3, 0.4, 0.5, 0.6, 0.7, 0.8, 0.9, and 1.0 mM. The antioxidant activity of GBSE was expressed as FRAP [mM Fe^{2+}/1 g GBSE].

Table 1. Analytical conditions for detection of GBSE composition.

Analyte	Column	Mobile Phase	Gradient	Flow Rate [mL/min]	HESI-Ionizator Mode	Collision Energy [eV]	Scan Mode
Flavonoids	Syncronis C18 (150 × 4.6 mm, 5 µm); Thermo Fisher Scientific Inc., Waltham, MA, USA	A–0.1% HCOOH in MeCN–H_2O (90:10, *v*/*v*) B–0.1% HCOOH in MeCN–H_2O (10:90, *v*/*v*)	0–3 min–70% B, 3–10 min–42.5% B, 10–11 min–17.5% B, 11–19 min–0% B, 19–25 min–80% B	0.7	Negative	32 24 27 31	SRM
Terpen trilactones	Syncronis C18 (150 × 4.6 mm, 5 µm); Thermo Fisher Scientific Inc., Waltham, MA, USA	0.1% HCOOH in MeCN-H_2O (50:50, *v*/*v*)	**Isocratic**	0.5	Negative	14 16 17 18 14	SRM
Ginkgolic Acid C13:0	Syncronis C18 (150 × 4.6 mm, 5 µm); Thermo Fisher Scientific Inc., Waltham, MA, USA	A-100% MeCN + 0.1% HCOOH B-10% MeCN + 0.1% HCOOH	0 min–30% B, 0–5 min–0 % B, 5–14 min–0% B, 14–25 min–30% B	1.0	Negative	23	SRM
Ginkgotoxin	Hypersil C18 (150 × 3 mm, 3 µm); Thermo Hypersil-Keystone Corp., Waltham, MA, USA	A-0.1% HCOOH in MeCN-H_2O (60:40, *v*/*v*) B–0.1% HCOOH in MeCN–H_2O (10:90, *v*/*v*)	**Reverse gradient** 0–1 min–80% B, 1–6 min–20% B, 6–15 min–10% B, 15–20 min–80% B	0.5	Positive	13	SRM

2.6. CUPRAC Assay

The CUPRAC analysis was performed according to Apak et al. [27] with modifications. The following solutions 10 mM $CuCl_2$ in ddH_2O, 1.0 M ammonium acetate buffer; pH 7.0, and 7.5 mM neocuproin (NC) in 96% ethanol were mixed in the ratio of 1:1:1. The tested substances were pipetted to Eppendorf tubes, and the volume was adjusted to 0.550 mL by adding H_2O. Then, 1.5 mL of the reaction mixture was added. After incubation at 50 °C for 20 min, absorbance was measured at 450 nm against blank (1.5 mL reaction mixture and 0.550 mL H_2O). The standard curve was prepared with Trolox at various concentrations ranging from 0.1 to 1.0 mM, and the results were expressed as CUPRAC (μM Trolox)/g GBSE).

2.7. Metal Chelating Assay

Metal chelation assay was applied after Venditti et al. [28]. A total of 0.2 mL of the sample solution, containing different concentrations of GBSE and reaction mixture, containing 0.74 mL of 0.1 M acetate buffer (pH 5.25), and 0.02 mL of 2 mM ferrous sulfate solution in 0.2 M hydrochloric acid were mixed for 10–15 s. Then, 0.04 mL ferrozine solution (5 mM) was added, and after incubation in the dark for 10 min at room temperature, the absorbance of the mixture at 562 nm was measured against the appropriate blank. The metal chelating capacity was determined using the following formula: Activity (%) = 100 (Ac − As)/(Ac), where Ac is the absorbance of the control and As is the absorbance of the sample solution. EDTA was used as a positive control, and the metal chelation capacity of GBSE was expressed as (mmol EDTA)/(g GBSE).

2.8. Statistics

All measurements were made in triplicate and the data in graphics were presented as mean ± standard deviation.

3. Results

3.1. Chromatographic Determination of GBSE Composition

The chromatographic analysis by LC-MS showed the presence of flavonoids (quercetin, isorhamnetin, and rutin) and terpenes (ginkgolides A, B, C, J, and bilobalide), as well as ginkgotoxin (4′-O-methylpyrodoxin), and ginkolic acid (Figure 1). The average levels of these bioactive constituents of the extracts are shown in Table 2.

Table 2. Concentration (μg per gram of dry matter) and analytical characteristics of analyzed bioactive compounds in GBSE.

	№	Compound Name	Formula	RT [min]	Molecular Mass	Q1 → Q3 (m/z)	Detected Amount [μg/g]
Formula	1.	Rutin	$C_{27}H_{30}O_{16}$	3.36	610.50	609.2 → 300.5	27.59
	2.	Quercetin	$C_{15}H_{10}O_7$	6.87	302.04	300.9 → 151.1	0.12
	3.	Kaempferol	$C_{15}H_{10}O_6$	8.02	286.05	285 → 185.1	ND *
	4.	Isorhamnetin	$C_{16}H_{12}O_7$	8.23	316.06	315 → 151	1.26
Terpenes	5.	Ginkgolide A	$C_{20}H_{24}O_9$	5.67	408.14	453 → 407	104.67
	6.	Ginkgolides B, J	$C_{20}H_{24}O_{10}$	5.66	424.14	423 → 367.5	221.44
	7.	Ginkgolide C	$C_{20}H_{24}O_{11}$	3.79	440.13	439 → 383.02	47.52
	8.	Bilobalide	$C_{15}H_{18}O_8$	5.25	326.10	324.9 → 163.0	103.69
	9.	Ginkgolic Acid C13:0	$C_{22}H_{34}O_3$	12.08	346.50	319.2 → 275.3	3.90
	10.	Ginkgotoxin	$C_9H_{13}NO_3$	7.11	183.20	183.97 → 152.04	125.05

* ND—not detected; Ginkgolides B and J are isomers and cannot be separated under the chromatographic conditions used [4].

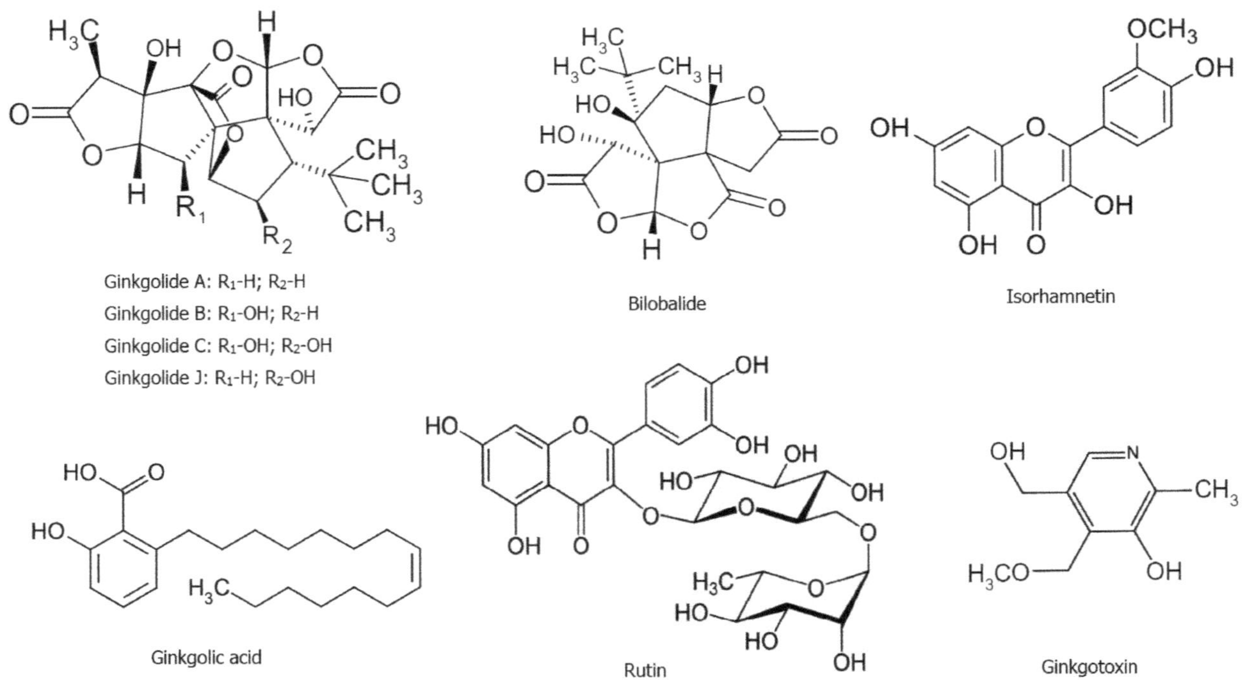

Figure 1. The main bioactive compounds in GBSE.

From the group of flavonoids, three representatives were identified, the largest amount being the glycoside flavonol rutin (quercetin-3-O-rutinoside) and a trace amount of its aglycone quercetin. The flavonol glycoside isorhamnetin was also detected (see Table 2 and chromatogram in Supplementary Material Figure S1).

Ginkgo-specific terpenes have been found to contain the largest amounts. The isomeric ginkgolides B and J have the highest content (221.44 μg/g), followed by ginkgolide A and bilobalide in almost equal concentrations (see Table 2 and chromatograms in Supplementary Material, Figure S2).

The toxic phenolic compounds, ginkgolic acid, and ginkgotoxin were also present in the studied GBSE see Table 2 and chromatograms in Supplementary Material, Figures S3 and S4). Ginkgolic acid isomers are lipophilic compounds, and after extraction with hexane, this amount was reduced to below 4 μg/g without significant changes in the levels of flavonoids and terpene trilactones. This extract was used thereafter in the analyses of the redox-modulating capacity of the studied GBSE.

3.2. *ABTS-Radical-Scavenging Properties*

The results obtained for the inhibition of ABTS radicals ($ABTS^{\bullet +}$) by the tested GBSE concentrations are presented in Figure 2. As can see in the graphic, the antiradical activity of the extract showed a strict dose–response relationship in the concentration range of 0.5–2.5 mg/mL.

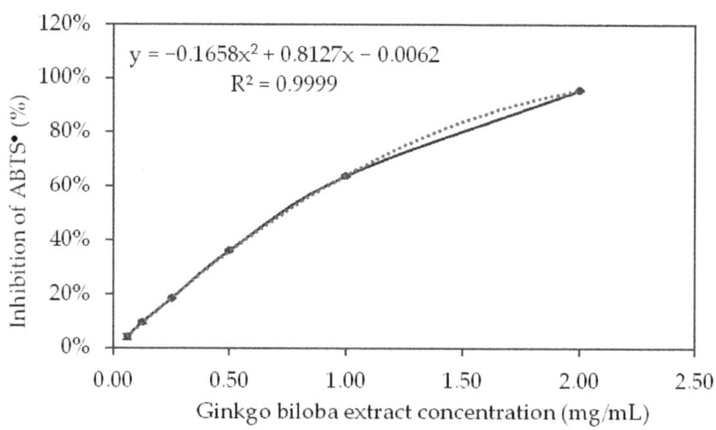

Figure 2. $ABTS^{\bullet +}$ inhibition curve from different concentrations of the *Ginkgo biloba* seed extract.

Based on the polynomial, which described the curve of inhibition of the $ABTS^{\bullet +}$ by GBSE, the IC_{50} of the GBSE was calculated:

$$\text{GBSE } IC_{50} = 728.2 \text{ mg/L}$$

Using the linear equation of the calibration curve of inhibition of the $ABTS^{\bullet +}$ by Trolox ($R^2 = 0.9998$) the concentration of Trolox (MW = 250.9), which leads to 50% inhibition of the $ABTS^{\bullet +}$, was calculated:

$$\text{Trolox } IC_{50} = 39.020\ \mu M = 9.7663 \text{ mg/L}$$

Trolox equivalent antioxidant capacity (TEAC%) of GBSE was computed as follow:

$$\text{TEAC\%} = 100 \times (\text{Trolox } IC_{50})/(\text{GBSE } IC_{50})$$

$$\text{GBSE (TEAC\%)} = 100 \times 9.7663/728.2 = 1.34\%$$

This experiment showed that the antioxidant capacity of GBSE determined by the ABTS method was equal to 1.34% of that of the same amount of Trolox.

3.3. DPPH-Radical-Scavenging Properties

The results of the DPPH radicals ($DPPH^{\bullet}$) inhibition by the different GBSE concentrations are presented in Figure 3. In the concentration range of 0–1.0 mg/mL, the radical-cleansing effect of GBSE increased to almost 70%, and the trend of the effect followed that of the reference substance Trolox. After this concentration, the effect of the extract was slightly lower than that of the standard.

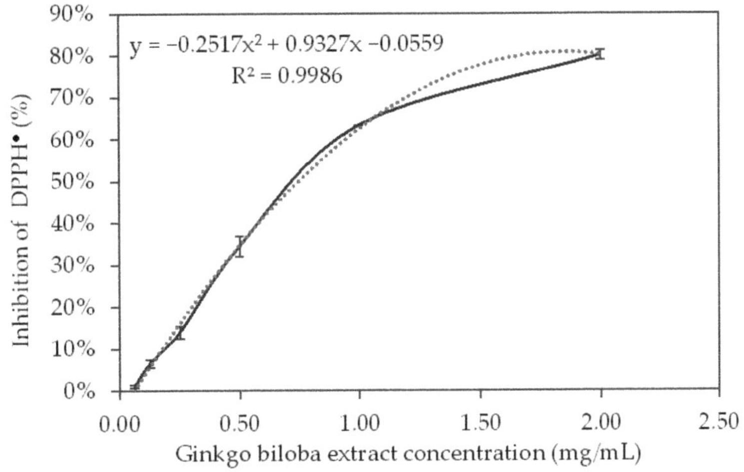

Figure 3. $DPPH^{\bullet}$ inhibition curve from different concentrations of the *Ginkgo biloba* seed extract.

Based on the polynomial, which described the inhibition curve of the $DPPH^{\bullet}$ ($R^2 = 0.9986$), the concentration of GBSE, which leads to 50% inhibition of the $DPPH^{\bullet}$, was calculated:

$$\text{GBSE } IC_{50} = 746\ \mu g/mL = 746\ mg/L$$

Based on the linear equation of the calibration curve of $DPPH^{\bullet}$ inhibition by Trolox ($R^2 = 0.9997$), we calculated the concentration of Trolox (MW = 250.9), which leads to 50% inhibition of the $DPPH^{\bullet}$:

$$\text{Trolox } IC_{50} = 17.28\ \mu M = 4.32\ mg/L$$

Trolox equivalent antioxidant capacity (TEAC%) of GBSE was calculated as follows:

$$\text{TEAC\%} = 100 \times (IC_{50} \text{ on Trolox})/(IC_{50} \text{ on GBSE})$$

$$\text{GBSE (TEAC\%)} = 100 \times 4.32/746.0 = 0.58\%$$

Thus, the experiment showed that the antioxidant capacity of GBSE determined by the DPPH method was equal to 0.58% of that of the same amount of Trolox.

3.4. FRAP Assay

The results for the degree of reduction of Fe^{3+} in complex with TPTZ to Fe^{2+} from GBSE are presented in Figure 4.

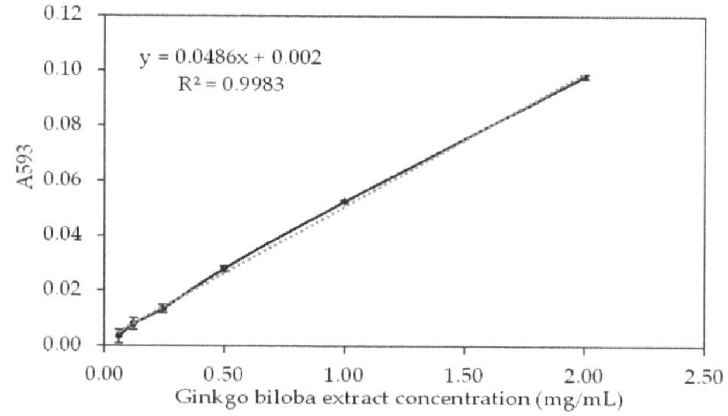

Figure 4. Reduction curve of Fe^{3+} in complex with TPTZ from different concentrations of the *Ginkgo biloba* seed extract.

The reduction curve of Fe^{3+} to Fe^{2+} in a complex with TPTZ (described by A_{593}) demonstrates a well-defined concentration dependence on the effect in the studied concentration range. The relation of the different concentrations of GBSE is described ($R^2 = 0.9983$) with the following linear correlation equation.

$$A_{593} = 0.0486x - 0.0002; \text{ where x is GBSE concentration (mg/mL)} \quad (1)$$

The calibration curve of the FRAP method with different concentrations of $FeSO_4$ is described with the following linear correlation equation:

$$A_{593} = 6.8883x - 0.0021;$$
$$\text{where x is } FeSO_4 \text{ concentration (mmol/L) } (R^2 = 0.9967)$$

From Equation (1), we calculated that the absorption of concentration of GBSE = 1.0 mg/mL is $A_{593} = 0.0484$, and by substituting A_{593} with this value in the above equation, we found that the effect of 1 mg/mL GBSE was equal to $FeSO_4$ having concentration of 7.33 mM.

3.5. Cupric Reducing Antioxidant Capacity—CUPRAC

The CUPRAC method is based on the absorbance measurement of Cu^{1+}-neocuproine (Nc) chelate formed as a result of the redox reaction of chain-breaking antioxidants with the CUPRAC reagent, Cu^{2+}-Nc, where absorbance is recorded at the maximal light-absorption wavelength of 450 nm [29]. The results for the degree of reduction of Cu^{2+} by GBSE, measured by chelation of the resulting Cu^{+} with neocuproine are presented in Figure 5.

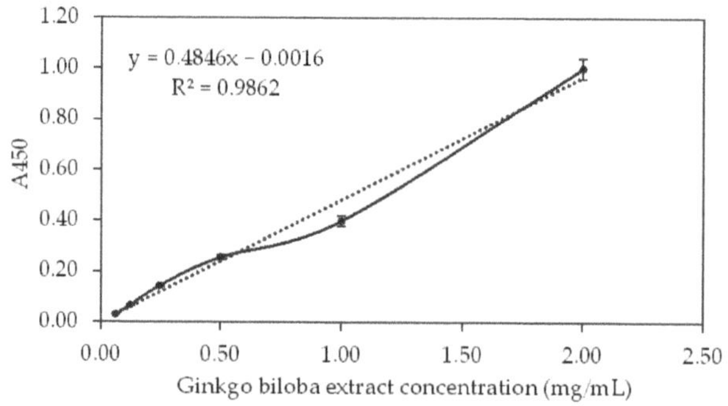

Figure 5. Neocuproine-Cu^{+} complex concentration resulting from Cu^{2+} reduction by different concentrations of the *Ginkgo biloba* seed extract.

Antioxidant index for GBSE using its Cu^{2+}/Cu^{+} reducing capability measured in the presence of neocuproine is described with the following linear equation ($R^2 = 0.9862$):

$$A_{450} = 0.04846x - 0.0016; \text{ where x is GBSE concentration (mg/mL)} \quad (2)$$

The calibration line obtained of the CUPRAC method with different concentrations of Trolox is described with the following linear correlation equation:

$$A_{450} = 0.0023x - 0.0124;$$
$$\text{where x is Trolox concentration } (\mu mol/L)\ (R^2 = 0.9981)$$

From Equation (2) it was found that at a concentration of GBSE = 1.0 mg/mL, the calculated absorption was $A_{450} = 0.483$. By substituting A_{450} in the above equation with the resulting value, we obtained that the effect of 1 mg/mL GBSE was equal to that which would be obtained at a concentration of Trolox = 215.39 μM, or CUPRAC = 215.39 (μmol Trolox)/(g GBSE).

3.6. Metal Chelation Capacity

The results for the metal-chelating properties of GBSE are presented in Figure 6.

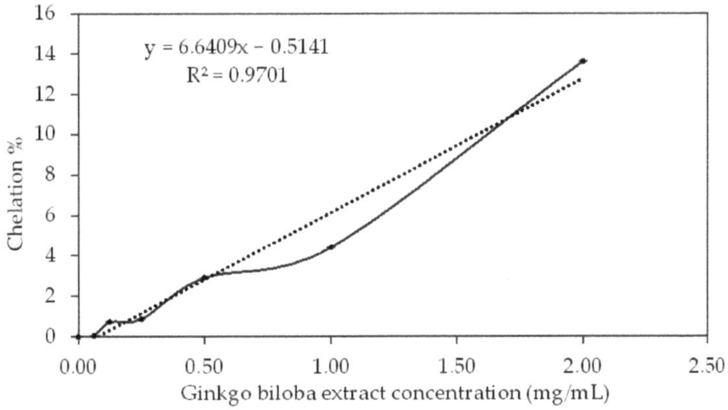

Figure 6. The curve of the chelation activity of iron ions from different concentrations of the *Ginkgo biloba* seed extract.

The degree of chelation of iron ions from different concentrations of GBSE is well described ($R^2 = 0.9701$) with the following linear equation:

$$\text{Chelation } \% = 6.6409x - 0.5141; \text{ where x is GBSE concentration (mg/mL)}$$

The obtained calibration line for metal-chelating capacity with different concentrations of EDTA is described ($R^2 = 0.9966$) with the following linear correlation equation:

$$\text{Metal-chelating capacity } (\%) = 457.4x - 2.5082;$$
$$\text{x is the EDTA concentration (mM)}$$

With a concentration of GBSE = 1.0 mg/mL, the calculated metal-chelating capacity was equal to 6.13%, and by substituting the above equation with this value, we could calculate that the chelating effect of 1 mg/mL GBSE was equal to EDTA at a concentration 0.0189 mM, from which we can calculate:

$$1 \text{ g/L GBSE} = 0.0189 \text{ mM EDTA}$$

$$(\text{mmol EDTA})/(\text{g GBSE}) = 0.0189$$

4. Discussion

In this study, we investigated the phytochemical profile and the redox modulating potential—radical-scavenging, metal-reducing, and metal-chelating properties of methanol extract from seeds of *G. biloba* growing in Bulgaria's Plovdiv region. The existing data have shown significant differences in the measured antioxidant activity of the *G. biloba* extracts depending on the used raw material, method of extraction, solvent, and region of growth even in the same country [30,31]. The antioxidant capacity of plant extracts is most often associated with the presence of secondary metabolites of the polyphenol group. Evidence of this is the good correlation found between the content of phenolic compounds and the antioxidant properties of extracts of different plant species, evaluated by applying various standardized antioxidant methods [32]; therefore, studies on the chromatography profile of the extracts and identification of the metabolites that could determine the antioxidant properties are needed in order to obtain standardized preparations with certain qualities.

The data of the chromatographic analysis of GBSE in this study showed the presence of two types of polyphenol representatives—flavonoids and terpenes. The established phytochemical composition of the methanol extract of seeds is similar to that of leaf extracts but differs in proportions [33]. In accordance with the literature data, terpenes (the triterpene ginkgolides A, B, C, and J, and the sesquiterpene bilobalide) prevail in the seed extract, while glycosylated flavonoids predominate in *G. biloba* leaf extracts [4]. In the standardized extract EGb 761® (Tanakan®), commercially available in European markets, flavonoids constitute 24% and terpenes constitute 6% [34]. It has been established that terpenes isolated from *G. biloba* leaf have a well-defined antioxidant activity with possibilities for therapeutic applications in pathologies with oxidative stress etiology [13], although it has been shown that ginkgolides have less ability to scavenge DPPH radicals than ginkgo flavonoids [35]. The good therapeutic effect of terpenes in neurodegenerative diseases is probably due to the activation of various signaling pathways to protect cells from oxidative stress injury, in addition to their direct antioxidant action [13,36]. Flavonoids are the major components identified in *Ginkgo biloba* leaf extracts (ranging from 23–35%) in comparison to terpenoids (ranging from 0.17–11.3% for bilobalide, ginkgolide A, ginkgolide B, and ginkgolide C) [2]. Of the flavonoids in this study, we detected the presence of rutin, quercetin, and isorhamnetin. The most abundant was rutin (quercetin-3-rhamnosyl glucoside) (Table 2). Assays for establishing the in vitro antioxidant activity of rutin have found that its DPPH radical scavenging activity is much stronger than that of butylated hydroxytoluene and is comparable to vitamin C, as well as it effectively inhibits lipid peroxidation [37]. Isorhamnetin and quercetin are among the most important ingredients in the *G. biloba* leaves, and are responsible for their therapeutic benefits [38,39]. Both have shown excellent antioxidant activity [40]. With an equivalent concentration, the activity of quercetin in scavenging either DPPH or ABTS radicals was stronger than the activity of isorhamnetin [41]. These stronger effects of quercetin are likely due to the presence of the phenolic hydroxyl group linked to the 3'Carbon atom in the molecule, while the 3'Carbon atom of isorhamnetin is linked with the methoxy group, and this is the only difference between them; however, isorhamnetin has demonstrated very similar to quercetin activity for inhibiting lipid peroxidation [41]. The phytochemical analysis of the tested GBSE herein indicated the presence of two toxic compounds—ginkgotoxin and ginkgolic acid (Table 2). Although data on content ginkgotoxin (4'-O-methylpyrodoxin) in different parts of Ginkgo tree are contradictory, it is considered that in seeds its amount is larger than in the leaves. Ginkgotoxin is also known as antivitamin B_6 and in certain quantities can cause stomach pain, nausea, irritability, epileptic convulsions, and death in very rare cases [42]. The European Pharmacopoeia has not yet standardized the safe levels of ginkgoxin in extracts. Concerning another toxic constituent of GBSE, ginkgolic acid, allergenic and genotoxic effects have been shown [43]. Initial analysis for ginkgolic acids in the GBSE tested in this study showed a concentration of 18 μg/g, which significantly exceeded the permissible levels set by the European Pharmacopoeia for leaf extract; however, ginkgolic acid isomers could be reduced without significant changes in the flavonoids and terpenes levels, as herein was performed.

The antioxidant activity of polyphenols involves different mechanisms: hydrogen atom transfer, single electron transfer, sequential proton loss electron transfer, and chelation of transition metals [44]. In this study, the widely accepted methods based on single-electron transfer and/or hydrogen atom transfer, such as ABTS, DPPH, FRAP, CUPRAC, as well as metal chelation tests, were applied. The obtained results demonstrated excellent antioxidant activity of the GBSE compared to leaf extracts when ABTS and FRAP methods were applied. The ABTS radical scavenging effect of the aqueous extract of Ginkgo biloba leaves obtained by the "reflux" method showed that at a final extract concentration of 100 μg/mL, 47.96 ± 0.40% of the ABTS radicals are neutralized [45]. This final concentration corresponds to the resulting concentration in our experiment since the sample with an initial concentration of 2.00 mg/mL was diluted 20-fold and became 100 μg/mL. At this concentration in our experiment, we found almost twice as much neutralization of ABTS radicals—95.6% ± 0.5%. Some data for the iron-reducing capacity of the leaf extracts of *Ginkgo biloba* obtained by the FRAP method showed that at a final extract concentration of 50 μg/mL, it was equivalent to 74.70 ± 1.71 μM $FeSO_4$ [45]. At the same concentration (the initial concentration of 1.00 mg/mL GSBE was diluted 20-fold in the sample), we found that iron-reducing capacity was ten times greater, equivalent to only 7.33 μM $FeSO_4$.

The antioxidant properties of the *Ginkgo biloba* leaf extracts studied by the DPPH method vary depending on the extraction procedure, growing region, and methods for calculation of the results. There are data demonstrating that leaf extract at a final concentration of 100 μg/mL and of 1000 μg/mL neutralized 58.47 ± 1.17% and 89.84 ± 2.11% of the DPPH radicals, respectively [45]. The above-mentioned concentrations correspond to the concentrations in our experiment (initial concentrations of 0.25 and 2.00 mg/mL with a two-fold dilution in the sample), at which we obtained 14.0 ± 0.88% and 80.1 ± 0.63% DPPH radical inhibition. The observed differences in the antioxidant effect at concentrations of 100 μg/mL in our experiment, compared to that at a similar concentration in the cited experiment, can be explained by the differences in the composition of the two extracts. Other authors [46] investigating the antioxidant properties of the methanol extracts of *Ginkgo biloba* leaves by the DPPH method found that at a final concentration of 150 μg/mL, 75.86% of DPPH radicals were neutralized. Actually, at a 150 μg/mL concentration, our extract would show only 7.8% neutralization of DPPH radicals. Ahmad et al. [47], in a comparative study of the antioxidant properties of methanolic and ethanolic extracts of *Ginkgo biloba* leaves by the DPPH method, found better antioxidant properties in methanol extract—82.8% compared to 76.7% for the ethanol extract, at the same concentration of 0.25 mg/mL. The methanol GBSE studied by us showed a close effect (80% inhibition) at four times higher concentration—2.0 mg/mL. When studying the antioxidant properties of leaf extracts of *Ginkgo biloba* from six regions of India by the DPPH method, the authors [30] obtained the lowest IC_{50} values for the extracts obtained by the "reflux" method (0.325 mg/mL) and the highest IC_{50} for the extracts obtained by the "shaker" method (6.5 mg/mL). The results for the extract obtained by the "reflux" method are comparable to the results for GBSE in our study—0.746 mg/mL. Values in the same range (max 1.545 ± 0.112 mg and min 0.394 ± 0.011 TEAC/g fresh matter) were measured for ethanol extracts of leaves from *Ginkgo biloba* growing in nine regions of Slovakia. The antioxidant activity assessed by the DPPH method could vary between extracts of the plant from different regions since the composition and biological activity of the compounds strongly depend on the agro-ecological conditions [31]. The significant difficulty in comparing the published results is the lack of a standard for determining the TEAC. TEAC is calculated by different methods, which cannot always be converted into each other.

The presence of free redox ions in biological systems, mainly Fe (II) and Cu (I), is also a possible source of reactive oxygen species generation and oxidative stress through the so-called Fenton reaction [23]; therefore, the use of compounds that may chelate these ions is also one of the antioxidant strategies. El-Beltagi et al. [46] investigated the metal-chelating effect of a methanolic extract of *Ginkgo biloba* leaves and found that at a concentration of 0.2 mg/mL, the extract chelated 32.2% of divalent iron ions. In our experiment, GBSE at a concentration of 0.25 mg/mL chelated only 0.9% of iron ions; however, it should be noted that the values given by the authors for the chelation of iron ions from the standard EDTA chelator differ significantly from those obtained by us. Thus, the authors find that at a concentration of 0.2 mg/mL (0.68 mM) EDTA chelates 51.2% of iron ions, and we obtain 99.3% chelation of iron ions at 0.5 mM EDTA [28]. In our opinion, it is necessary to establish standard methods for laboratory measurement and calculation of the antioxidant and other pharmacological properties of plant extracts for proper comparison and outlining of the expected therapeutic results. We could not compare the measured copper-reducing effect of GBSE because we did not find similar data in the literature.

These findings indicate that, along with the typical healing benefits of *Ginkgo biloba* seeds outlined by traditional Chinese medicine (pulmonary disorders, bladder inflammation, and enuresis), GBSE has antioxidant properties as a free radical scavenger and may be active in preventing free-radical-driven pathological reactions.

5. Conclusions

Recently, we have witnessed a resurgence of global interest in herbal medicines. More and more researchers are turning to the study of herbal products in healthcare; however, the concentrations of plant bioactive substances vary depending on a number of factors. In this study, we examined *Ginkgo biloba* seed extract, which is less used in practice but has the same biologically active substances as identified in the leaves. Our research focuses on the possibility of using this product in pathological conditions related to oxidative stress. In this study, we applied a panel of methods to investigate the redox-modulating activities of *Ginkgo biloba* seed extract containing complex organic compounds, including some with antioxidant potential.

The studied methanolic extract of *Ginkgo biloba* seeds was rich in flavonoids such as rutin and isorhamnetin and terpenes such as ginkgolides A, B, C, J, and bilobalide. Terpenes predominated in seed extract compared to flavonoids. Two toxic compounds have been identified in the crude extract—ginkgotoxin and ginkgolic acid. Their concentration can be reduced without reducing flavonoid and terpene levels. The studied Ginkgo biloba seed extract showed excellent properties for neutralizing ABTS and DPPH radicals, and the values found were comparable to those obtained by other authors with extracts of Ginkgo biloba leaves and nuts. The effects of iron reduction and iron chelation were significantly weaker than the published results for leaf extracts; however, in our opinion, these results are a promising start. Further studies of seed extracts in in vivo models of various pathological conditions are needed to fully determine their bioactivity and properties in order to understand the full phenomena of their healing potential.

Supplementary Materials: The following supporting information can be downloaded at: https://www.mdpi.com/article/10.3390/life12060878/s1, Figure S1. Chromatograms of flavonoids standards: Kaempferol (1A), Quercetin (2A), Isorhamnetin (3A), Rutin (4A), and of flavonoids in GBSE: Kaempferol (1B), Quercetin (2B), Isorhamnetin (3B), and Rutin (4B); Figure S2. Chromatograms of terpenes standards: Bilobalide (1A), Ginkgolides B and J (2A), Ginkgolide C (3A), Ginkgolide A (4A), and of terpenes in GBSE: Bilobalide (1B), Ginkgolides B and J (2B), Ginkgolide C (3B), and Ginkgolide A (4B); Figure S3. Chromatogram of Ginkgolic Acid standard (A) and of Ginkgolic Acid in GBSE (B); Figure S4. Chromatogram of Ginkgotoxin standard (A) and of Ginkgotoxin in GBSE (B).

Author Contributions: Conceptualization, L.P., A.A., M.A. and M.M.; methodology, A.A., M.A., A.G. and T.T.; software, L.P.; validation, L.P., A.A. and M.A.; formal analysis, T.T., A.G. and E.T.; investigation, T.T., A.G. and E.T.; resources, A.A. and M.A.; data curation, L.P., A.A., M.A. and M.M.; writing—original draft preparation, L.P., A.A. and M.M.; writing—review and editing, A.A. and M.M.; visualization, L.P., T.T., M.A. and A.G.; supervision, A.A.; project administration, A.A.; funding acquisition, A.A. and M.M. All authors have read and agreed to the published version of the manuscript.

Funding: This work was supported by the Bulgarian Ministry of Education and Science (Grant D01-217/30.11.2018 and agreements DO1-323/18.12.2019, DO1-358/17.12.2020, and DO1-278/03.12.2021) under the National Research Programme "Innovative Low-Toxic Bioactive Systems for Precision Medicine (BioActiveMed)" approved by DCM # 658/14.09.2018.

Institutional Review Board Statement: Not applicable.

Informed Consent Statement: Not applicable.

Data Availability Statement: The authors confirm that the data supporting the findings of this study are available within the article. Row data that support the findings of this study are available from the corresponding author upon reasonable request.

Conflicts of Interest: The authors declare no conflict of interest. The funders had no role in the design of the study; in the collection, analyses, or interpretation of data; in the writing of the manuscript, or in the decision to publish the results.

References

1. Ivanova, V.; Natcheva, L.; Gercheva, P. Use of *Ginko biloba*, L. in green system of the town of Plovdiv. In Proceedings of the Scientific Works of the Union of Scientists in Bulgaria-Plovdiv, Plovdiv, Bulgaria, 5–6 November 2015; pp. 241–244.
2. Zhang, Z.; Chen, S.; Mei, H.; Xuan, J.; Guo, X.; Couch, L.; Mei, N. *Ginkgo biloba* leaf extract induces DNA damage by inhibiting topoisomerase II activity in human hepatic cells. *Sci. Rep.* **2015**, *5*, 14633. [CrossRef] [PubMed]
3. Goh, L.M.; Barlow, P.J. Antioxidant capacity in Ginkgo Biloba. *Food Res. Int.* **2002**, *35*, 815–820. [CrossRef]
4. Tomova, T.; Doncheva, N.; Mihaylova, A.; Kostadinov, I.; Peychev, L.; Argirova, M. An experimental study on phytochemical composition and memory enhancing effect of *Ginkgo biloba* seed extract. *Folia Med.* **2021**, *63*, 203–212. [CrossRef] [PubMed]
5. Wonther, K.; Randlov, C.; Rein, E.; Mehlsen, J. Effects of Ginkgo boloba extracts on cognitive function and blood presure in elderly subjects. *Curr. Ther. Res.* **1998**, *59*, 881–888. [CrossRef]
6. Diamond, B.J.; Shiflett, S.C.; Feiwel, N.; Matheis, R.J.; Noskin, O.; Richards, J.A.; Schoenberger, N.E. *Ginkgo biloba* extract: Mechanisms and clinical indications. *Arch. Phys. Med. Rehabil.* **2000**, *81*, 668–678. [CrossRef]
7. Yuan, Q.; Wang, C.W.; Shi, J.; Lin, Z.X. Effects of *Ginkgo biloba* on dementia: An overview of systematic reviews. *J. Ethnopharmacol.* **2017**, *195*, 1–9. [CrossRef] [PubMed]
8. Zeng, K.; Li, M.; Hu, J.; Mahaman, Y.A.; Bao, J.; Huang, F.; Wang, X. *Ginkgo biloba* extract EGb761 attenuates Hyperhomocysteinemia-induced AD like tau hyperphosphorylation and cognitive impairment in rats. *Curr. Alzheimer Res.* **2018**, *15*, 89–99. [CrossRef]
9. Tanaka, K.; S-Galduroz, R.F.; Gobbi, L.T.B.; Galduróz, J.C.F. *Ginkgo biloba* extract in an animal model of Parkinson's disease: A systematic review. *Curr. Neuropharmacol.* **2013**, *11*, 430–435. [CrossRef]
10. Kumar Singh, S.; Barreto, G.E.; Aliev, G.; Echeverria, V. *Ginkgo biloba* as an alternative medicine in the treatment of anxiety in dementia and other psychiatric disorders. *Curr. Drug Metab.* **2017**, *18*, 112–119. [CrossRef]
11. Tian, J.; Liu, Y.; Chen, K. *Ginkgo biloba* extract in vascular protection: Molecular mechanisms and clinical applications. *Curr. Vasc. Pharmacol.* **2017**, *15*, 532–548. [CrossRef]
12. Zhou, W.; Chai, H.; Lin, P.H.; Lumsden, A.B.; Yao, Q.; Chen, C. Clinical use and molecular mechanisms of action of extract of *Ginkgo biloba* leaves in cardiovascular diseases. *Cardiovasc. Drug Rev.* **2004**, *22*, 309–319. [CrossRef] [PubMed]
13. Liu, Q.; Jin, Z.; Xu, Z.; Yang, H.; Li, L.; Li, G.; Xiao, W. Antioxidant effects of ginkgolides and bilobalide against cerebral ischemia injury by activating the Akt/Nrf2 pathway in vitro and in vivo. *Cell Stress Chaperones* **2019**, *24*, 441–452. [CrossRef] [PubMed]
14. Yoshikawa, T.; Naito, Y.; Kondo, M. *Ginkgo biloba* leaf extract: Review of biological actions and clinical applications. *Antioxid. Redox Signal.* **1999**, *1*, 469–480. [CrossRef] [PubMed]
15. Mohanta, T.K.; Tamboli, Y.; Zubaidha, P.K. Phytochemical and medicinal importance of *Ginkgo biloba* L. *Nat. Prod. Res.* **2014**, *28*, 746–752. [CrossRef] [PubMed]
16. Liu, L.; Wang, Y.; Zhang, J.; Wang, S. Advances in the chemical constituents and chemical analysis of *Ginkgo biloba* leaf, extract, and phytopharmaceuticals. *J. Pharm. Biomed. Anal.* **2021**, *193*, 113704. [CrossRef] [PubMed]
17. Avula, B.; Sagi, S.; Gafner, S.; Upton, R.; Wang, Y.H.; Wang, M.; Khan, I.A. Identification of *Ginkgo biloba* supplements adulteration using high performance thin layer chromatography and ultra high performance liquid chromatography-diode array detector-quadrupole time of flight-mass spectrometry. *Anal. Bioanal. Chem.* **2015**, *407*, 7733–7746. [CrossRef]
18. Santos-Sánchez, N.F.; Salas-Coronado, R.; Villanueva-Cañongo, C.; Hernández-Carlos, B. Antioxidant compounds and their antioxidant mechanism. *Antioxidants* **2019**, *10*, 1–29.
19. Di Domenico, F.; Barone, E.; Perluigi, M.; Butterfield, D.A. Strategy to reduce free radical species in Alzheimer's disease: An update of selected antioxidants. *Expert Rev. Neurother.* **2015**, *15*, 19–40. [CrossRef]
20. Simons, B.D.; Clevers, H. Strategies for homeostatic stem cell self-renewal in adult tissues. *Cell* **2011**, *145*, 851–862. [CrossRef]
21. Wang, K.; Zhang, T.; Dong, Q.; Nice, E.C.; Huang, C.; Wei, Y. Redox homeostasis: The linchpin in stem cell self-renewal and differentiation. *Cell Death Dis.* **2013**, *4*, e537. [CrossRef]
22. Schlesier, K.; Harwat, M.; Böhm, V.; Bitsch, R. Assessment of antioxidant activity by using different in vitro methods. *Free Radic. Res.* **2002**, *36*, 177–187. [CrossRef] [PubMed]

23. Halliwell, B.; Gutteridge, J.M. *Free Radicals in Biology and Medicine*; Oxford University Press: Oxford, MS, USA, 2015.
24. Re, R.; Pellegrini, N.; Proteggente, A.; Pannala, A.; Yang, M.; Rice-Evans, C. Antioxidant activity applying an improved ABTS radical cation decolorization assay. *Free. Radic. Biol. Med.* **1999**, *26*, 1231–1237. [CrossRef]
25. Brand-Williams, W.; Cuvelier, M.E.; Berset, C. Use of a free radical method to evaluate antioxidant activity. *LWT—Food Sci. Technol.* **1995**, *28*, 25–30. [CrossRef]
26. Benzie, I.F.F.; Strain, J.J. The ferric reducing ability of plasma (FRAP) as a measure of 'antioxidant power': The FRAP assay. *Anal. Biochem.* **1996**, *239*, 70–76. [CrossRef] [PubMed]
27. Apak, R.; Güçlü, K.; Özyürek, M.; Karademir, S.E. Novel total antioxidant capacity index for dietary polyphenols and vitamins C and E, using their cupric ion reducing capability in the presence of neocuproine: CUPRAC method. *J. Agric. Food Chem.* **2004**, *52*, 7970–7981. [CrossRef] [PubMed]
28. Venditti, E.; Bacchetti, T.; Tiano, L.; Carloni, P.; Greci, L.; Damiani, E. Hot vs. cold water steeping of different teas: Do they affect antioxidant activity? *Food Chem.* **2010**, *119*, 1597–1604. [CrossRef]
29. Özyürek, M.; Güçlü, K.; Tütem, E.; Başkan, K.S.; Erçağ, E.; Çelik, S.E.; Apak, R. A comprehensive review of CUPRAC methodology. *Anal. Methods* **2011**, *3*, 2439–2453. [CrossRef]
30. Sati, P.; Dhyani, P.; Bhatt, I.D.; Pandey, A. *Ginkgo biloba* flavonoid glycosides in antimicrobial perspective with reference to extraction method. *J. Tradit. Complementary Med.* **2019**, *9*, 15–23. [CrossRef]
31. Ražná, K.; Sawinska, Z.; Ivanišová, E.; Vukovic, N.; Terentjeva, M.; Stričík, M.; Kačániová, M. Properties of *Ginkgo biloba* L.: Antioxidant characterization, antimicrobial activities, and genomic microRNA based marker fingerprints. *Int. J. Mol. Sci.* **2020**, *21*, 3087. [CrossRef]
32. Tzanova, M.; Grozeva, N.; Gerdzhikova, M.; Argirova, M.; Pavlov, D.; Terzieva, S. Flavonoid content and antioxidant activity of Betonica bulgarica Degen et Neič. *Bulg. Chem. Commun.* **2018**, *50*, 90–97.
33. van Beek, T.A.; Montoro, P. Chemical analysis and quality control of *Ginkgo biloba* leaves, extracts, and phytopharmaceuticals. *J. Chromatogr. A* **2009**, *1216*, 2002–2032. [CrossRef] [PubMed]
34. McKeage, K.; Lyseng-Williamson, K.A. *Ginkgo biloba* extract EGb 761®in the symptomatic treatment of mild-to-moderate dementia: A profile of its use. *Drugs Ther. Perspect.* **2018**, *34*, 358–366. [CrossRef] [PubMed]
35. Guo, Y.; Mao, M.; Li, Q.; Yu, X.; Zhou, L. Extracts of Ginkgo flavonoids and ginkgolides improve cerebral ischaemia-reperfusion injury through the PI3K/Akt/Nrf2 signalling pathway and multicomponent in vivo processes. *Phytomedicine* **2022**, *99*, 154028. [CrossRef] [PubMed]
36. Omar, S.H. *Ginkgolides and Neuroprotective Effects*; Springer: Berlin/Heidelberg, Germany, 2013.
37. Yang, J.; Guo, J.; Yuan, J. In vitro antioxidant properties of rutin. *LWT-Food Sci. Technol.* **2008**, *41*, 1060–1066. [CrossRef]
38. Gong, G.; Guan, Y.Y.; Zhang, Z.L.; Rahman, K.; Wang, S.J.; Zhou, S.; Zhang, H. Isorhamnetin: A review of pharmacological effects. *Biomed. Pharmacother.* **2020**, *128*, 110301. [CrossRef] [PubMed]
39. Xu, D.; Hu, M.J.; Wang, Y.Q.; Cui, Y.L. Antioxidant activities of quercetin and its complexes for medicinal application. *Molecules* **2019**, *24*, 1123. [CrossRef] [PubMed]
40. Xiao, Z.; He, L.; Hou, X.; Wei, J.; Ma, X.; Gao, Z.; Yue, T. Relationships between structure and antioxidant capacity and activity of glycosylated flavonols. *Foods* **2021**, *10*, 849. [CrossRef]
41. Zuo, A.; Yanying, Y.; Li, J.; Binbin, X.; Xiongying, Y.; Yan, Q.; Shuwen, C. Study on the relation of structure and antioxidant activity of isorhamnetin, quercetin, phloretin, silybin and phloretin isonicotinyl hydrazone. *Free Radic. Antioxid.* **2011**, *1*, 39–47. [CrossRef]
42. Wada, K.; Ishigaki, S.; Ueda, K.; Take, Y.; Sasaki, K.; Sakata, M.; Haga, M. Studies on the Constitution of Edible and Medicinal Plants. I.: Isolation and Identification of 4-O-Methylpyridoxine, Toxic Principle from the Seed of *Ginkgo biloba* L. *Chem. Pharm. Bull.* **1988**, *36*, 1779–1782. [CrossRef]
43. Ahlemeyer, B.; Selke, D.; Schaper, C.; Klumpp, S.; Krieglstein, J. Ginkgolic acids induce neuronal death and activate protein phosphatase type-2C. *Eur. J. Pharmacol.* **2001**, *430*, 1–7. [CrossRef]
44. Zeb, A. Concept, mechanism, and applications of phenolic antioxidants in foods. *J. Food Biochem.* **2020**, *44*, e13394. [CrossRef] [PubMed]
45. Park, J.S. Analysis of Antioxidant Efficacy of Ginkgo Biloba Leaves and Acer Palmatum Leaves. *Turk. J. Comput. Math. Educ. (TURCOMAT)* **2021**, *12*, 698–703. [CrossRef]
46. El-Beltagi, H.S.; Badawi, M.H. Comparison of antioxidant and antimicrobial properties for *Ginkgo biloba* and rosemary (*Rosmarinus officinalis* L.) from Egypt. *Not. Bot. Horti Agrobot. Cluj-Napoca* **2013**, *41*, 126–135. [CrossRef]
47. Ahmad, N.; Fazal, H.; Abbasi, B.H.; Farooq, S. Efficient free radical scavenging activity of *Ginkgo biloba*, Stevia rebaudiana and Parthenium hysterophorous leaves through DPPH (2, 2-diphenyl-1-picrylhydrazyl). *Int. J. Phytomed.* **2010**, *2*, 231–239.

life

Article

Biogas Production Potential of Thermophilic Anaerobic Biodegradation of Organic Waste by a Microbial Consortium Identified with Metagenomics

Lyudmila Kabaivanova *, **Penka Petrova**, **Venelin Hubenov and Ivan Simeonov**

The "Stephan Angeloff" Institute of Microbiology, Bulgarian Academy of Sciences, Acad G. Bonchev Str., Bl. 26, 1113 Sofia, Bulgaria; pepipetrova@yahoo.com (P.P.); vhubenov7@gmail.com (V.H.); issim@abv.bg (I.S.)
* Correspondence: lkabaivanova@yahoo.com; Tel.: +359-2-979-3167

Abstract: Anaerobic digestion (AD) is a widespread biological process treating organic waste for green energy production. In this study, wheat straw and corn stalks without any harsh preliminary treatment were collected as a renewable source to be employed in a laboratory-scale digester to produce biogas/biomethane. Processes parameters of temperature, pH, total solids, volatile solid, concentration of volatile fatty acids (VFA), and cellulose concentration, were followed. The volume of biogas produced was measured. The impact of organic loading was stated, showing that the process at 55 °C tolerated a higher substrate load, up to 45 g/L. Further substrate increase did not lead to biogas accumulation increase, probably due to inhibition or mass transfer limitations. After a 12-day anaerobic digestion process, cumulative volumes of biogas yields were 4.78 L for 1 L of the bioreactor working volume with substrate loading 30 g/L of wheat straw, 7.39 L for 40 g/L and 8.22 L for 45 g/L. The degree of biodegradation was calculated to be 68.9%, 74% and 72%, respectively. A fast, effective process for biogas production was developed from native wheat straw, with the highest quantity of daily biogas production occurring between day 2 and day 5. Biomethane concentration in the biogas was 60%. An analysis of bacterial diversity by metagenomics revealed that more than one third of bacteria belonged to class *Clostridia* (32.9%), followed by *Bacteroidia* (21.5%), *Betaproteobacteria* (11.2%), *Gammaproteobacteria* (6.1%), and *Alphaproteobacteria* (5%). The most prominent genera among them were *Proteiniphilum*, *Proteiniborus*, and *Pseudomonas*. Archaeal share was 1.37% of the microflora in the thermophilic bioreactor, as the genera *Methanocorpusculum*, *Methanobacterium*, *Methanomassiliicoccus*, *Methanoculleus*, and *Methanosarcina* were the most abundant. A knowledge of the microbiome residing in the anaerobic digester can be further used for the development of more effective processes in conjunction with theidentified consortium.

Keywords: biogas; renewable energy sources; anaerobic biodegradation; thermophilic conditions; metagenomics

Citation: Kabaivanova, L.; Petrova, P.; Hubenov, V.; Simeonov, I. Biogas Production Potential of Thermophilic Anaerobic Biodegradation of Organic Waste by a Microbial Consortium Identified with Metagenomics. *Life* **2022**, *12*, 702. https://doi.org/10.3390/life12050702

Academic Editor: Milka Mileva

Received: 18 April 2022
Accepted: 6 May 2022
Published: 8 May 2022

1. Introduction

Today's civilization faces growing energy and environmental problems related to the depletion of fossil energy sources [1]. Yet society's energy needs continue to increase due to fast economic growth and, therefore, much attention needs to be paid to the utilization of renewable resources for the production of valuable products and energy [2]. Renewable energy sources are part of the earth's physical structure and are constantly being renewed by natural means so they cannot deplete [3]. The use of biofuels is highly beneficial from an ecological, economic and strategic point of view, as it significantly reduces CO_2 emissions to the atmosphere [4]. Biomass energy is generated or produced by living or once-lived organisms. Biomass is assumed to be a renewable material because plants continue their growth after being harvested, thus the available stock remains without decreasing in addition to the incorporated carbon quantity over time. Biomass energy can be produced from plants, trees, crop residues, and other agricultural waste [5].

Globally, increased biogas production will favor energy supply through the use of renewable alternatives [6]. Lignocellulosic biomass is one of the most abundant renewable resources not only in the world, but also Bulgaria. One method of utilization is its involvement in anaerobic degradation accomplished by microorganisms. Anaerobic digestion technology uses microorganisms to utilize waste and produce methane, which could serve as a source of clean renewable energy. Temperature and substrate composition are among the main factors affecting the performance and stability of anaerobic digestion processes [7,8]. Anaerobic digestion of lignocellulosic biomass provides an excellent opportunity to convert abundantly available bioresources into energy [9]. Straw is an agricultural residue as a result of the production of rapeseed or sunflowers. It is a promising lignocellulosic substrate with a beneficial greenhouse gas balance during biofuel production [10]. Wheat straw is a complex material. Mechanical fractionation and grinding can represent a promising and efficient pretreatment of raw lignocellulosic materials for bioconversion without the addition of water and unnecessary effluent production [11]. Wheat straw must undergo several steps for conversion into biogas. The consecutive pathways include hydrolysis to produce simpler and soluble materials, acetogenesis to produce VFA, and methanogenesis to produce biogas/biomethane [12]. According to the national long-term program to encourage the use of biomass (Sofia, Bulgaria), the total quantity of wheat straw in Bulgaria is 2,714,500 t/year, of which 542,900 t/year remains unused and available [13]. Anaerobic digestion for biogas/biomethane production is an approach with environmental benefit, combining waste disposal with energy production [14]. This technology favors nutrient recycling and increases the stability of the energy supply [15]. Conducting the process of anaerobic digestion under thermophilic conditions is known to strongly affect the performance of biogas digesters leading to increased hydrolysis rates. Elevated temperature accelerates biochemical processes, enabling faster degradation and higher biogas yields from a wide variety of substrates, compared with mesophilic anaerobic digestion [16]. There is an insufficient availability of data in the literature comparing mesophilic and thermophilic treatments. There is a report on spent animal bedding consisting of feces and straw [17]. Böske et al. [18] used a continuous up-flow anaerobic solid-state reactor to treat spent horse bedding, whereas Gómez et al. [19] used a dry batch system to digest spent cow bedding. Under thermophilic conditions, the first authors observed a higher methane yield than at mesophilic temperature, while the second authors reported a lower methane yield at increased temperature. Therefore, the operating parameters of anaerobic digesters and the methods of acceleration and optimization used to improve process efficiency remain of utmost importance. Further intensive research must be carried out in the field of renewable energy generation to solve such a global dilemma.

The aim of this study is to evaluate the biogas/biomethane production potential of a biotechnological process in a laboratory-scale bioreactor, together with identification of the participating microbes, influenced by the high temperature conditions.

2. Materials and Methods

2.1. Bioreactor Design and Anaerobic Biodegradation Performance

The design consisted of a metal bioreactor (BR) with a total volume of 20 dm^3 with a glass hatch, enabling monitoring of the level of culture fluid. Corresponding nozzles and silicone hoses were used as an inlet and outlet. During the experiments, the working volume was 14 dm^3. The BR was equipped with a regulator for automatic temperature regulation in mesophilic (37 ± 0.5 °C) and thermophilic (55 ± 0.5 °C) modes, and a stirring system provided by a DC electric motor at a constant stirring of about 100 rpm. The inoculum used for the mesophilic process was taken from a working pilot-scale methane generating bioreactor, situated in our lab. Bacterial communities for the thermophilic process were taken from a thermophilic bioreactor, containing wheat straw, working with liquid phase recirculation. After feeding, a purge with nitrogen gas was performed to ensure an anaerobic environment. Anaerobic cultivation techniques and equipment were used. The laboratory-scale bioreactor is schematically presented in Figure 1.

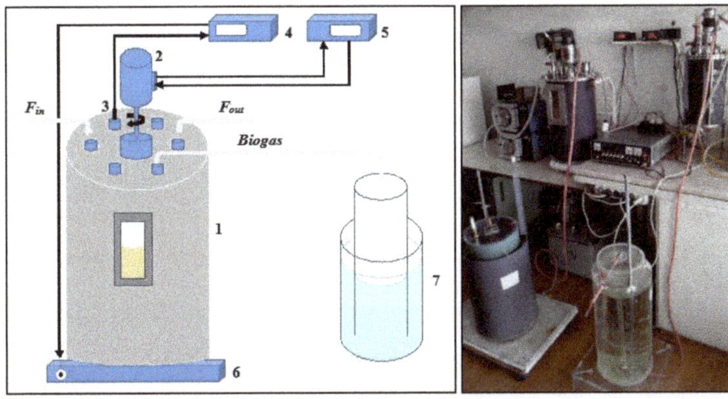

Figure 1. Bioreactor set-up diagram (**left**) and original experimental setup (**right**): 1. bioreactor; 2. DC-motor for stirrer; 3. Pt-100 temperature probe; 4. regulator and process indicator for temperature control; 5. regulator and process indicator for stirring speed; 6. heating device; 7. gas holder (water displacement principle).

Wheat straw and corn stalks were tested as substrates for the anaerobic digestion process. Prior to feeding, the substrates were mechanically pre-treated using a knife mill to decrease the particle size. The final particle size of the substrate was $\leq$2 mm. The air-dried portions were fed into the bioreactor manually in the following manner: about 9 L of bioreactor content was withdrawn and the solid fraction was separated from the liquid fraction using a $\approx$1 mm pore size sieve. The new portion of wheat straw or corn stalks was added to the liquid fraction from the previous step, and the total solids content was recalculated to correspond to 0.5% to 4.5% total solids (w/v). We assumed, using this process, that not only could the digestate water content be recycled, but also a bigger portion of microorganisms saved, together with preserving the buffer capacity of digestate. After the addition of this mixture to the bioreactor, we added the appropriate water quantity to reach the total working volume (14 L), usually about 2.0 liters of distilled water.

2.2. Analytical Methods

The volume of biogas produced (daily and cumulative) was measured with a graduated gas holder by method of water displacement. The composition of the biogas was determined by a specialized gas analyzer (Dräger GmbH, Lubeck, Germany) equipped with infrared sensors for the determination of methane and carbon dioxide, and term conductivity sensors for hydrogen sulfide and oxygen.

The concentration of VFA was analyzed using a Thermo Scientific Focus GC (Thermo Scientific Inc., Waltham, MA, USA) gas chromatograph, equipped with a TG-WAXMS A column (L: 30 m, I.D.: 0.25 mm, film: 0.25 mm), split/splitless injector and FID. A 1mL sample aliquot was acidified using 37% H_3PO_4 until pH reached 2.0, followed by centrifugation at 13,000 rpm/10 min using a Hermle Z 326K centrifuge (HERMLE Labortechnik GmbH, Gosheim, Germany). From every sample, an aliquot of 100 μL was withdrawn and mixed with an equivalent amount of internal standard solution (3,3-dimethylbutyric acid), mixed well on vortex, and 1 μL was then manually injected with a microliter chromatograph syringe.

An assay for residual cellulose concentration was performed according to the photometric method proposed by Updegraff [20]. The appropriate amount of sample from the bioreactor was centrifuged at 3000 g/10 min. Briefly, the solid fraction was processed in the following manner: it was washed with distilled water and then treated with an acetic-nitric reagent for 30 min in a boiling water bath in screw-type glass tubes. After cooling to room temperature, the samples were centrifuged and the solid fraction was preserved and washed with distilled water. The resulting white precipitate (containing mainly cellulose) was dissolved in 67% H_2SO_4. The resulting solution was diluted with distilled water and aliquots were taken, then cooled in an ice bath, mixed with anthrone reagent, and heated in a boiling water bath for 16 min. After cooling down to room temperature, the absorbance of samples was measured at 620 nm using a Jenway 6305 spectrophotometer (Bibby Scientific, Stone, Staffordshire, UK).

Total solids were measured by dehydration of a certain volume of culture liquid at 105 °C, while volatile solids were measured by burning at 575 °C [21].

Biodegradation degree (BD) was calculated according to the following formula:

$$BD = [(ODMi - ODMo)/ODMi] \times 100$$

where BD is the biodegradation degree, ODMi is the input organic dry matter, and ODMo is the output organic dry matter.

Elemental analysis with an automatic analyzer EuroEA 3000 was carried out for elements determination and C/N ratio.

Light microscopy observations were performed after Gram staining, and visualized on a Levenhuk D870T 8M (Levenhuk LTD, Tampa, FL, USA).

2.3. Metagenome's Sequencing and Bioinformatics Analysis

Metagenome library construction and sequencing were performed by Macrogen Inc. (Seoul, South Korea). For library construction, total DNA was extracted from a sample using GeneMATRIX Bacterial & Yeast Genomic DNA Purification Kit (EUR_X, Gdańsk, Poland). The preparation of the 16S metagenomic sequencing library for bacteria was performed using primer pair, which targeted the V3–V4 region (Macrogen primer set), while the archaeal metagenomic library was constructed using the primers 519F_Arch 5′CAGCMGCCGCGGTAA3′ and 806R_Arch 5′GGACTACVSGGGTATCTAAT3′ [22]. Both libraries were analyzed with a Herculase II Fusion DNA Polymerase Nextera XT Index Kit V2. The sequencing (Illumina platform) was conducted with a reading length of 301 bp and FastQC quality control. The assembly results showed that the quality-filtered data contained around 1,626,035,712 total bases, and 5,402,112 read counts for each sample. The percentage of Q20 quality reads was 94.52%.

3. Results and Discussion

Biotechnological processes for anaerobic digestion at 35 °C and 55 °C were performed consecutively in a laboratory-scale bioreactor. Initial experiments followed the biomethane yields for two substrates, namely, corn stalks and wheat straw as waste substrates with no chemical or biological substrate pretreatment applied, unlike as suggested by previous researchers [23,24]. Different types of pretreatment resulted, in most cases, in high investment costs. Substrates were only milled in a chopper/knife mill. Previously, for the delignification of lignocellulosic substrates, various physical, chemical, and biological pretreatment techniques were performed [25,26]. This was required due to the lignocellulosic substrates' complexity, as the typical composition of lignocellulosic materials comprises 10–25% lignin, 40–50% cellulose, and 5–30% hemicellulose [27]. The novelty of our study was that we used untreated wheat straw, relying only on the increased temperature during the process to favor and enhance substrate swelling, decreasing the integrity, increasing the accessibility, and hence the biodegradability.

Two temperature regimes (35 °C and 55 °C) and two substrates were initially applied to estimate and compare the effect. For both substrates subjected to biodegradation at equal loading of 5 g/L, the volumes of biogas released were greater at 55 °C (Figure 2).

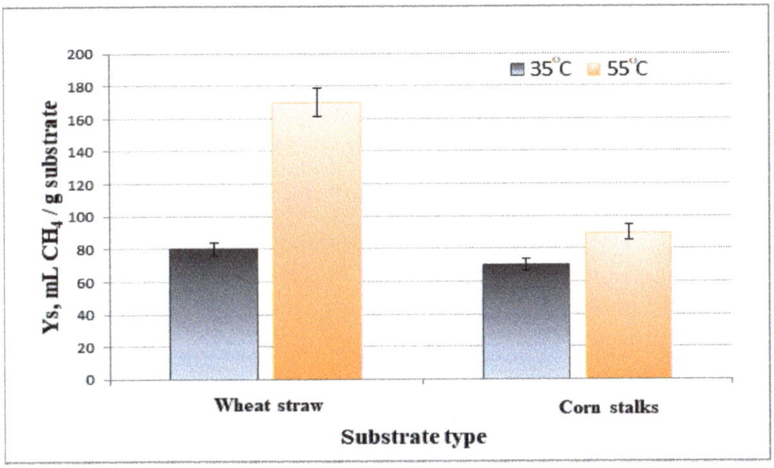

Figure 2. Comparison of the yield coefficients (Y_S) for two substrate types at 35 °C and 55 °C at a constant loading rate of 5 g/L.

The yield coefficient was calculated. Ys is the amount of product obtained/the amount of introduced substrate. As presented in Figure 2, the increase in temperature during the anaerobic digestion process led to a significant increase in biomethane production when applying wheat straw as a substrate (52.8%) and a slight increase when using corn stalks as a substrate (21.9%). This may have been due to the higher lignin content in corn stalks (Table 1).

Table 1. Comparison of some components of wheat straw and corn stalks.

Parameter	Wheat Straw	Corn Stalks
TS, %	93.1 ± 0.05	95.0 ± 0.05
VS, %	88.4 ± 0.05	89.8 ± 0.05
Total nitrogen, g/L	1.1 ± 0.05	0.92 ± 0.05
Proteins, g/L	6.5 ± 0.05	4.0 ± 0.05
Cellulose, % VS	32–38 ± 0.05	26–37 ± 0.05
Hemicellulose, % VS	21–28 ± 0.05	22–29 ± 0.05
Lignin, % VS	15–20 ± 0.05	17–23 ± 0.05

With an aim to break down the three-dimensional morphologic structures, modify the degree of polymerization, and increase the surface area and pore size in these complex substrates, the effect of elevated temperature is to exert its influence on one or more structural features and thus on biomass digestibility, which varies with the changes in the surrounding conditions [28]. In detail, Sorensen et al. suggested thermal acceleration of the reaction steps in four purified cellulases over a small range of temperature (10–50 °C) and substrate loads (0–100 g/L) [29]. At mesophilic conditions, the AD of lignocellulosic wastes occurs at low rates and is almost impossible without thermal, chemical or biological pretreatments [30], while at elevated temperature, these processes take place faster and with a higher biodegradation degree [31]. A possible explanation may be the better swelling of the lignocellulosic material at the higher temperature, and consequently the greater availability of the substrates for the microbial cellulase enzyme systems. Other researchers have also pointed out that thermophilic reactor performance is better, compared with mesophilic reactor performance [32].

Initial experiments involved two substrates; however, the better performance of wheat straw under the tested conditions led us to continue with that substrate, also bearing in mind that wheat straw represents the most abundant biomass available in the European Union for use in bioenergy production.

Biogas volume, and composition produced, were recorded daily to evaluate the process performance. The daily biogas yield using native wheat straw 5 g/L was evaluated (Figure 3).

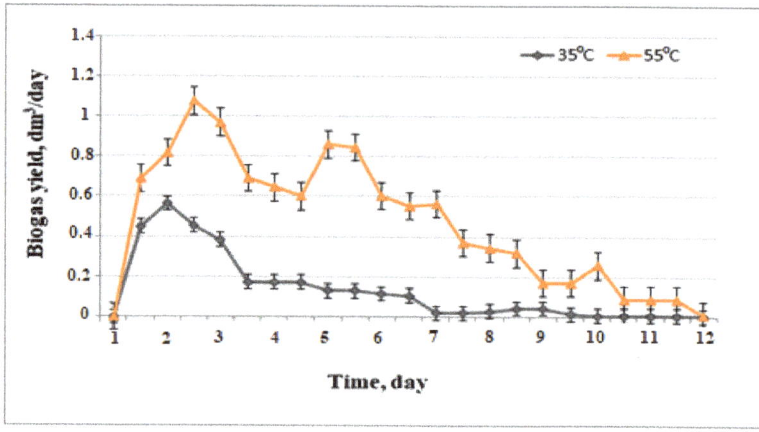

Figure 3. Comparison of biogas yield obtained under mesophilic and thermophilic conditions.

A sharp difference was observed in biogas production at the two temperature regimes in favor of the thermophilic conditions, so the experiments were continued at 55 °C using wheat straw as a substrate. Moreover, a high initial biogas production rate in the thermophilic reactor was found. Thermophilic digestion was evaluated in the same way by Suhartini et al. [33] as giving higher biogas and methane productivity than mesophilic, and was able to operate in a stable manner, whereas mesophilic digestion showed signs of instability.

The next step included estimation of the impact of loading on biogas production (Figure 4).

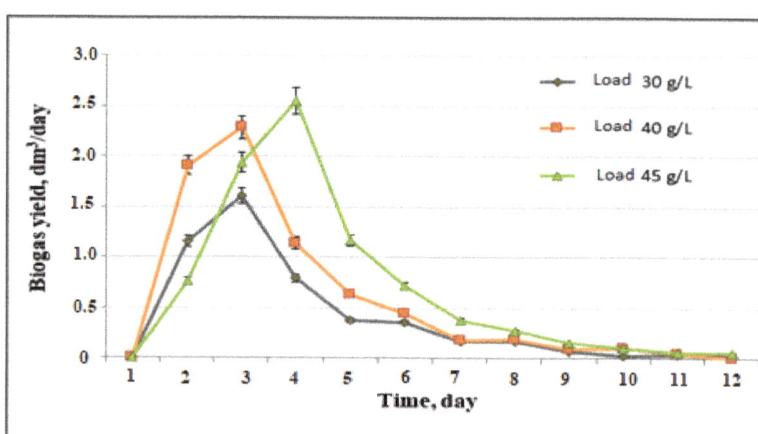

Figure 4. Biogas yield according to substrate loading.

As seen in Figure 4, the biogas yield became higher with the increase in substrate loading. After completion of a 12-day anaerobic digestion process, the cumulative volume of biogas yield was 4.78 L for 1 L of the bioreactor working volume with a substrate loading of 30 g/L of wheat straw, 7.39 L for 40 g/L, and 8.22 L for 45 g/L. The degree of biodegradation was calculated to be 68.9%, 74%, and 72%, respectively.

Simultaneous with the biogas/biomethane yield increase, the quantity of residual cellulose decreased (Figure 5). The highest percent of methane measured was 60% (Figure 5). Similar results were obtained by Poh and Chong [34], who depicted an anaerobic thermophilic system of biodegradation, obtaining 64% of methane production using palm effluent.

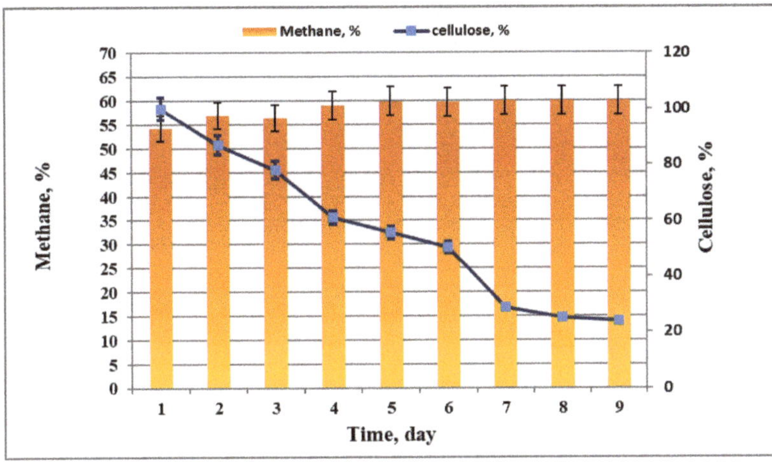

Figure 5. Percentage of methane measured in the biogas as a result of cellulose biodegradation.

The pH during the process of biogas production was in the range of 7.58 to 7.2 (Figure 6). With the increase in VFA concentration (Figure 7) on day 2, the pH slightly dropped. This result suggests that changes in the distribution of fermentative products could cause variation in pH and influence the process itself. The acidification in the initial phase was relatively low, possibly due to the buffering capacity of the inoculum used.

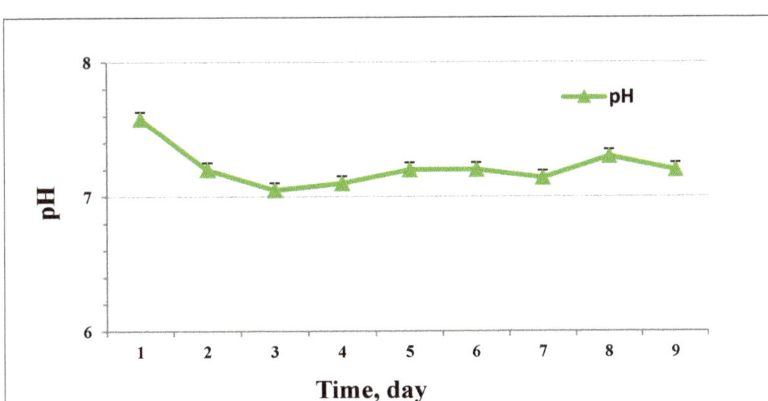

Figure 6. pH values during the process with 30 g/L loading.

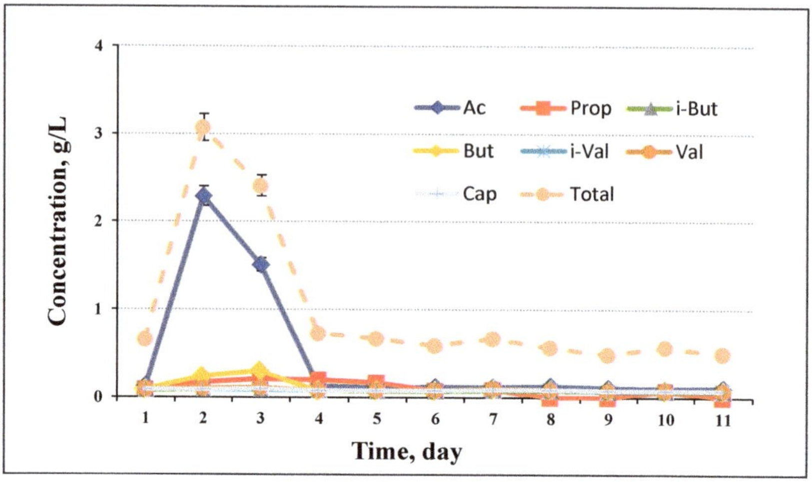

Figure 7. Profile of VFA during the anaerobic digestion process at 30 g/L loading.

Sometimes, in digesters with low buffering capacity, changes in pH, can lead to an imbalance in the process [35]. The profile of VFA, major metabolic product in anaerobic biodegradation, was investigated. Biodegradation of cellulose leads to accumulation in the medium, mainly of acetate, followed by propionate and butyrate. In our experiments, acetate was 80% of other VFA detected (Figure 7). Speece et al. [36] stated that the average VFA concentration correlated with digester temperature, increasing from approximately 400 mg/L to more than 1000 mg/L as the temperature increased from 53 °C to 58 °C, and dropped as temperature decreased back to 53 °C. Aitken et al. [37] also reported that effluent propionate concentrations were relatively high in thermophilic digesters which operated from 51 °C to 55 °C.

The distribution profile of VFA at the other two loadings followed a similar trend. Acetate and propionate were the major VFA produced, followed by butyrate. The results in Figure 6 show that the process was not overloaded. The total VFA content was about 1.2 g/L, decreasing quickly on the second day. This demonstrated that the methane producing process was stable, and only a small amount of straw components were converted to liquid products which remained in the liquid phase.

The role of the C/N ratio is well known and reported in the literature. Most literature sources do not indicate the exact optimal value, instead an optimal range, as the C/N ratio should belong to the interval (15:1 to 30:1). Other authors have determined the ratio for anaerobic degradation of organic waste to be between 20 and 35 [38]. Practically, with all three loads, similar values were observed. Therefore, for the studied processes, the C/N ratio was in the optimal range (Table 2).

Table 2. Elemental analysis of final solid fraction.

Samples	Element (%)		
Substrate	**Carbon, C**	**Nitrogen, N**	**C/N Ratio**
Loading 30 g/L	42.18	2.31	18.26
Loading 40 g/L	42.41	2.27	18.68
Loading 45 g/L	43.06	2.29	18.80

Low C/N ratio is regarded as an important factor limiting anaerobic digestion [39]. When a certain substrate has a very low C/N ratio, co-digestion with carbon-rich co-substrates is recommended, which appeared not necessary in our case. The main factors of anaerobic biodegradation of wheat straw proved that they strongly affected the performance of biogas producing bioreactors, leading to increased hydrolysis rates of thermophilic digesters. Their optimization becomes crucial for the results sought. In this study, the highest measured cumulative biogas yield was reached (8.22 L for a liter of BR working volume) at 45 g/L loading. The highest biodegradation was achieved at 40 g/L substrate loading (74%). However, conducting the processes at 55 °C led to a faster and higher degradation rate in all cases. Despite that lignocellulosic substrates are difficult to decompose, thermophilic AD showed that it can enable a higher organic loading. The accomplished process at 55 °C revealed that thermophilic systems were able to treat this type of agricultural waste at high loads. An increased methane yield was achieved for a shorter period. Accelerated metabolic rates and biogas yields from a wide variety of substrates at 50–70 °C greater than mesophilic anaerobic digestion (30–42 °C) were reported by Weiland [15].

Decreasing the viscosity, and hence the need for energy for vigorous stirring, are other important issues [40]. Therefore, we turned our attention to how much energy we would obtain from the methane obtained from the biogas mixture. It is well known that methane production using wheat straw is one of the best applications of agricultural waste [41]. In addition, biogas/methane is produced using anaerobic degradation of lignocellulosic biomass, which has a higher energy efficiency compared with ethanol production from wheat straw [42]. In addition, wheat was selected as it is the main crop of many nations; residues from agricultural activities in the European Union exceed 200 million tons every year and mostly constitute cereal straw [43].

Before discussing the energy efficiency of the studied process with only mechanical pretreatment, it is worth noting that it was carried out at laboratory-scale in a 14 L batch reactor. This methodology was useful and effective for the determination of optimal substrate pretreatment conditions and loading, and can serve as the basis for further detailed studies to be performed on full-scale devices, where the energy balance will be different in a positive way.

The highest energy yield was reached at a load of 40 g/L. We obtained less energy at 45 g/L per unit of substrate (Figure 8). The difference in the obtained energy between 40 g/L and 45 g/L load was smaller than between 30 g/L and 40 g/L (Figure 9).

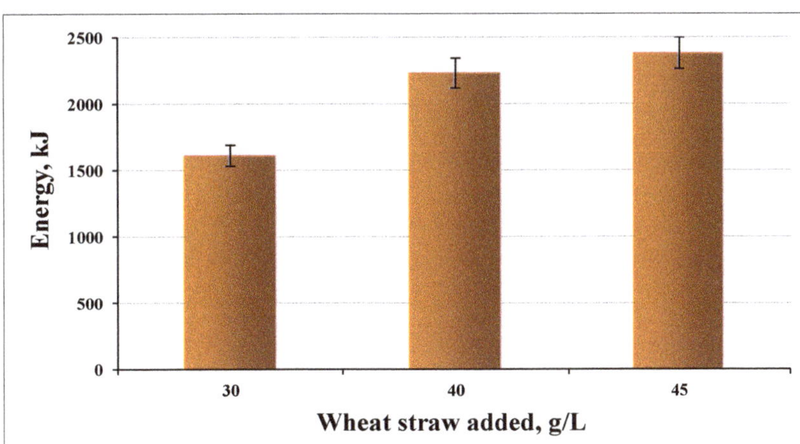

Figure 8. Energy yield at different substrate loadings.

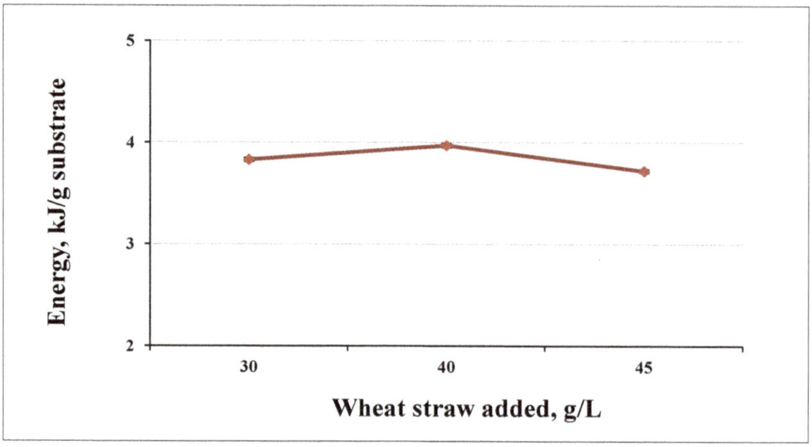

Figure 9. Energy yields at different loading for 1 g of substrate.

Another important feature of the high temperature conditions was keeping away pathogens. Operation under thermophilic conditions also reduces the need for external sanitation. High temperature conditions can positively affect the pathogen elimination in the resultant digestate [44]. Figure 10 reveals a light microscopy image, obtained from our thermophilic methanogenic digester operating at 55 °C, where short rod-shaped and coccoid forms were observed. The participating microbial community had a complex structure and probably used synergetic mechanisms in the performance of anaerobic biodegradation of the strong and complex substrate, as native wheat straw was converted to energy production.

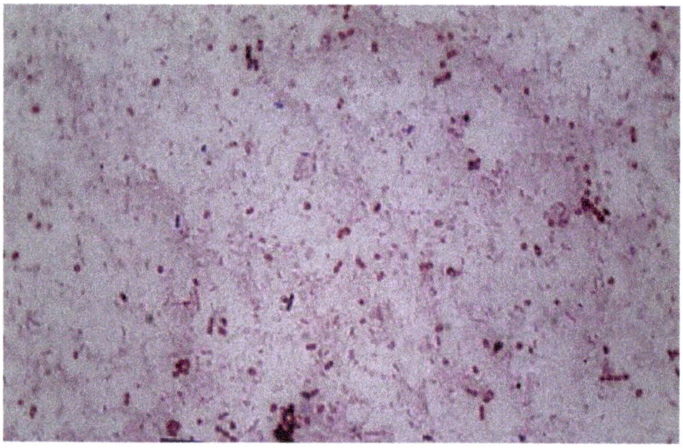

Figure 10. Light microscopy images obtained from the thermophilic digester (magnification ×1000).

In nature, cellulose, lignocellulose, and lignin are major sources of renewable plant biomass and, therefore, their recycling is indispensable for the carbon cycle. The synergistic action of a variety of microorganisms is needed for recycling lignocellulosic materials. The AD process consists of four sequential biochemical steps: hydrolysis by hydrolytic bacteria, acidogenesis by acidogenic bacteria, acetogenesis by acetogenic bacteria, and methanogenesis by methanogenic archaea. Most essential for the last step of methanogenesis are both acetotrophic and hydrogenotrophic methanogens, but reports about their roles during this phase of the process are very limited. Microorganisms usually utilize the available carbon and nitrogen sources, which are necessarily channeled towards microbial proliferation to produce biogas and energy [45]. Temperature plays the most crucial role in the digestion rate, especially the rates of hydrolysis and methanogenesis, and determines the variety of microorganisms [46]. Metagenomics enables the comprehensive analysis of microbiomes by identifying all species participating in this complex process.

In our study, the metagenome investigation of the biodiversity in the thermophilic anaerobic digester showed that bacteria strongly prevailed over archaea, comprising 98.63% of the microbial content (Figure 11). The most abundant phyla were *Firmicutes* (38.39%), *Proteobacteria* (24.58%), *Bacteroidetes* (22.66%), *Spirochaetes* (3.92%), *Synergistetes* (2.72%), *Chloroflexi* (1.11%), *Coprothermobacterota* (0.89%), and *Actinobacteria* (0.71%).

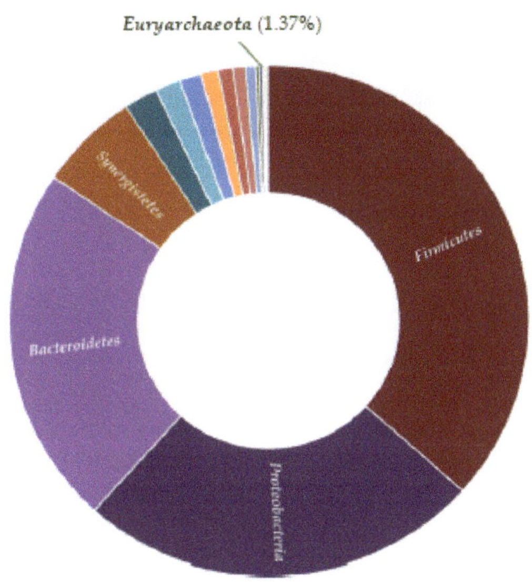

Figure 11. Microbial diversity in the thermophilic bioreactor (the main bacterial and archaeal phyla).

Among bacterial classes, the most abundant were Clostridia, Bacteroidia, Betaproteobacteria, Gammaproteobacteria, and Alphaproteobacteria, followed by Spirochaetia (Figure 12).

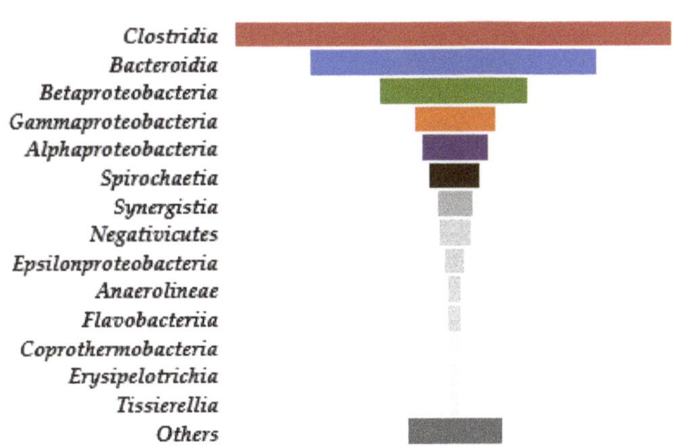

Figure 12. Microbial diversity in the thermophilic bioreactor.

Utilizing a wide range of substrates and tolerating high concentrations of VFA and alcohols, *Clostridia* were the most abundant bacteria, followed by *Bacteroidetes*, which are largely involved in biomass degradation and are indispensable in the prevention of bioreactor acidosis [47]. The prevailing bacterial genera were *Proteiniphilum* (12.25%), *Proteiniborus* (8.40%), *Pseudomonas* (5.97%), *Advenella* (4.26%), *Treponema* (3.68%), *Gracilibacter* (3.03%), *Parabacteroides* (2.94%), *Variimorphobacter* (2.84%), *Comamonas* (2.72%), *Anaerobacterium* (1.81%), *Ruminiclostridium* (1.78%), *Acetomicrobium* (1.74%), and *Thermoclostridium* (1.65%). The corresponding species are shown in Figure 13.

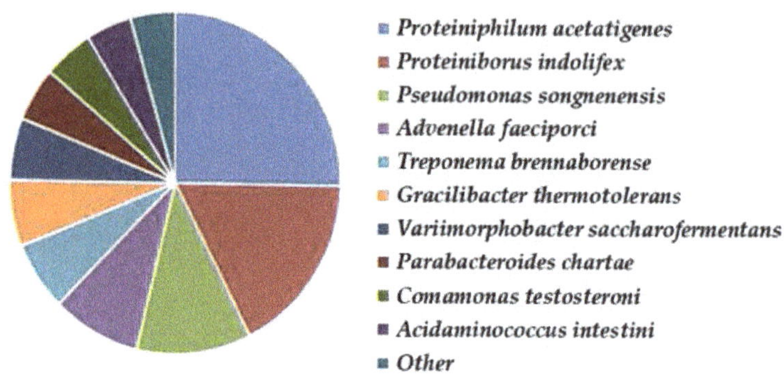

Figure 13. Bacterial diversity in the thermophilic bioreactor for biogas production.

Archaeal share was 1.37% of the microflora in the thermophilic bioreactor, as representatives of the genera *Methanocorpusculum*, *Methanobacterium*, *Methanoculleus*, *Methanosarcina*, *Methanomassiliicoccus*, *Methanosarcina*, and *Methanoregula* were the most abundant (Figure 14). All are members of the thermophilic phylum *Euryarchaeota*.

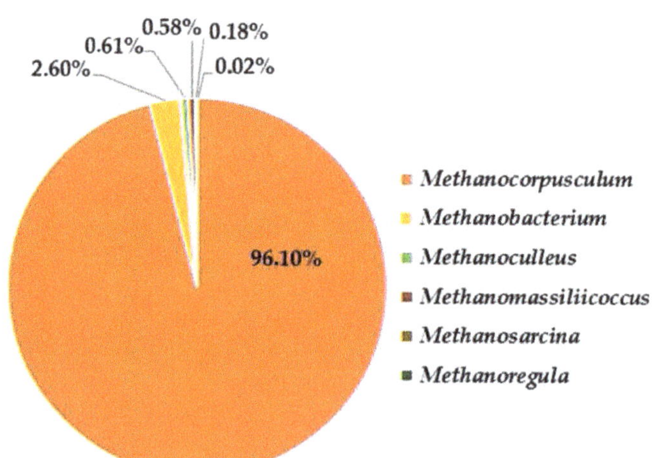

Figure 14. Archaeal diversity in the thermophilic bioreactor for biogas production.

Methanocorpusculum is coccoid in shape, has a temperature optimum between 30 and 40 °C, and is able to reduce CO_2 to methane using hydrogen, formate, or alcohol. The species *Methanocorpusculum aggregans* and *Methanobacterium formicicum* (Figure 15) are rather mesophilic archaeons [48], and hydrogenotrophic methanogens, which are reported here for the first time in a thermophilic bioreactor. The presence of *Methanobacterium* (0.04%) suggested that acetotrophic methanogenesis also takes part in the bioreactor.

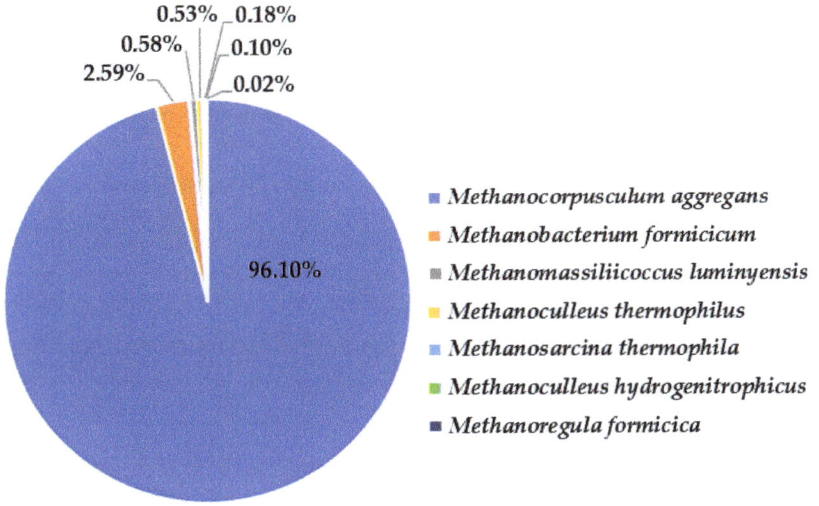

Figure 15. Archaeal species in the thermophilic bioreactor for biogas production.

Reporting the presence of *Methanobacteriales* (*M. formicium*), our results are consistent with previous studies on H_2 methane production [49–51]. As suggested by other authors, this order of methanogens has a selective advantage over other hydrogenotrophic methanogens in ex situ bio-methanation [52]. Thermophilic conditions can also favor coccoid hydrogenotrophic methanogens, such as *Methanosarcina* [53]. Lee et al. [54] proved that in conditions below 65 °C, microbes affiliated with methanogens dominated the population, while at higher temperatures acidogenic bacteria prevailed.

Stage-specific bacterial and archaeal populations were found to reside in thermophilic or mesophilic AD bioreactors [55]. Several metagenomic studies revealed high functional redundancy in AD bioreactors [56], explained by taxonomic variations under certain changing operational conditions, while biogas production remained relatively stable. Such a phenomenon could also explain the recovery of AD bioreactors without additional inoculation. However, the functional potential of metagenomes in AD bioreactors depends strongly on the feedstocks. Microbial communities in vegetable and fruit residues digesters have been studied [57]. This process could continue in a circle, as the resultant waste digestate after the anaerobic digestion process could serve as a biofertilizer or a cultivation medium for microalgae [58]. In this way, lowering the production cost of microalgae increases their potential for producing high-value products [59]. It was also reported that in some cases the nutrients from the digestate were more useful than other fertilizers, and digestate application and incorporation into the soil before planting could be recommended.

4. Conclusions

Biotechnological exploitation of lignocellulosic wastes is promising for sustainable and environmentally-friendly energy production, because of the abundant availability of these renewable sources.

Wheat straw, as agricultural waste, was involved in anaerobic digestion at elevated temperature, and biogas was obtained with biomethane as an energy carrier for further fuel production.

Conducting the process at 55 °C led to faster and higher hydrolysis rates and an increased biodegradation degree, achieved in a stable system that enabled higher organic loading.

A microbial consortium is the main tool for efficient utilization of complex waste substrates. Metagenomics was applied in the community structure elucidation, proving its complexity and probable synergetic mechanisms in the performance of anaerobic biodegradation. The most important genera were *Proteiniphilum*, *Proteiniborus*, and *Pseudomonas*, while the archaeal share of the microflora in the thermophilic bioreactor included the genera *Methanocorpusculum*, *Methanobacterium*, *Methanomassiliicoccus*, *Methanoculleus* and *Methanosarcina*. Better understanding of natural microbial communities can make their future application in energy production more felicitous.

Author Contributions: Conceptualization, project administration, L.K.; methodology, investigation V.H., P.P. and I.S.; writing—original draft preparation and editing, P.P. and L.K. All authors have read and agreed to the published version of the manuscript.

Funding: This research was funded by the BNSF, grant number KP-06-IP-China/3. The APC was funded by the same grant.

Institutional Review Board Statement: Not applicable.

Informed Consent Statement: Not applicable.

Data Availability Statement: Not applicable.

Conflicts of Interest: The authors declare no conflict of interest.

References

1. Hööka, M.; Tang, X. Depletion of fossil fuels and anthropogenic climate change—A review. *Energy Policy* **2013**, *52*, 797–809. [CrossRef]
2. Demirbas, A. Competitive liquid biofuels from biomass. *Appl. Energy* **2011**, *88*, 17–28. [CrossRef]
3. Kralova, I.; Sjöblom, J. Biofuels–Renewable Energy Sources: A Review. *J. Dispers. Sci. Technol.* **2010**, *31*, 409–425. [CrossRef]
4. Moghaddam, E.A.; Ahlgren, S.; Hulteberg, C.; Nordberg, Å. Energy balance and global warming potential of biogas-based fuels from a life cycle perspective. *Fuel Proc. Technol.* **2015**, *132*, 74–82. [CrossRef]
5. Saghir, M.; Zafar, S.; Tahir, A.; Ouadi, M.; Siddique, B.; Hornung, A. Unlocking the potential of biomass energy in Pakistan. *Front. Energy Res.* **2019**, *7*, 24. [CrossRef]
6. Dahlgren, S. Biogas-based fuels as renewable energy in the transport sector: An overview of the potential of using CBG, LBG and other vehicle fuels produced from biogas. *Biofuels* **2020**, *13*, 587–599. [CrossRef]
7. Ziganshin, A.M.; Liebetrau, J.; Pröter, J.; Kleinsteuber, S. Microbial community structure and dynamics during anaerobic digestion of various agricultural waste materials. *Appl. Microbiol. Biotechnol.* **2013**, *97*, 5161–5174. [CrossRef]
8. Labatut, R.A.; Angenent, L.T.; Scott, N.R. Conventional mesophilic vs. thermophilic anaerobic digestion: A trade-off between performance and stability? *Water Res.* **2014**, *15*, 249–258. [CrossRef]
9. Sawatdeenarunat, C.; Surendra, K.C.; Takara, D.; Oechsner, H.; Khanal, S.K. Anaerobic digestion of lignocellulosic biomass: Challenges and opportunities. *Bioresour. Technol.* **2015**, *178*, 178–186. [CrossRef]
10. Passoth, V.; Sandgren, M. Biofuel production from straw hydrolysates: Current achievements and perspectives. *Appl. Microbiol. Biotechnol.* **2019**, *103*, 5105–5116. [CrossRef]
11. Barakat, A.; Chuetor, S.; Monlau, F.; Solhy, A.; Rouau, X. Eco-friendly dry chemo-mechanical pretreatments of lignocellulosic biomass: Impact on energy and yield of the enzymatic hydrolysis. *Appl. Energy* **2014**, *113*, 97–105. [CrossRef]
12. Meegoda, J.N.; Li, B.; Patel, K.; Wang, L.B. A Review of the Processes, Parameters, and Optimization of Anaerobic Digestion. *Int. J. Environ. Res. Public Health* **2018**, *15*, 2224. [CrossRef] [PubMed]
13. Woody Biomass for Energy. NGO Concerns and Recommendations. Available online: http//Biomass_Programme_EN.pdf (accessed on 16 December 2021).
14. Singh, A.; Kuila, A.; Adak, S.; Bishai, M.; Banerjee, R. Utilization of vegetable wastes for bioenergy generation. *Agric. Res.* **2012**, *1*, 213–222. [CrossRef]
15. Weiland, P. Biogas production: Current state and perspective. *Appl. Microbiol. Biotechnol.* **2010**, *85*, 849–860. [CrossRef]
16. Moset, V.; Poulsen, M.; Wahid, R.; Hojberg, O.; Moller, H.B. Mesophilic versus thermophilic anaerobic digestion of cattle manure: Methane productivity and microbial ecology. *Microb. Biotechnol.* **2015**, *8*, 787–800. [CrossRef]
17. Riggio, S.; Hernandéz-Shek, M.A.; Torrijos, M.; Vives, G.; Esposito, G.; van Hullebusch, E.D.; Steyer, J.P.; Escudié, R. Comparison of the mesophilic and thermophilic anaerobic digestion of spent cow bedding in leach-bed reactors. *Bioresour. Technol.* **2017**, *234*, 466–471. [CrossRef]
18. Böske, J.; Wirth, B.; Garlipp, F.; Mumme, J.; Van den Weghe, H. Upflow anaerobic solidstate (UASS) digestion of horse manure: Thermophilic vs. mesophilic performance. *Bioresour. Technol.* **2015**, *175*, 8–16. [CrossRef]
19. Gómez, X.; Blanco, D.; Lobato, A.; Calleja, A.; Martínez-Núñez, F.; Martin-Villacorta, J. Digestion of cattle manure under mesophilic and thermophilic conditions: Characterization of organic matter applying thermal analysis and ^{1}H NMR. *Biodegradation* **2011**, *22*, 623–635. [CrossRef]
20. Updegraff, D.M. Semimicro determination of cellulose in biological materials. *Anal. Biochem.* **1969**, *3*, 420–424. [CrossRef]

21. American Public Health Association. *Standard Methods for the Examination of Waste and Wastewater APHA*; APHA: Washington, DC, USA, 2005.
22. Pausan, M.R.; Csorba, C.; Singer, G.; Till, H.; Schöpf, V.; Santigli, E.; Klug, B.; Högenauer, C.; Blohs, M.; Moissl-Eichinger, C. Exploring the Archaeome: Detection of Archaeal Signatures in the Human Body. *Front. Microbiol.* **2019**, *10*, 2796. [CrossRef]
23. Novakovic, J.; Kontogianni, N.; Barampouti, E.M. Towards upscaling the valorization of wheat straw residues: Alkaline pretreatment using sodium hydroxide, enzymatic hydrolysis and biogas production. *Environ. Sci. Pollut. Res.* **2021**, *28*, 24486–24498. [CrossRef] [PubMed]
24. Andersen, L.F.; Parsin, S.; Lüdtke, O. Biogas production from straw—The challenge feedstock pretreatment. *Biomass Convers. Biorefinery* **2020**, *12*, 379–402. [CrossRef]
25. Simeonov, I.; Denchev, D.; Kabaivanova, L.; Kroumova, E.; Chorukova, E.; Hubenov, V.; Mihailova, S. Different types of pre-treatment of lignocellulosic wastes for methane production. *Bulg. Chem. Commun.* **2017**, *49*, 430–435.
26. Hubenov, V.; Carcioch, R.A.; Ivanova, J.; Vasileva, I.; Dimitrov, K.; Simeonov, I.; Kabaivanova, L. Biomethane production using ultrasound pre-treated maize stalks with subsequent microalgae cultivation. *Biotechnol. Biotechnol. Equip.* **2020**, *34*, 800–809. [CrossRef]
27. Anwar, Z.; Gulfraz, M.; Irshad, M. Agro-industrial lignocellulosic biomass a key to unlock the future bio-energy: A brief review. *J. Radiat. Res. Appl. Sci.* **2014**, *7*, 163–173. [CrossRef]
28. Zhu, L.; O'Dwyer, J.; Chang, V.; Granda, C.; Holtzapple, M. Structural features affecting biomass enzymatic digestibility. *Bioresour. Technol.* **2008**, *99*, 3817–3828. [CrossRef]
29. Sorensen, T.; Cruys-Bagger, N.; Windahl, M.; Badino, S.; Borch, K.; Westh, P. Temperature Effects on Kinetic Parameters and Substrate Affinity of Cel7A Cellobiohydrolases. *J. Biol. Chem.* **2015**, *290*, 22193–22202. [CrossRef]
30. Ning, X.; Shixun, L.; Fengxue, X.; Jie, Z.; Honghua, J.; Jiming, X.; Min, J.; Weiliang, D. Biomethane Production From Lignocellulose: Biomass Recalcitrance and Its Impacts on Anaerobic Digestion. *Front. Bioeng. Biotechnol.* **2019**, *7*, 191. [CrossRef]
31. Tu, M.; Pan, X.; Saddler, J.N. Adsorption of cellulase on cellulolytic enzyme lignin from lodgepole pine. *J. Agric. Food. Chem.* **2009**, *57*, 7771–7778. [CrossRef]
32. Shi, X.; Guo, X.; Zuo, J.; Wang, Y.; Zhang, M. A comparative study of thermophilic and mesophilic anaerobic co-digestion of food waste and wheat straw: Process stability and microbial community structure shifts. *Waste Manag.* **2018**, *75*, 261–269. [CrossRef]
33. Suhartini, S.; Heavena, S.; Banks, C.J. Comparison of mesophilic and thermophilic anaerobic digestion of sugar beet pulp: Performance, dewaterability and foam control. *Bioresour. Technol.* **2014**, *152*, 202–211. [CrossRef] [PubMed]
34. Poh, P.E.; Chong, M.F. Biomethanation of Palm Oil Mill Effluent (POME) with a thermophilic mixed culture cultivated using POME as a substrate. *Chem. Eng. J.* **2010**, *164*, 146–154. [CrossRef]
35. Murto, M.; Björnsson, L.; Mattiasson, B. Impact of food industrial waste on anaerobic co-digestion of sewage sludge and pig manure. *J. Environ. Manag.* **2004**, *70*, 101–107. [CrossRef] [PubMed]
36. Speece, R.E.; Boonyakitsombut, S.; Kim, M.; Azbar, N.; Ursillo, P. *Overview of Anaerobic Treatment: Thermophilic and Propionate Implications Water Environment Research*; Wiley: Hoboken, NJ, USA, 2006; Volume 78, pp. 460–473. Available online: https://www.jstor.org/stable/25053534 (accessed on 17 April 2022).
37. Aitken, M.D.; Walters, G.W.; Crunk, P.L.; Willis, J.L.; Farrell, J.B.; Schafer, P.L.; Arnet, C.; Turner, B.G. Laboratory Evaluation of Thermophilic-Anaerobic Digestion to Produce Class A Biosolids. 1. Stabilization Performance of a Continuous-Flow Reactor at Low Residence Time. *Water Environ. Res.* **2005**, *77*, 3019–3027. [CrossRef] [PubMed]
38. Cerón-Vivas, A.; Cáceres, K.T.; Rincón, A.; Cajigas, Á.A. Influence of pH and the C/N ratio on the biogas production of wastewater. *Rev. Fac. De Ing. Univ. De Antioq.* **2019**, *92*, 70–79. [CrossRef]
39. Khalid, A.; Arshad, M.; Anjum, M.; Mahmood, T.; Dawson, L. The anaerobic digestion of solid organic waste. *Waste Manag.* **2011**, *31*, 1737–1744. [CrossRef] [PubMed]
40. Grim, J.; Malmros, P.; Schnürer, A.; Nordberg, Å. Comparison of pasteurization and integrated thermophilic sanitation at a full-scale biogas plant—Heat demand and biogas production. *Energy* **2015**, *79*, 419–427. [CrossRef]
41. Weitemeyer, S.; Kleinhans, D.; Vogt, T.; Agert, C. Integration of Renewable Energy Sources in future power systems: The role of storage. *Renew. Energy* **2015**, *75*, 14–20. [CrossRef]
42. De Wit, M.; Faaij, A. European biomass resource potential and costs. *Biomass Bioenergy* **2010**, *34*, 188–202. [CrossRef]
43. Seruga, P.; Krzywonos, M.; Palusza, Z.; Urbanowska, A.; Pawlak-Kruczek, H.; Niedzwiecki, Ł.; Pinkowska, H. Pathogen Reduction Potential in Anaerobic Digestion of Organic Fraction of Municipal Solid Waste and Food Waste. *Molecules* **2020**, *25*, 275. [CrossRef]
44. Börjesson, P.; Mattiasson, B. Biogas as a resource-efficient vehicle fuel. *Trends Biotechnol.* **2008**, *26*, 7–13. [CrossRef] [PubMed]
45. Kaldis, F.; Cysneiros, D.; Day, J.; Karatzas, K.-A.G. Chatzifragkou A, Anaerobic Digestion of Steam-Exploded Wheat Straw and Co-Digestion Strategies for Enhanced Biogas Production. *Appl. Sci.* **2020**, *10*, 8284. [CrossRef]
46. Anukam, A.; Mohammadi, A.; Naqvi, M.; Granström, K. A Review of the Chemistry of Anaerobic Digestion: Methods of Accelerating and Optimizing Process Efficiency. *Processes* **2019**, *7*, 504. [CrossRef]
47. Bertucci, M.; Calusinska, M.; Goux, X.; Rouland-Lefèvre, C.; Untereiner, B.; Ferrer, P.; Gerin, P.A.; Delfosse, P. Carbohydrate Hydrolytic Potential and Redundancy of an Anaerobic Digestion Microbiome Exposed to Acidosis, as Uncovered by Metagenomics. *Appl. Environ. Microbiol.* **2019**, *85*, e00895-19. [CrossRef]

48. Jiang, J.; Liu, J.; Wang, C.; Wang, C.; Zhao, X.; Yin, F.; Zhang, W. Long-Term Biocatalytic Methanation of Hydrogen and Carbon Dioxide Based on Closed Nutrient Recycling and Microbial Community Succession. *SSRN eJournal* **2022**, 25. [CrossRef]
49. Darcy, T.J.; Sandman, K.; Reeve, J.N. *Methanobacterium formicicum*, a mesophilic methanogen, contains three HFo histones. *J. Bacteriol.* **1995**, *177*, 858–860. [CrossRef]
50. Kougias, P.G.; Treu, L.; Benavente, D.P.; Boe, K.; Campanaro, S.; Angelidaki, I. Ex-situ biogas upgrading and enhancement in different reactor systems. *Bioresour. Technol.* **2017**, *225*, 429–437. [CrossRef]
51. Rachbauer, L.; Beyer, R.; Bochmann, G.; Fuchs, W. Characteristics of adapted hydrogenotrophic community during biomethanation. *Sci. Total Environ.* **2017**, *595*, 912–919. [CrossRef]
52. Savvas, S.; Donnelly, J.; Patterson, T.; Chong, Z.S.; Esteves, S.R. Biological methanation of CO_2 in a novel biofilm plug-flow reactor: A high rate and low parasitic energy process. *Appl. Energy* **2017**, *202*, 238–247. [CrossRef]
53. Demirel, B.; Scherer, P. The roles of acetotrophic and hydrogenotrophic methanogens during anaerobic conversion of biomass to methane: A review. *Rev. Environ. Sci. Biotechnol.* **2008**, *7*, 173–190. [CrossRef]
54. Lee, M.; Hidaka, T.; Tsuno, H. Effect of temperature on performance and microbial diversity in hyperthermophilic digester system fed with kitchen garbage. *Bioresour. Technol.* **2008**, *99*, 6852–6860. [CrossRef] [PubMed]
55. Kabaivanova, L.; Hubenov, V.; Dimitrova, L.; Simeonov, I.; Wang, H.; Petrova, P. Archaeal and Bacterial Content in a Two-Stage Anaerobic System for Efficient Energy Production from Agricultural Wastes. *Molecules* **2022**, *27*, 1512. [CrossRef] [PubMed]
56. Yang, Y.; Yu, K.; Xia, Y.; Lau, F.; Tang, D.; Fung, W.; Fang, H.; Zhang, T. Metagenomic analysis of sludge from full-scale anaerobic digesters operated in municipal wastewater treatment plants. *Appl. Microbiol. Biotechnol.* **2014**, *98*, 5709–5718. [CrossRef] [PubMed]
57. Ivanova, J.; Vasileva, I.; Kabaivanova, L. Enhancement of Algal Biomass Accumulation Using Undiluted Anaerobic Digestate. *Int. J. Pharma Med. Biol. Sci.* **2020**, *9*, 111–116. [CrossRef]
58. Vasileva, I.; Ivanova, J.; Angelova, L. Urea from waste waters—Perspective nitrogen and carbon source for green algae Scenedesmus sp. cultivation. *J. Int. Sci. Publ.* **2016**, *10*, 311–319.
59. Lošák, T.; Hlušek, J.; Bělíková, H.; Vítězová, M.; Vítěz, T.; Antonkiewicz, J. What is more suitable for kohlrabi fertilization—Digestate or mineral fertilizers? *Acta Univ. Agric. Silvic. Mendel. Brun.* **2015**, *63*, 787–791. [CrossRef]

life

MDPI

Article

Application of Sewage Sludge in a Rice (*Oryza sativa* L.)-Wheat (*Triticum aestivum* L.) System Influences the Growth, Yield, Quality and Heavy Metals Accumulation of Rice and Wheat in the Northern Gangetic Alluvial Plain

Surendra Singh Jatav [1], Satish Kumar Singh [1,*], Manoj Parihar [2], Amnah Mohammed Alsuhaibani [3], Ahmed Gaber [4] and Akbar Hossain [5,*]

[1] Department of Soil Science and Agricultural Chemistry, Institute of Agricultural Sciences, Banaras Hindu University, Varanasi 221005, Uttar Pradesh, India; surendra.jatav1@bhu.ac.in
[2] ICAR-Vivekananda Parvatiya Krishi Anusandhan Sansthan, Almora 263601, Uttarakhand, India; manojbhu7@gmail.com
[3] Department of Physical Sport Science, College of Education, Princess Nourah bint Abdulrahman University, P.O. Box 84428, Riyadh 11671, Saudi Arabia; amalsuhaibani@pnu.edu.sa
[4] Department of Biology, College of Science, Taif University, P.O. Box 11099, Taif 21944, Saudi Arabia; a.gaber@tu.edu.sa
[5] Department of Agronomy, Bangladesh Wheat and Maize Research Institute, Dinajpur 5200, Bangladesh
* Correspondence: sksingh_1965@rediffmail.com (S.K.S.); akbar.hossain@bwmri.gov.bd (A.H.)

check for updates

Citation: Jatav, S.S.; Singh, S.K.; Parihar, M.; Alsuhaibani, A.M.; Gaber, A.; Hossain, A. Application of Sewage Sludge in a Rice (*Oryza sativa* L.)-Wheat (*Triticum aestivum* L.) System Influences the Growth, Yield, Quality and Heavy Metals Accumulation of Rice and Wheat in the Northern Gangetic Alluvial Plain. *Life* **2022**, *12*, 484. https://doi.org/10.3390/life12040484

Academic Editor: Milka Mileva

Received: 4 March 2022
Accepted: 25 March 2022
Published: 27 March 2022

Abstract: For a sustainable and profitable agriculture production system, balanced and integrated use of nutrients is a key strategy. In addition, partial replacement of chemical fertilizers with organics ones reduces both environmental concerns and economic costs and provides greater soil health benefits. With this hypothesis, an experiment was conducted to assess the yield and economic benefits of a rice-wheat cropping system (RWCS) as influenced by the joint application of sewage sludge (SSL) and fertilizer. The treatments comprised: without fertilizer or SSL; 100% recommended dose of fertilizers (RDF); 100% RDF + 20 Mg ha^{-1} SSL; 100% RDF + 30 Mg ha^{-1} SSL; 50% RDF + 20 Mg ha^{-1} SSL; 60% RDF + 20 Mg ha^{-1} SSL; 70% RDF + 20 Mg ha^{-1} SSL; 50% RDF + 30 Mg ha^{-1} SSL; 60% RDF + 30 Mg ha^{-1} SSL and 70% RDF + 30 Mg ha^{-1} SSL. The experiment was laid out in a randomized block design with three replications. The result of our study indicate that the highest percent increase in mean plant height i.e., ~14.85 and ~13.90, and grain yield i.e., ~8.10 and ~18.90 for rice and wheat, respectively, were recorded under 100% RDF + 30 Mg SSL ha^{-1} treatment compared to 100% RDF, while 70% RDF + 20 Mg ha^{-1} SSL produced a statistically equivalent grain yield of 100% RDF in RWCS. The application of 20 and 30 Mg SSL ha^{-1} along with recommended or reduced fertilizer dose, significantly increased the heavy metal content in plant and soil systems above that of 100% RDF, but this enhancement was found within permissible limits. Moreover, the reduced use of SSL i.e., 20 Mg SSL ha^{-1}, resulted in lower heavy metal content in grain and soil than did the 30 Mg ha^{-1} SSL treatment, but significantly higher than in the absolute control or 100% RDF treatment. In summary, the use of 20 Mg ha^{-1} SSL along with 70% RDF provided a safer, profitable and sustainable option in a rice-wheat cropping system in the middle Ganegatic alluvial plain.

Keywords: plant height; yield; protein content; rice-wheat cropping system; sewage sludge

1. Introduction

A rice-wheat cropping system (RWCS) is the main cropping pattern occupying 24 million hectares (Mha) of cultivated land in the Indo-Gangetic Plains (IGP) in South Asian subtropics and China [1]. This covers an area of 13.5 Mha in the IGP. Out of this 10, 2.2, 0.8 and 0.5 Mha lie in India, Pakistan, Bangladesh, and Nepal, respectively, and the remainder of 10.5 Mha is found in China [1–3]. Therefore, it was judged suitable to use with a test crop in the present study. In RWCS, nutrient removal occurs more than replenishment with the application of chemical fertilizer [4]. This situation is anticipated to worsen in the future, as more food needs to be produced to feed an ever-increasing population. As a result, the use of organic amendments with chemical fertilizers is to be encouraged in order to maintain soil fertility for sustainable agriculture [5]. In sustainable agricultural, the use of traditionally applied inorganic fertilizers cannot be over-emphasized because of the high fertilizer cost and their negative impact on the soil environment. Therefore, the substitution of available organic wastes is required [6,7]. The long-term and continuous uptake of nutrients from soils without adding organic manure leads to land degradation. The intensive application of inorganic fertilizers also decreases soil quality due to salt accumulation (Cl^- and SO_4^{2-}) in the rhizosphere zone [8]. Hence, there is an imperative need to select suitable organic manure for replacement or reduce inorganic fertilizer doses in the RWCS.

Different organic manure and waste occur in nature, but due to continuous urbanization, a vast amount of sewage sludge (SSL) is being produced which can also be used as manure for improving agricultural production and to mitigate environmental concerns [9] with economic feasibility [10]. Sewage sludge is a heterogeneous mixture of undigested organic materials such as cellulose, plant residues, oil or faecal material, and inorganic materials [11]. In developing countries, the number of sewage treatment plants is increasing due to growing urbanization and development. In India, around 100,000 million tons (Mt) SSL or soil waste is generated annually from 59 cities [12]. Sewage sludge (semi-solid material) is produced during the sewage treatment process [6]. Application of SSL in soil improves the availability of nutrients, soil water retaining capacity, soil structure, and porosity [13–16], and maintains organic matter [17,18] thereby reducing the need for synthetic fertilizer [19]. Earlier studies on SSL with chemical fertilizer application have shown improved growth, yield and yield of the crop, [20] as well as better macronutrient and micronutrient status [21] and amplification of microbial counts in the soil [7]. In agriculture, the combined use of chemical fertiliser with SSL as a source of nutrients improves soil fertility and farm profitability [22–24]. Integrated use of chemical fertilizers along with SSL has shown better yield performance, improvement in mineralizable nitrogen (N) and microbial biomass [12,25]. Several researchers also reported that the sludge can be used as an amendment option for degraded land, which improves soil physical properties, i.e., bulk density, micro aggregate, water retention, porosity, and hydraulic conductivity [26] as compared to inorganic fertilizer due to its carbon enriched nature. However, in addition, sludge has causes some undesirable modifications, such as a decline in pH, rise in salinity, and heavy metal content in soil [16]. Thus, its agricultural application requires monitoring to avoid the risk of heavy metal contamination in the soil and plant system [27].

The novelty of the present study is the integrated and balanced utilization of SSL and inorganic fertilizer for sustainable growth of crops. Sewage sludge has low-cost and easily available in urban areas and could substitute for farmyard manure (FYM). The current study was conducted (i) to determine the effect of joint application of SSL with chemical fertilizer on growth and yield of rice and wheat, (ii) to determine the effect of SSL with chemical fertilizer on the protein content of rice and wheat and (iii) to evaluate the effect of joint application of SSL with chemical fertilizer on bioaccumulation of heavy metals in RWCS. It is hypothesized that the joint application of SSL and fertilizer in a proper combination may positively influence the growth, yield and quality of RWCS in the middle Ganegatic alluvial plain.

2. Materials and Methods

2.1. Experimental Site

An experiment with two cropping cycles of rice (*Oryza sativa*; Arize 6444)–wheat (*Triticum aestivum*; HD 2967) was completed in 2015–2016 (I-rice and I-wheat) and 2016–2017 (II-rice and II-wheat). The present investigation comprised the next two cycles of rice-wheat set up in 2017–2018 (III-rice and III-Wheat) and 2018–2019 (IV-Rice and IV-wheat) without disturbing the field design of the previous experiment at the Agricultural Research Farm, Banaras Hindu University, Varanasi (UP), India. This farm is situated in the Northern Gangetic Alluvial (Inceptisol) Plain (128.93 m asl; latitude 25°19′ N, and longitude 83° E) (Figure 1).

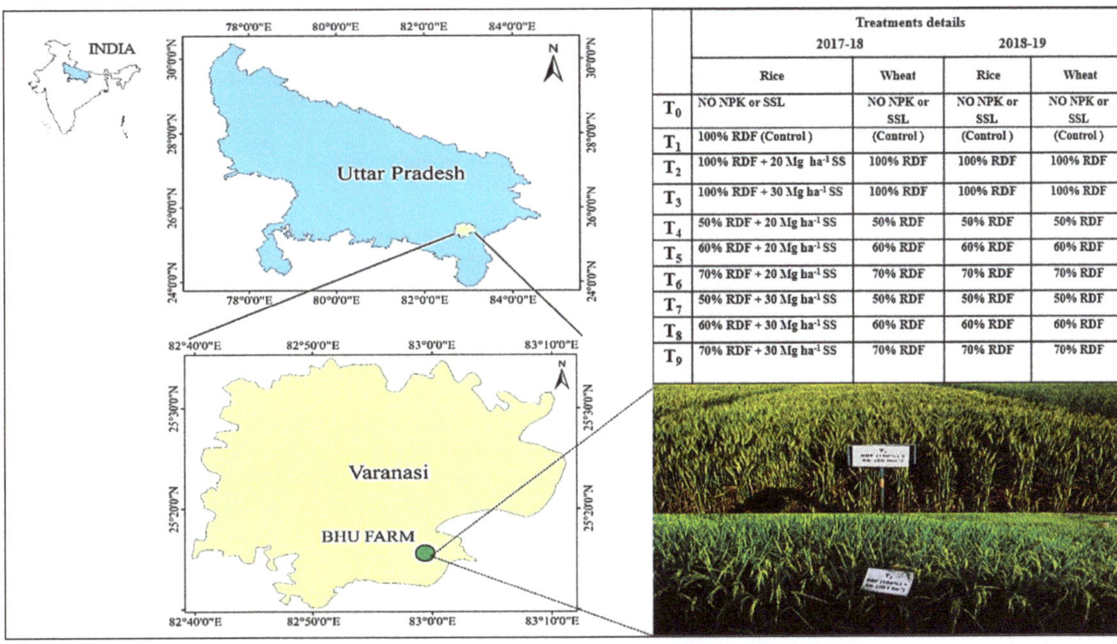

	Treatments details			
	2017-18		2018-19	
	Rice	Wheat	Rice	Wheat
T_0	NO NPK or SSL	NO NPK or SSL	NO NPK or SSL	NO NPK or SSL
T_1	100% RDF (Control)	(Control)	(Control)	(Control)
T_2	100% RDF + 20 Mg ha^{-1} SS	100% RDF	100% RDF	100% RDF
T_3	100% RDF + 30 Mg ha^{-1} SS	100% RDF	100% RDF	100% RDF
T_4	50% RDF + 20 Mg ha^{-1} SS	50% RDF	50% RDF	50% RDF
T_5	60% RDF + 20 Mg ha^{-1} SS	60% RDF	60% RDF	60% RDF
T_6	70% RDF + 20 Mg ha^{-1} SS	70% RDF	70% RDF	70% RDF
T_7	50% RDF + 30 Mg ha^{-1} SS	50% RDF	50% RDF	50% RDF
T_8	60% RDF + 30 Mg ha^{-1} SS	60% RDF	60% RDF	60% RDF
T_9	70% RDF + 30 Mg ha^{-1} SS	70% RDF	70% RDF	70% RDF

Figure 1. Location of the experimental site, layout and experimental view.

2.2. Weather and Soil Condition

The region has semi-arid to sub-humid climatic conditions. Annual mean rainfall received during the experimentation was 727.75 mm and 1121.10 mm between 2017–2018 and 2018–2019, respectively, and 75% of this amount was received from June to September (Figure 2).

The experimental soil was alkaline in nature (pH 8.24), non-saline (EC 0.15 dS m^{-1}), low in organic carbon content (4.60 g kg^{-1}), low in available N (141.72 kg ha^{-1}), medium in available P (17.42 kg ha $^{-1}$), medium in available K (132.74 kg ha^{-1}) and medium in sulfur content (14.65 mg kg^{-1}). The DTPA-extractable Fe, Cu, Zn, Mn Pb, Cd, Cr, and Ni contents in the initial soil were 42.65, 2.17, 1.02, 11.41, 0.55, 2.12, 9.24 and 6.79 mg kg^{-1}, respectively.

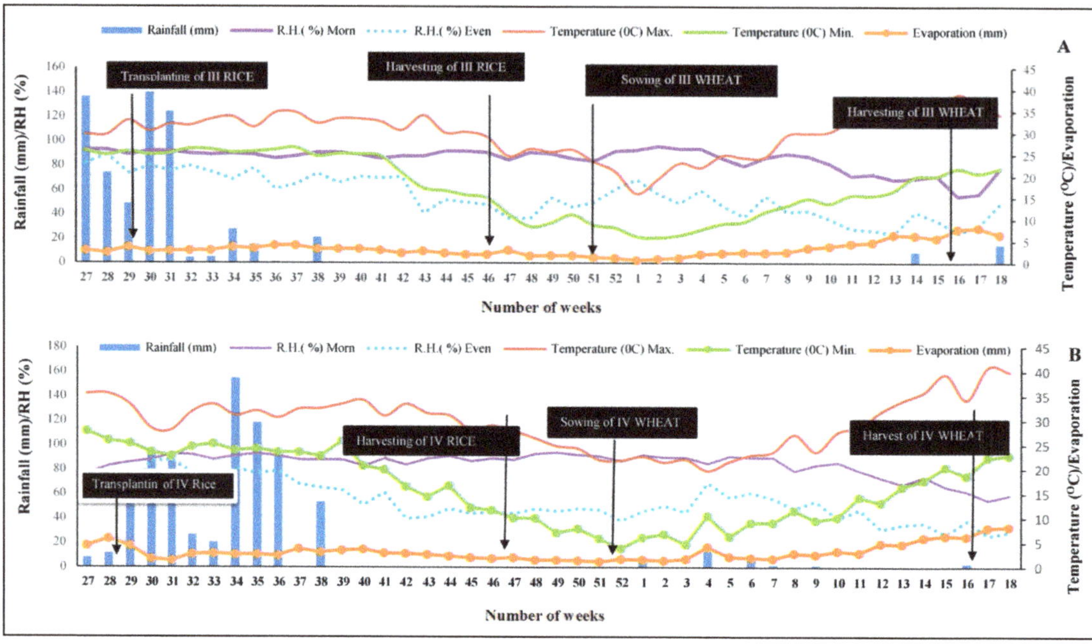

Figure 2. Meteorological data during the experiment in 2017–2018 (**A**) and 2018–2019 (**B**).

2.3. Characteristics of Sewage Sludge

SSL of domestic origin was collected from a Sewage Treatment Plant (STP) in Bhagwanpur, Varanasi, in the month of May 2017. For further analysis, a composite sample was ground and passed through a 2 mm sieve and stored in a polythene bag. The SSL used in the experiment had pH 7.02, EC: 3.25 dS m^{-1}, organic carbon: 7.98%, total N: 1.85%, total P: 1.40% and total K: 1.20%. According to the Council of the European Communities [28], the maximum permissible limits (MPLs) for potentially toxic elements such as Zn, Cu, Cd, Pb, Ni and Cr in sludge used in agricultural soils are 2500, 1000, 20, 750, 300 and 750 mg kg^{-1}, respectively. The sludge used for the study contained 200, 247, 8, 52, 17 and 44 mg kg^{-1} of Zn, Cu, Cd, Pb, Ni and Cr, respectively. Thus, all the heavy metals were within the MPL.

2.4. Experimental Design and Treatments

The experiment was conducted in a randomized block design with different recommended doses of fertilizers (RDF), i.e., 150 (N), 75 (P_2O_5) and 75 (K_2O) kg ha^{-1} for rice, and 120 (N), 60 (P_2O_5) and 60 (K_2O) kg ha^{-1} for wheat, and SSL levels which were replicated in triplicate. Treatments in the present study were as follows: $T_{0:}$ (no NPK or SSL); T_1 100% RDF, T_2 (T_1 + SSL 20 Mg ha^{-1}); T_3 (T_1 + SSL 30 Mg ha^{-1}); T_4 (50% RDF + SSL 20 Mg ha^{-1}); T_5 (60% RDF + SSL 20 Mg ha^{-1}); T_6 (70% RDF + SSL 20 Mg ha^{-1}); T_7 (50% RDF + SSL 30 Mg ha^{-1}); T_8 (60% RDF + SSL 30 Mg ha^{-1}) and T_9 (70% RDF + SSL 30 Mg ha^{-1}). A half dose of N and a full dose of P_2O_5 and K_2O were applied at the time of transplanting/sowing of the crops, while the remaining N fertilizer was applied in two equal parts at 30 and 60 days after transplanting or days after sowing (DAT/DAS). The RDF was applied in both the crop and season as per the mentioned treatments. However, SSL (dry weight basis) was applied only once and was spread in the various plots as per treatments and thoroughly mixed with soil one week before the start of the third cycle of rice-wheat.

2.5. Data and Collection Procedures

Randomly, five plants from each plot were selected and labelled. The height of both rice and wheat plants was measured using a meter-rod from the base to the tip of the topmost leaf of the plant at 30, 60, and 90 DAT/DAS and the harvest stage, and then averaged. The leaf greenness of the plants (SPAD chlorophyll value) was measured at 30, 60, and 90 DAT/DAS in the uppermost fully expanded leaf using a SPAD-502 (SPAD-502 Plus Konica Minolta). Five representative panicles from rice and the ear from wheat were sampled and grain number in each was recorded. Length (cm) of the panicle/ear was measured from the base of the topmost spikelet. From the yield of the net plot of each experimental unit, 1000 grains were counted and their weight was recorded. The harvest index (HI) was calculated using the following formula:

$$\text{Harvest Index} = \frac{\text{Grain yield}\left(\text{kg ha}^{-1}\right)}{\text{Biological yield}\left(\text{kg ha}^{-1}\right)} \times 100$$

2.6. Plant, Soil and Sewage Sludge Analyses

Rice and wheat grain samples were washed sequentially in detergent solution (0.2% liquid), 0.1 N HCl solution and deionized water then dried at 65 °C until a constant was weight achieved. Nitrogen concentration was determined by digestion (H_2SO_4), distillation and a titrimetric method using a standard Kjeldahl Auto analyzer (DISTYL-EM; Pelican, CIT Nagar, Chennai, Tamil Nadu) procedure [29]. Grain samples were digested in a di-acid mixture (HNO_3:$HClO_4$::3:1 v/v) for the estimation of Cd, Cr, Ni and Pb using an atomic absorption spectrophotometer (Agilent FS-240, 5301 Stevens Creek Blvd, Santa Clara, CA, USA) as per the procedure outlined by [30].After processing, soil samples were analyzed for soil reaction (pH) and electrical conductivity (EC) [31], organic carbon [32], available nitrogen by the alkaline potassium permanganate method [33], available phosphorus by spectrophotometry [34], available potassium by flame photometry [35], and available sulphur by a turbidimetric method [36], and DTPA extractable Zn, Cu, Mn, Fe, Pb, Cd, Cr, and Ni [37] were analyzed by atomic absorption spectrophotometry (AAS) (Agilent FS-240). Total N, P and K content in SSL was analyzed by the methods outlined by [38]. Total heavy metals (Cd, Cr, Ni and Pb) in SSL were analyzed by an aqua regia digestion procedure which consist of digesting SSL samples digested on a hot plate with a mixture of HCl and HNO_3 (3:1 v/v) [39] followed by analysis using AAS (Agilent FS-240) as per the procedure outlined by [40]. After completion of the experiment (IV-wheat), total heavy metal (Pb, Cd, Cr, and Ni) content in post-harvest soil was determined by AAS using aqua regia (HCl:HNO_3::3:1 v/v) digestion [39]. The certified reference standards (CRS) for Pd (5190–8287), Cd (5190–8270), Cr (5190–8275), and Ni (5190–8298) were purchased from Agilent, 5301 Stevens Creek Blvd, Santa Clara, CA 95051, USA. To control analytical precision, quality control check samples were taken from materials with parameters of known value and set at concentrations near the midpoint of the calibration range. The recovery rate of Pd, Cd, Cr, and Ni were 97.2, 98.5, 96.4, and 98.8%, respectively.

The protein content in grain was calculated by multiplying N (%) in the grain of rice and wheat by a factor of 6.25 [41].

2.7. Statistical Data Analysis

The data were statistically analysed using one-way analysis of variance (ANOVA) in SPSS Inc., Chicago Ver. 22. Duncan's multiple range test (DMRT) was used to test the significance of the difference between the treatments at the 5% level. Figures were drawn using Sigma plot 12.5 software and Microsoft Excel 2016.

3. Results and Discussion

3.1. Effect of Joint Application of Sewage Sludge and Fertilizers on the Growth of Rice and Wheat

The effect of joint application of SSL with chemical fertilizer (CF) at different growth stages (30, 60, 90 DAT/DAS and harvest stage) of rice and wheat crop is presented in Tables 1 and 2. The greatest plant height was measured in treatment T_3, i.e., the combination of 30 Mg ha^{-1} SSL + 100% RDF (104.22 cm) followed by T_2, i.e., 20 Mg ha^{-1} SSL + 100% RDF (102.41 cm) at the harvest stage of III-rice and these treatments having significantly increased height compared to 100% RDF. However, treatment T_2 i.e., 20 Mg ha^{-1} SSL + 100% RDF and T_3 i.e., 30 Mg ha^{-1} SSL + 100% RDF were found to be statistically similar with respect to plant height at the harvest stage. A similar trend was noticed in III-wheat. During 2018–2019, the maximum plant height for IV-rice (102.32) and IV-wheat (103.80) at harvest stage was recorded in treatments T_3 (30 Mg ha^{-1} SSL + 100% RDF), significantly greater than T_1 (100% RDF). However, T_3 (30 Mg ha^{-1} SSL + 100% RDF) was statistically at par with treatment T_2 (20 Mg ha^{-1} SSL + 100% RDF) in IV-rice and IV-wheat. In IV-wheat crop, application of 20 Mg ha^{-1} SSL + 100% RDF (T_2) and 30 Mg ha^{-1} SSL + 100% RDF (T_3) significantly increased plant height over 100% RDF (Tables 1 and 2).

Table 1. Impact of sewage sludge use with fertilizers on plant height of rice and wheat.

Treatments	**Plant Height (cm) at 30 DAT/DAS**			
	2017–2018		**2018–2019**	
	III-Rice	**III-Wheat**	**IV-Rice**	**IV-Wheat**
T_0 (WF)	45.49 ± 1.44 d	18.75 ± 2.56 c	42.45 ± 2.32 e	16.48 ± 2.44 d
T_1 (RDF 100)	75.60 ± 0.68 ab	30.05 ± 1.81 ab	76.54 ± 0.94 ab	30.97 ± 0.95 abc
T_2 (RDF 100 + SSL 20)	79.31 ± 1.51 a	33.17 ± 2.32 ab	78.91 ± 1.70 a	32.11 ± 1.06 ab
T_3 (RDF 100 + SSL 30)	81.49 ± 2.12 a	36.87 ± 1.53 a	79.85 ± 2.38 a	35.10 ± 1.75 a
T_4 (RDF 50 + SSL 20)	64.69 ± 0.97 c	26.55 ± 1.50 bc	63.42 ± 1.76 d	24.39 ± 1.73 c
T_5 (RDF 60 + SSL 20)	67.51 ± 0.90 bc	27.50 ± 1.32 b	65.90 ± 1.20 cd	25.90 ± 1.54 bc
T_6 (RDF 70 + SSL 20)	70.43 ± 2.53 bc	28.96 ± 0.78 ab	68.25 ± 0.71 bcd	27.30 ± 0.68 bc
T_7 (RDF 50 + SSL 30)	66.98 ± 0.33 bc	28.21 ± 2.59 b	64.57 ± 0.33 cd	26.78 ± 1.07 bc
T_8 (RDF 60 + SSL 30)	70.54 ± 2.20 bc	30.76 ± 0.74 ab	69.60 ± 2.23 bcd	29.36 ± 1.32 abc
T_9 (RDF 70 + SSL 30)	73.80 ± 2.83 ab	31.14 ± 1.47 ab	72.02 ± 1.75 abc	30.38 ± 1.41 abc
Significance level	**	**	**	**
Treatments	Plant Height (cm) at 60 DAT/DAS			
T_0 (WF)	56.94 ± 2.57 e	36.78 ± 0.51 e	55.57 ± 2.5 e	34.91 ± 1.35 d
T_1 (RDF 100)	82.28 ± 2.11 bcd	61.38 ± 2.33 cd	83.07 ± 1.96 abcd	62.95 ± 1.22 ab
T_2 (RDF 100 + SSL 20)	89.63 ± 0.30 ab	70.42 ± 2.03 ab	87.77 ± 2.3 ab	67.891.15 a
T_3 (RDF 100 + SSL 30)	92.33 ± 1.58 a	73.41 ± 1.99 a	89.62 ± 0.86 a	69.46 ± 1.57 a
T_4 (RDF 50 + SSL 20)	74.63 ± 2.09 d	56.23 ± 2.54 d	73.54 ± 1.56 d	52.86 ± 2.02 c
T_5 (RDF 60 + SSL 20)	77.77 ± 0.73 cd	57.67 ± 1.24 cd	77.03 ± 2.10 cd	55.58 ± 1.60 bc
T_6 (RDF 70 + SSL 20)	79.19 ± 1.79 cd	61.95 ± 2.04 bcd	78.07 ± 1.58 bcd	58.42 ± 1.73 bc
T_7 (RDF 50 + SSL 30)	78.08 ± 0.58 cd	59.58 ± 1.95 cd	77.02 ± 2.45 cd	57.06 ± 1.45 bc
T_8 (RDF 60 + SSL 30)	81.03 ± 1.46 bcd	64.09 ± 0.48 bcd	79.98 ± 2.36 abcd	61.27 ± 1.87 abc
T_9 (RDF 70 + SSL 30)	84.41 ± 2.47 abc	66.26 ± 0.62 abc	84.05 ± 0.98 abc	63.78 ± 2.77 ab
Significance level	**	**	**	**

Mean values within the same column having the same letters differ non-significantly ($p \leq 0.01$), while different letters indicate significant difference ($p \leq 0.01$). Mean (±SE) was taken from three replicates for each treatment. **, indicates significant at 1% level of probability.

Table 2. Impact of sewage sludge use with fertilizers on plant height of rice and wheat.

Treatments	Plant Height (cm) at 90 DAT/DAS			
	2017–2018		2018–2019	
	III-Rice	III-Wheat	IV-Rice	IV-Wheat
T_0 (WF)	67.51 ± 2.85 c	62.26 ± 1.13 d	62.58 ± 1.23 d	60.62 ± 2.98 c
T_1 (RDF 100)	92.28 ± 6.52 b	89.90 ± 0.97 bc	93.84 ± 2.04 bc	92.61 ± 1.26 ab
T_2 (RDF 100 + SSL 20)	105.47 ± 1.05 a	101.15 ± 2.61 a	104.91 ± 2.66 ab	100.03 ± 4.73 a
T_3 (RDF 100 + SSL 30)	108.02 ± 1.71 a	103.66 ± 1.63 a	106.87 ± 2.39 a	101.17 ± 1.44 a
T_4 (RDF 50 + SSL 20)	87.80 ± 0.90 b	85.27 ± 2.31 c	85.84 ± 2.30 c	83.10 ± 1.83 b
T_5 (RDF 60 + SSL 20)	88.50 ± 0.68 b	87.21 ± 0.05 bc	86.40 ± 2.57 c	84.77 ± 2.47 b
T_6 (RDF 70 + SSL 20)	90.15 ± 1.3 b	89.09 ± 1.30 bc	88.83 ± 2.35 c	85.90 ± 1.24 b
T_7 (RDF 50 + SSL 30)	90.95 ± 2.80 b	92.29 ± 0.65 bc	90.11 ± 2.38 c	90.51 ± 1.69 ab
T_8 (RDF 60 + SSL 30)	96.03 ± 1.39 ab	92.97 ± 0.72 b	95.09 ± b3.02 c	91.26 ± 1.63 ab
T_9 (RDF 70 + SSL 30)	97.48 ± 0.95 ab	93.75 ± 0.42 b	97.01 ± 1.85 abc	93.31 ± 1.41 ab
Significance level	**	**	**	**
Treatments	Plant Height (cm) at Harvest			
T_0 (WF)	64.95 ± 0.41 d	61.59 ± 1.98 d	61.83 ± 1.98 c	64.86 ± 3.53 e
T_1 (RDF 100)	90.13 ± 0.89 bc	89.10 ± 0.81 c	89.69 ± 0.81 b	92.76 ± 1.72 bc
T_2 (RDF 100 + SSL 20)	102.41 ± 1.94 a	99.64 ± 0.79 ab	99.27 ± 0.79 a	101.06 ± 1.81 a
T_3 (RDF 100 + SSL 30)	104.22 ± 1.55 a	102.55 ± 2.65 a	102.32 ± 2.65 a	103.80 ± 2.12 a
T_4 (RDF 50 + SSL 20)	86.30 ± 1.59 c	84.96 ± 2.42 c	83.22 ± 2.24 b	84.61 ± 0.84 d
T_5 (RDF 60 + SSL 20)	88.08 ± 1.27 c	85.46 ± 3.76 c	84.62 ± 3.73 b	85.79 ± 2.09 d
T_6 (RDF 70 + SSL 20)	88.95 ± 1.85 bc	88.62 ± 2.98 c	87.23 ± 2.90 b	87.54 ± 1.12 cd
T_7 (RDF 50 + SSL 30)	89.38 ± 3.37 bc	91.90 ± 3.43 bc	86.40 ± 3.43 b	93.20 ± 2.35 bc
T_8 (RDF 60 + SSL 30)	93.10 ± 3.74 bc	92.34 ± 1.55 bc	91.01 ± 1.55 b	94.11 ± 2.34 bc
T_9 (RDF 70 + SSL 30)	95.49 ± 1.84 b	93.22 ± 1.24 bc	91.11 ± 1.24 b	97.18 ± 2.57 ab
Significance level	**	**	**	**

Mean values within the same column having alike alphabets differ non-significantly ($p \leq 0.05$), while different alphabets show a significant difference ($p \leq 0.05$). Mean (±SE) was taken from three replicates for each treatment. **, indicates significant at 1% level of probability.

There was a significant reduction in plant height in T_4 (50% RDF + 20 Mg ha^{-1} SSL) and T_5 (60% RDF + 20 Mg ha^{-1} SSL), whereas T_6 (70% RDF + 20 Mg ha^{-1} SSL), T_7 (50% RDF + 30 Mg ha^{-1} SSL), T_8 (60% RDF + 30 Mg ha^{-1} SSL) and T_9 (70% RDF + 30 Mg ha^{-1} SSL) were at par with T_1 (100% RDF). The plant heights in treatment T_2 (100% RDF + 20 Mg ha^{-1} SSL) and T_3 (100% RDF + 30 Mg ha^{-1} SSL) at harvest were 13.62 and 15.63% higher than T_1 (100% RDF) in III-rice crop, and in IV-rice the same treatment had respective increases by 10.70 and 14.08% over T_1 (100% RDF).

In the case of III-wheat, the respective increases in T_2 (100% RDF + 20 Mg ha^{-1} SSL) and T_3 (100% RDF + 30 Mg ha^{-1} SSL) were 11.82 and 15.09%, and for IV-wheat, 8.94 and 11.90% over T_1 (100% RDF). It was observed that at all the growth stages (30, 60 and 90 DAT/DAS), T_3 (30 Mg ha^{-1} SSL + 100% RDF) and T_2 (20 Mg ha^{-1} SSL + 100% RDF) showed the highest plant heights in rice and wheat crops (Tables 1 and 2).

It is well-known that applying SSL with CF to croplands can enhance plant height. According to Latare et al. [21] combining SSL with chemical fertilizers may improve soil fertility and increase the availability of nitrogen and trace elements to plants, thus indirectly enhanced plant development. The addition of SSL with CF enhances the direct availability of N and P from chemical fertilizers, and indirect or slow-release from SSL, which results in increased leaf area and higher dry matter accumulation [42,43]. Thus, the improvement of soil fertility associated with the application of SSL and CF would have supported improved rice and wheat plant growth. Similarly, Zhang et al. [44] revealed a significant increase in rice plant height by greater soil fertility and nutrient status after applying SSL amendments. Our findings also resemble the work of Rehman and Qayyum [45], who reported a significant influence of SSL compost on crop productivity and biomass accumulation in rice and wheat crops.

3.2. Leaf Greenness (SPAD) at Different Growth Stages of Rice and Wheat

The data presented in Figure 3 show a significant increase of leaf greenness (chlorophyll content) in rice and wheat due to the joint application of SSL and CF in both years.

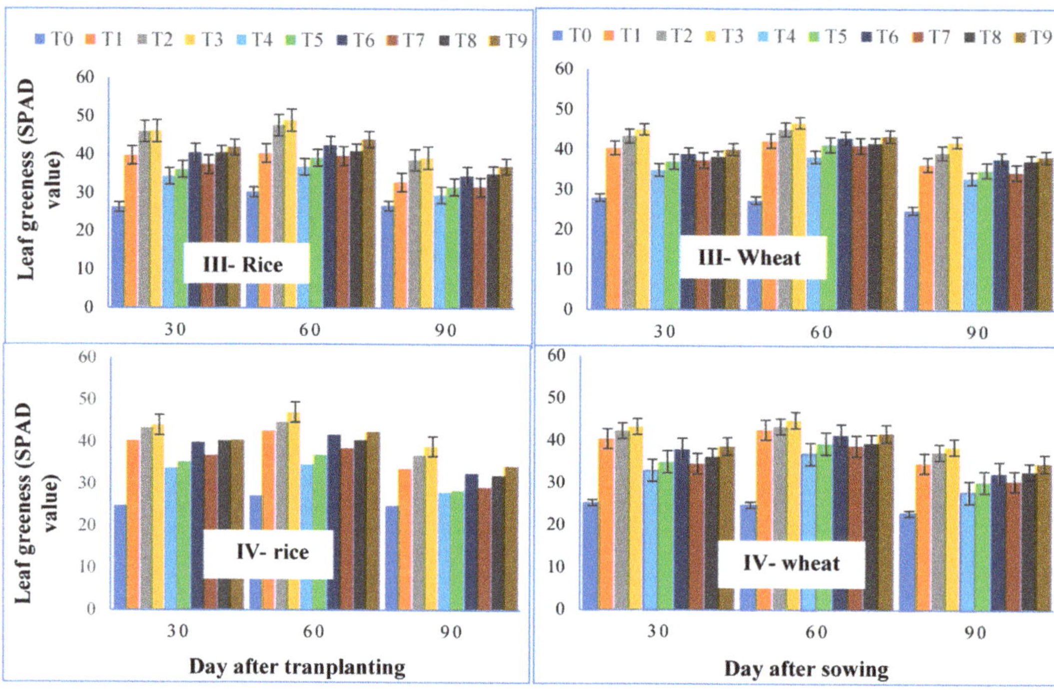

Figure 3. Impact of sewage sludge use with fertilizer on leaf greenness (SPAD value). The bars represent the means ± SE of three replicates.

Maximum leaf greenness was measured in T_3 (100% RDF + 30 Mg ha^{-1} SSL) followed by T_2 (100% RDF + 20 Mg ha^{-1} SSL), and the minimum was in T_0 (without SSL and CF). Treatment T_3 (100% RDF + 30 Mg ha^{-1} SSL) and T_2 (100% RDF + 20 Mg ha^{-1} SSL) had significantly increased leaf greenness over T_1 (100% RDF) in III-rice, whereas these treatments were statistically similar with 100% RDF (T_1) in III-Wheat.

However, the treatments from T_4 toT_9 showed a non-significant difference of T_1 in terms of leaf greenness in III-rice and III-wheat in the year 2017–2018. Similarly, the treatments T_2 (100% RDF + 20 Mg ha^{-1} SSL) and T_3 (100% RDF + 30 Mg ha^{-1} SSL) were found statistically at par with each other in III -rice and III-wheat. During 2018–2019, the leaf greenness ranged from 24.84 to 44.01 and 25.13 to 43.27 SPAD in IV-Rice and IV-Wheat, respectively. The maximum leaf greenness in IV-Rice and IV-wheat, during both years was seen in T_3, i.e., the combination of 30 Mg ha^{-1} SSL + 100% RDF at 30 DAT/DAS.

Although, it was noticed that the T_3 (100% RDF + 30 Mg ha^{-1} SSL), treatment was statistically at par with T_1 (100% RDF), T_2 (100% RDF + 20 Mg ha^{-1} SSL), T_6 (70% RDF + 20 Mg ha^{-1} SSL), T_8 (60% RDF + 30 Mg ha^{-1} SSL), and T_9 (100% RDF + 30 Mg ha^{-1} SSL), it was significantly superior over the rest of the treatments in terms of leaf greenness in IV-rice and IV-wheat at 30 DAT during the year 2018–2019. An almost similar trend was observed with the leaf greenness recorded at 60 and 90 DAT/DAS. AT 60 DAT/DAS, a slight increase in plant leaf greenness was noticed compared to observations at 30 DAT/DAS, whereas at 90 DAT, a decrease was noticed compared to 30 and 60 DAT/DAS. Chlorophyll content (SPAD) directly influences the photosynthetic rate of plants. The increase in assimilatory pigments content in leaves was observed when crops were grown in SSL-amended soil. Romani and Beltarre [46] found that repeated 7 years of treatment with SSL (3.7 Mg ha^{-1}) resulted in a significant increase in chlorophyll content (SPAD index). Latare et al. [43] reported that at 30 days after transplanting and sowing (DAT/DAS) in rice and wheat, leaf greenness index did not increase but increased significantly at 60 and 90 DAT/DAS. This might be because Fe, Mg, and Mn contents in the SSL, are liberated after decomposition of SSL and remain directly associated with chlorophyll synthesis [47,48].

3.3. *Effect of Sewage Sludge and Fertilizers on Yield Attributes of Rice and Wheat*

The panicle/ear length (cm) ranged between 18.26–37.61 and 12.99–35.95 with mean values of 29.49 and 27.15 in III-Rice and IV-rice, respectively whereas the corresponding value in III-wheat and IV-wheat varied between 6.09–15.54 and 5.90–15.16 with a mean value of 11.89 and 11.28 (Table 3). It was observed that for the rice crop, the significantly highest length of the panicle (37.61 and 35.95 cm) was recorded with T_3 (100% RDF + 30 Mg ha^{-1} SSL) followed by T_2 (34.61 and 32.87 cm) during both years, respectively. An almost similar trend was noticed during both years of wheat experimentation. In III-Rice, a significant increase was recorded with T_2 (18.16%) and T_3 (28.41%) concerning panicle length. Similarly, the ear length of the wheat crop (III and IV) increased significantly in T_3 (29.50 and 20.03%) over 100% RDF (Table 3). The result show that the application of SSL with CF increased ear/panicle length. This is due to the role of N in flowering, fruiting, and crop maturation, as well as seed formation. Latare et al. and Jamil et al. [21,49] reported an increase in spike length of wheat with different doses of SSL compared to the non-treated plot. A similar trend was observed in the wheat crop [50]. Zhang et al. [51] reinforced the results, finding that an adequate supply of organic wastes along with NPK fertilizer improves the yield attribute. Thus, combined application of SSL with chemical fertilizer in different levels appears beneficial with respect to yield attributes without showing any toxic effects on plants. The number of grains per panicle/ear (Table 3) varied from 54.44 to 154 and 15.36 to 41.02 in III-rice and III-wheat (2017–2018), respectively. Application of 30 Mg ha^{-1} SSL + 100% RDF resulted in the highest number of grains per panicle i.e., 154.41 and 41.02 in III-rice and III-wheat, respectively.

During 2018–2019, the maximum grains per panicle/ear for IV-rice (146.147) and IV-wheat (39.64) were measured in treatments T_3 (30 Mg ha^{-1} SSL + 100% RDF). This was significantly superior to T_1 in IV-rice but statistically similar in IV-wheat (Table 3). A significant reduction was noticed in grains per panicle in T_4 (50% RDF + 20 Mg ha^{-1} SSL) and T_5 (60% RDF + 20 Mg ha^{-1} SSL), whereas T_6 (70% RDF + 20 Mg ha^{-1} SSL), T_7 (50% RDF + 30 Mg ha^{-1} SSL), T_8 (60% RDF + 30 Mg ha^{-1} SSL), and T_9 (70% RDF + 30 Mg ha^{-1} SSL) were at par with T_1 (100% RDF) in IV-rice. While, IV-wheat showed a marked reduction in T_4 (50% RDF + 20 Mg ha^{-1} SSL) and T_5 (60% RDF + 20 Mg ha^{-1} SSL), T_6 (70% RDF + 20 Mg ha^{-1} SSL), T_7 (60% RDF + 30 Mg ha^{-1} SSL), T_8 (70% RDF + 30 Mg ha^{-1} SSL), and T_9 (70% RDF + 30 Mg ha^{-1} SSL) were statistically at par with respect to grains per panicle. T_2 (RDF 100% + SSL 20 Mg ha^{-1}) was 22.06% higher than T_1 (RDF 100%) in the III-rice crop, and in the case of IV-rice, the same treatment showed a 12.04% increase over T_1 (RDF 100%).

Table 3. Impact of sewage sludge use with fertilizers on panicle/ear length and grain per panicle or ear of rice and wheat.

Treatments	Panicle/Ear Length (cm)			
	2017–2018		2018–2019	
	III-Rice	III-Wheat	IV-Rice	IV-Wheat
T_0 (WF)	18.26 ± 1.34 e	6.09 ± 1.01 c	12.99 ± 2.11 e	5.90 ± 0.54 e
T_1 (RDF 100)	29.29 ± 1.36 cd	12.00 ± 1.05 b	30.36 ± 2.14 bc	12.63 ± 0.85 bc
T_2 (RDF 100 + SSL 20)	34.61 ± 1.14 ab	13.24 ± 0.87 ab	32.87 ± 0.77 ab	13.08 ± 0.074 ab
T_3 (RDF 100 + SSL 30)	37.61 ± 1.98 a	15.54 ± 1.12 a	35.95 ± 0.90 a	15.16 ± 0.62 a
T_4 (RDF 50 + SSL 20)	25.27 ± 1.83 d	11.22 ± 0.85 b	24.30 ± 1.65 d	9.88 ± 0.55 d
T_5 (RDF 60 + SSL 20)	27.10 ± 1.63 cd	11.50 ± 0.43 b	26.71 ± 1.35 cd	10.01 ± 0.41 de
T_6 (RDF 70 + SSL 20)	31.48 ± 1.02 bc	11.97 ± 0.65 b	30.71 ± 1.61 bc	10.64 ± 0.25 cde
T_7 (RDF 50 + SSL 30)	28.04 ± 1.12 cd	11.45 ± 0.67 b	26.44 ± 1.11 cd	11.05 ± 1.23 bcde
T_8 (RDF 60 + SSL 30)	31.00 ± 2.83 bc	12.47 ± 0.64 b	29.93 ± 0.52 bc	12.19 ± 1.28 bcde
T_9 (RDF 70 + SSL 30)	32.25 ± 1.03 bc	12.71 ± 0.26 b	31.43 ± 0.86 b	12.27 ± 0.20 bcd
Significance level	**	**	**	**
	Grains/Panicle/Ear			
T_0 (WF)	54.44 ± 2.80 e	15.36 ± 0.49 f	53.04 ± 3.53 g	14.38 ± 1.07 e
T_1 (RDF 100)	122.99 ± 8.11 d	33.37 ± 0.63 cde	125.71 ± 5.89 cde	35.79 ± 2.70 ab
T_2 (RDF 100 + SSL 20)	150.16 ± 3.47 ab	38.98 ± 0.13 ab	140.85 ± 5.32 ab	37.55 ± 1.23 ab
T_3 (RDF 100 + SSL 30)	154.41 ± 2.23 a	41.02 ± 1.35 a	146.17 ± 3.39 a	39.64 ± 1.44 a
T_4 (RDF 50 + SSL 20)	114.04 ± 3.30 d	31.73 ± 1.84 e	108.57 ± 4.90 f	28.84 ± 1.29 d
T_5 (RDF 60 + SSL 20)	120.75 ± 1.06 d	32.48 ± 1.99 de	116.19 e ± 0.51 f	29.73 ± 1.51 cd
T_6 (RDF 70 + SSL 20)	139.42 ± b2.90 c	35.39 ± 1.03 bcde	130.57 ± 0.45 bcd	32.94 ± 2.26 bcd
T_7 (RDF 50 + SSL 30)	125.26 ± 2.47 d	33.25 ± 1.83 cde	120.72 ± 2.54 de	30.30 ± 0.37 cd
T_8 (RDF 60 + SSL 30)	136.84 ± 2.15 c	36.10 ± 0.54 bcd	130.40 ± 3.97 bcd	34.14 ± 0.30 bc
T_9 (RDF 70 + SSL 30)	142.07 ± 1.55 bc	37.05 ± 1.07 bc	133.27 ± 2.90 bc	34.42 ± 1.24 bc
Significance level	**	**	**	**

Mean values within the same column having the same letter differ non-significantly ($p \leq 0.01$), while different letters indicate a significant difference ($p \leq 0.01$). Mean (±SE) was taken from three replicates for each treatment. **, indicates significant at 1% level of probability.

In the case of III-wheat, T_2 was 16.81% greater than T_1 (100% RDF), whereas this increase was only 4.93% in the IV-wheat crop. Tamrabet et al. [52] found an increased number of grains spike^{-1} of wheat after treatment with 20, 30, and 40 Mg ha^{-1} SSL. Moreover, SSL seemed to be more beneficial l the crop than inorganic fertilizer. The SSL treatment statistically improved spike fertility and plant biomass at the heading and maturity stage. By applying SSL and fertilizer, the yield of both crops was significantly improved significantly in all the treatments compared to no fertilizer (Figure 4).

In both years, in rice and wheat crops, a significantly higher grain yield was documented in T_3, with the combination of 30 Mg ha^{-1} SSL + 100% RDF, followed by T_2 with 20 Mg ha^{-1} SSL + 100% RDF, compared to other treatments, whereas the lowest yield was recorded in WF (T_0).

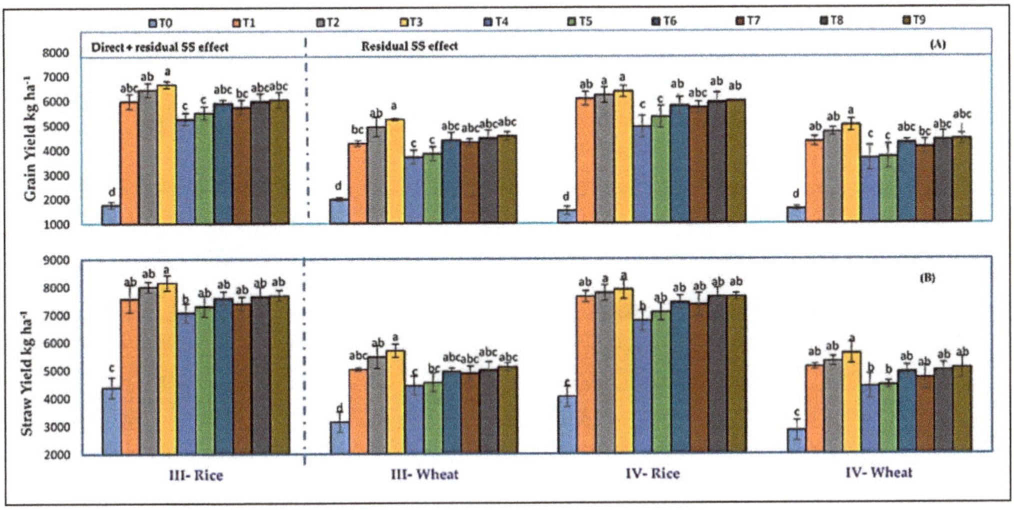

Figure 4. Impact of sewage sludge use with fertilizer on grain yield (**A**) and straw yield (**B**). Data (mean ± SE) followed by the same letter differ non-significantly ($p \leq 0.05$), while different letters indicate a significant difference ($p \leq 0.05$).

There was no significant difference within the rest of the treatments except T_4 (50% RDF + 20 Mg ha^{-1} SSL) and T_5 (60% RDF + 20 Mg ha^{-1} SSL) compared to only 100% RDF (T_1) in IV-rice and IV-Wheat. The grain yield in treatment T_2 (100% RDF + 20 Mg ha^{-1} SSL) and T_3 (100% RDF + 30 Mg ha^{-1} SSL) of III-rice was higher by 7.75% and 11.42%, respectively, compared to treatment T_1 where 100% RDF was applied. In the case of IV-rice, the grain yield of treatments T_2 (100% RDF + 20 Mg ha^{-1} SSL) and T_3 (100% RDF + 30 Mg ha^{-1} SSL) was higher by 2.44% and 4.83%, respectively, than T_1 (100% RDF). With III-wheat, the grain yield of treatment T_2 (20 Mg ha^{-1} SSL + 100% RDF) and T_3 (30 Mg ha^{-1} SSL + 100% RDF) showed a respective increase of 15.55% and 22.75% over T_1 (100% RDF), whereas, in the case of IV-wheat, treatment T_2 (20 Mg ha^{-1} SSL + 100% RDF) and T_3 (30 Mg ha^{-1} SSL + 100% RDF) showed only 9.12% and 15.20% yield increment over T_1 (100% RDF).

During 2017–2018, in III-rice, application of 100% RDF resulted in statistically similar grain yield in all other treatments except T_0 (WF). However, the yield of treatment T_1, i.e., 100% RDF, was at par with 20 Mg ha^{-1} SSL when applied with reduced doses of CF (T_4, T_5 and T_6), and also with 30 Mg ha^{-1} SSL with a reduced dose of CF (T_7, T_8 and T_9). This provides the option of reducing the dose of RDF up to 50% when applied with SSL. A similar trend was observed for the grain yield of III–wheat. During 2018–2019, grain yield of IV-Rice in T_1 (100% RDF) was statistically similar to T_2 (20 Mg ha^{-1} SSL + 100% RDF), T_3 (30 Mg ha^{-1} SSL + 100% RDF), T_6 (20 Mg ha^{-1} SSL + 70% RDF), T_7 (30 Mg ha^{-1} SSL + 50% RDF), T_8 (30 Mg ha^{-1} SSL + 60% RDF) and T_9 (30 Mg ha^{-1} SSL + 70% RDF). However, a significant reduction in grain yield was noticed in T_4 (20 Mg ha^{-1} SSL + 50% RDF) and T_5 (20 Mg ha^{-1} SSL + 60% RDF). An almost similar yield trend was observed for IV wheat. It is evident that providing only 50% RDF with 20 Mg ha^{-1} SSL resulted in yields similar to 100% RDF for the first two crops. However, in subsequent years, i.e., IV-rice and IV-wheat, due to a decrease in the residual effect of SSL (applied in III-rice), the amount of chemical fertilizer had to be increased to 70% RDF with SSL (20 Mg ha^{-1} SSL + 70% RDF) to obtain similar grain yield to that of 100% RDF. There is a strong relationship between yield attributes and yield, particularly with respect to grain number in the panicle/ear. It was noted that the joint application of SSL with chemical fertilizer treatments increased different yield indicators, such as effective tillers and the weight of 1000 grains, thus producing higher grain yield (Figure 4). It was found that the use of SSL in RWCS had the potential to substitute half the dose of fertilizers. The yield increment could be explained by the fact that SSL as a source of organic matter contains various nutrients (macro and micro) and provides them to crops slowly after their decomposition [53]. Thus, improved number of grains per panicle/ear, panicle/ear length, and tillers of rice and wheat were positively correlated with joint application of SSL and chemical fertilizer during both years. The results of the present study are supported by Rehman and Qayyum [45], who noted that SSL treatment increased the growth and yield of rice and wheat, which might be due to higher uptake of water and nutrients by plants.

Data depicted in Figure 4 show that joint application of chemical fertilizer and SSL produced significantly higher straw yield than without fertilization (T_0) but was at par with T_1 (100% RDF) in both years. Among all the treatments, T_3 (100% RDF + 30 Mg ha^{-1} SSL) had the highest straw yield of the rice crop (8150 and 7896 kg ha^{-1}) and wheat crop (5695 and 5599 kg ha^{-1}), respectively, during the course of the experiments. In III-rice, the straw yield of treatments T_2 (100% RDF + 20 Mg ha^{-1} SSL), T_3 (100% RDF + 30 Mg ha^{-1} SSL), and T_9 (70% RDF + 30 Mg ha^{-1} SSL) increased by 4.07, 5.98, and 0.07%, respectively, compared to treatment T_1 where 100% RDF was applied. In the case of IV-rice, treatment T_2 (100% RDF + 20 Mg ha^{-1} SSL) and T_3 (100% RDF + 30 Mg ha^{-1} SSL) produced 1.10 and 2.64% higher straw yield compared to the T_1. With the III-wheat crop, straw yield of treatment T_2 (20 Mg ha^{-1} SSL + 100% RDF), T_3 (30 Mg ha^{-1} SSL + 100% RDF) and T_9 (30 Mg ha^{-1} SSL + 70% RDF) showed 8.83, 13.18 and 1.58% increments over 100% RDF (T_1), whereas, in the case of IV-wheat, only treatment T_2 (20 Mg ha^{-1} SSL + 100% RDF) and T_3 (30 Mg ha^{-1} SSL + 100% RDF) showed a positive increment in straw yield over 100% RDF (T_1). Greater leaf chlorophyll contents improve photosynthetic rate, which results in higher crop biomass and yield. The joint application of SSL with chemical fertilizer improves nutrient availability to the plants which improves their root development, the number of tillers, leaves count and ultimately higher straw production. Similar results were reported by Jamil et al. [49] and Al-Mustafa et al. [54].

From two years of pooled experimental data, it was found that grain and straw yield of rice and wheat considerably increased or decreased compared to 100% RDF (T_1) (Figure 5). Application of 30 Mg ha^{-1} SSL + 100% RDF (T_3) had maximum enhancement of grain yield in rice (8.1%) over 100% RDF (T_1) followed by the 20 Mg ha^{-1} SSL + 100% RDF plot (5.1%). Application of a reduced dose of CF i.e., 50, 60, and 70% of RDF, along with 20 or 30 Mg ha^{-1} SSL i.e., T_4 (50% RDF + 20 Mg ha^{-1} SSL), T_5 (60% RDF + 20 Mg ha^{-1} SSL), T_6 (70% RDF + 20 Mg ha^{-1} SSL), T_7 (50% RDF + 30 Mg ha^{-1} SSL), T_8 (60% RDF + 30 Mg ha^{-1} SSL) and T_9 (70% RDF + 30 Mg ha^{-1} SSL) resulted in 15, 10, 2.9, 4.9 1.6 and 0.3% reductions in rice grain yield over T_1, respectively. However, the greatest decrease was seen in the T_0 treatment (72.9%).

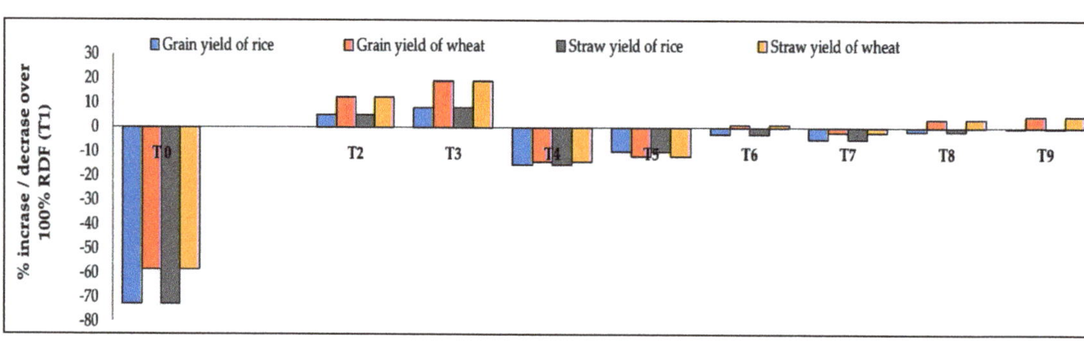

Figure 5. Impact of sewage sludge use with fertilizers on percent increase or decrease in grain and straw yield of rice and wheat (two years of pooled data).

Concerning wheat, the highest increase in grain production over 100% RDF (T_1) was recorded in T_3 (30 Mg ha^{-1} SSL + 100% RDF; 18.9%) followed by 20 Mg ha^{-1} SSL + 100% RDF (T_2), 20 Mg ha^{-1} SSL + 70% RDF (T_6), 30 Mg ha^{-1} SSL + 60% RDF (T_8; 3.3%) and 30 Mg ha^{-1} SSL + 70% RDF (T_9; 4.5%). An almost parallel trend was recorded with respect to the straw yield of rice and wheat. The highest increment in straw yield of rice (4.51%) and wheat (11.3%) over T_1 (100% RDF).was seen in T_3 (30 Mg ha^{-1} SSL + 100% RDF). Application of SSL with CF increased the grain and straw of rice and wheat yield percentage compared to 100% RDF due to slow release of nutrients from SSL throughout the period of crop growth, this finding being in accordance with the results of [22]. Application of any kind of fertilizer resulted in a greater response in wheat crop compared to rice due to different cultivation practices of rice and wheat. Yadav et al. [55], Gami et al. [56] and Bhatt et al. [57] stated that long-term integrated use of manure and fertilizers practiced in RWCS produces higher wheat yield than rice over only chemically fertilized plots.

3.4. Sewage Sludge and Fertilizers Influence the Harvest Index and Test Weight of Rice and Wheat

Data pertaining to HI as affected by the joint application of SSL and fertilizer as shown in Table 4. Among all the treatments, T_3 (100% RDF + 30 Mg ha^{-1} SSL) produced the highest HI in rice (45.00 and 44.67) and wheat crops (47.97 and 47.19) during both years. Significantly lower HI was recorded in T_0 in the rice crop (28.83 and 27.15) and in the wheat crop (38.66 and 36.12) during both years. Latare et al. [21], stated that an SSL treatment improved HI significantly in wheat but was non-significant in rice. This might be due to additional nutrients available to residual grown wheat from SSL-altered soil.

Table 4. Impact of sewage sludge use with fertilizers on harvest index and 1000 grain weight.

Treatment	Harvest Index (%)			
	2017–2018		2018–2019	
	III-Rice	III-Wheat	IV-Rice	IV-Wheat
T_0 (WF)	28.83 ± 2.83 b	38.66 ± 0.76 b	27.15 ± 0.51 b	36.12 ± 1.73 b
T_1 (RDF 100)	43.83 ± 2.72 a	45.85 ± 1.22 a	44.12 ± 0.53 a	45.86 ± 1.39 ab
T_2 (RDF 100 + SSL 20)	44.58 ± 0.78 a	47.37 ± 2.67 a	44.44 ± 1.88 a	47.13 ± 0.27 a
T_3 (RDF 100 + SSL 30)	45.00 ± 0.41 a	47.97 ± 1.25 a	44.67 ± 0.86 a	47.19 ± 1.35 a
T_4 (RDF 50 + SSL 20)	42.68 ± 0.73 a	45.31 ± 1.60 a	42.09 ± 2.55 a	45.27 ± 3.87 ab
T_5 (RDF 60 + SSL 20)	43.04 ± 1.77 a	45.82 ± 0.77 a	42.86 ± 0.78 a	45.43 ± 0.27 ab
T_6 (RDF 70 + SSL 20)	43.74 ± 1.07 a	46.92 ± 0.92 a	43.73 ± 1.45 a	46.61 ± 1.19 a
T_7 (RDF 50 + SSL 30)	43.63 ± 0.50 a	47.00 ± 1.10 a	43.73 ± 0.34 a	46.66 ± 2.18 a
T_8 (RDF 60 + SSL 30)	43.83 ± 2.14 a	47.11 ± 0.69 a	43.73 ± 2.06 a	46.86 ± 3.10 a
T_9 (RDF 70 + SSL 30)	43.93 ± 0.72 a	47.19 ± 0.50 a	43.90 ± 1.10 a	46.94 ± 0.62 a
Significance level	**	**	**	**
	1000 Grain Weight (g)			
T_0 (WF)	15.62 ± 1.85 d	28.89 ± 0.63 a	12.95 ± 0.34 d	28.20 ± 1.28 a
T_1 (RDF 100)	21.75 ± 1.50 bc	32.35 ± 0.84 a	21.84 ± 1.11 abc	31.69 ± 0.73 a
T_2 (RDF 100 + SSL 20)	25.37 ± 0.49 a	32.93 ± 0.62 a	24.55 ± 1.82 ab	32.19 ± 0.55 a
T_3 (RDF 100 + SSL 30)	25.88 ± 0.24 a	33.86 ± 0.29 a	24.80 ± 1.29 a	33.18 ± 1.75 a
T_4 (RDF 50 + SSL 20)	20.62 ± 0.50 c	31.23 ± 0.75 a	20.21 ± 0.45 c	30.57 ± 1.08 a
T_5 (RDF 60 + SSL 20)	20.75 ± 1.20 c	31.23 ± 0.81 a	20.37 ± 0.56 c	30.27 ± 0.83 a
T_6 (RDF 70 + SSL 20)	22.90 ± 0.49 abc	31.53 ± 0.42 a	21.16 ± 0.81 bc	30.64 ± 0.27 a
T_7 (RDF 50 + SSL 30)	21.81 ± 1.21 bc	31.53 ± 0.60 a	20.39 ± 1.09 c	30.47 ± 0.10 a
T_8 (RDF 60 + SSL 30)	24.30 ± 0.26 ab	32.80 ± 1.12 a	22.65 ± 1.14 abc	32.15 ± 1.37 a
T_9 (RDF 70 + SSL 30)	24.87 ± 0.33 ab	32.41 ± 0.60 a	24.08 ± 1.26 ab	31.84 ± 0.34 a
Significance level	**	**	**	**

Mean values within the same column having the same letters differ non-significantly ($p \leq 0.01$), while different letters indicate a significant difference ($p \leq 0.01$). Mean (±SE) was taken from three replicates for each treatment. **, indicates significant at 1% level of probability.

The data with respect to 1000 grain weight (test weight) of rice and wheat are shown in Table 4. Among all the treatments, T_3 (100% RDF + 30 Mg ha^{-1} SSL) had the highest test weight in the rice (25.88 and 25.80) and wheat crop (33.86 and 33.18) during 2017–2018 and 2018–2019, respectively. In the III-Rice crop, T_2 (100% RDF + 20 Mg ha^{-1} SSL) and T_3 (100% RDF + 30 Mg ha^{-1} SSL) had 16.61% and 18.97% greater test weights in comparison to T_1 (100% RDF). In the case of the IV-rice crop, T_2 (100% RDF + 20 Mg ha^{-1} SSL) and T_3 (100% RDF + 30 Mg ha^{-1} SSL) produced, respectively, 12.41% and 13.54% higher test weights over T_1 (100% RDF), whereas, in the case of the III-wheat crop, the test weight in T_2 (100% RDF + 20 Mg ha^{-1} SSL) and T_3 (100% RDF + 30 Mg ha^{-1} SSL) was 9.10% and 12.99% higher than T_1 (100% RDF), respectively. In the IV-wheat crop, T_2 (20 Mg ha^{-1} SSL + 100% RDF) and T_3 (30 Mg ha^{-1} SSL + 60% RDF) showed 2.50 and 6.61% higher test weights over 100% RDF (T_1), respectively. The SSL-amended soil favoured vegetative growth, development, and maturity of rice and wheat. This can be explained by optimal uptake of trace, micronutrients, and macronutrients by crop plants that support various biochemical and physiological processes, resulting in higher vigour of seeds. Jamil et al. [49] reported a significant rise in 1000 grain weight of wheat with the use of SSL compared to control. Barbarick et al. [58] and Elsokkary et al. [59] also reported that SSL-based nutrient management improved the productivity of crop plants, including 1000 grain weight.

3.5. Sewage Sludge and Fertilizers Influence the Nitrogen (%) and Protein (%) of Rice and Wheat

Application of SSL and fertilizer significantly influenced the nitrogen concentration and protein content of rice and wheat crops during both years (Table 5).

Table 5. Impact of sewage sludge use with fertilizers on nitrogen concentration and protein content.

Treatments	Nitrogen Concentration (%)			
	2017–2018		2018–2019	
	III-Rice	III-Wheat	IV-Rice	IV-Wheat
T_0 (WF)	1.28 ± 0.084 c	1.69 ± 0.09 f	1.26 ± 0.07 c	1.33 ± 0.05 e
T_1 (RDF 100)	1.54 ± 0.11 b	1.90 ± 0.07 cde	1.50 ± 0.10 b	1.95 ± 0.00 bc
T_2 (RDF 100 + SSL 20)	1.65 ± 0.08 a	2.44 ± 0.07 ab	1.58 ± 0.09 ab	2.15 ± 0.22 ab
T_3 (RDF 100 + SSL 30)	1.78 ± 0.06 a	2.69 ± 0.7 a	1.66 ± 0.08 a	2.26 ± 0.03 a
T_4 (RDF 50 + SSL 20)	1.43 ± 0.07 bc	1.72 ± 0.03 ef	1.39 ± 0.04 bc	1.66 ± 0.03 d
T_5 (RDF 60 + SSL 20)	1.51 ± 0.02 b	1.79 ± 0.03 bc	1.40 ± 0.07 bc	1.72 ± c0.03 d
T_6 (RDF 70 + SSL 20)	1.55 ± 0.52 b	1.88 ± 0.04 cde	1.47 ± 0.05 bc	1.80 ± 0.04 cd
T_7 (RDF 50 + SSL 30)	1.55 ± 0.00 b	1.89 ± 0.03 cde	1.48 ± 0.04 bc	1.82 ± 0.04 cd
T_8 (RDF 60 + SSL 30)	1.56 ± 0.08 b	1.96 ± 0.02 cd	1.51 ± 0.059 b	1.85 ± 0.02 cd
T_9 (RDF 70 + SSL 30)	1.59 ± 0.02 b	2.04 ± 0.00 c	1.54 ± 0.02 b	1.95 ± 0.07 bc
Significance level	**	**	**	**
	Protein Content (%)			
T_0 (WF)	8.03 ± 0.51 c	10.59 ± 0.58 f	7.90 ± 0.48 c	8.31 ± 0.35 e
T_1 (RDF 100)	9.63 ± 0.71 b	11.85 ± 0.44 cde	9.37 ± 0.66 ab	10.23 ± 0.02 d
T_2 (RDF 100 + SSL 20)	10.29 ± 0.50 ab	15.23 ± 0.46 b	9.86 ± 0.57 ab	13.42 ± 1.39 ab
T_3 (RDF 100 + SSL 30)	11.15 ± 0.41 a	16.81 ± 0.46 a	10.35 ± 0.51 a	14.13 ± 0.23 a
T_4 (RDF 50 + SSL 20)	8.94 ± 0.43 bc	10.75 ± 0.22 ef	8.71 ± 0.30 bc	10.35 ± 0.19 d
T_5 (RDF 60 + SSL 20)	9.42 ± 0.16 b	11.19 ± 0.20 def	9.03 ± 0.44 abc	10.78 ± 0.23 cd
T_6 (RDF 70 + SSL 20)	9.71 ± 0.32 b	11.75 ± 0.27 cde	9.17 ± 0.35 abc	11.26 ± 0.27 cd
T_7 (RDF 50 + SSL 30)	9.69 ± 0.32 b	11.79 ± 0.20 cde	9.25 ± 0.25 abc	11.39 ± 0.25 cd
T_8 (RDF 60 + SSL 30)	9.73 ± 0.49 b	12.25 ± 0.11 cd	9.46 ± 0.37 ab	11.59 ± 0.17 cd
T_9 (RDF 70 + SSL 30)	10.04 ± 0.18 b	12.75 ± 0.04 c	9.63 ± 0.12 ab	12.17 ± 0.48 bc
Significance level	**	**	**	**

Mean values within the same column having the same letter differ non-significantly ($p \leq 0.01$), while different letters indicate a significant difference ($p \leq 0.01$). Mean (±SE) was taken from three replicates for each treatment. **, indicates significant at 1% level of probability.

A significantly higher concentration of N and protein content of rice (1.78 and 1.66%, 11.15 and 10.35%) and wheat (2.69 and 2.66, 16.81 and 14.13%) were recorded with 100% RDF + 30 Mg ha^{-1} SSL during 2017–2018 and 2018–2019, respectively, whereas the lowest was recorded in the non-fertilized plot. The treatments (T_4–T_9), which received a reduced dose of CF with SSL, were statistically at par with 100% RDF up to IV-Rice but in the case of IV-wheat, a significant reduction was noticed with treatments T_4 (50% RDF + 20 Mg ha^{-1} SSL) and T_5 (60% RDF + 20 Mg ha^{-1} SSL) with respect to N concentration (Table 5).

Protein content in treatments that received a reduced dose of CF with SSL, i.e., T_4 to T_9, were statistically similar to the 100% RDF treatment during the study. The addition of SSL with CF increased the grain N concentration, because both are a good source of N and resulted in a greater supply of N to the plant [60]. This effect was observed in SSL and RDF-fertilized plots in comparison to nonfertilized plots. The N in the soil helps the decomposition process and, as a result, the rate of decomposition in RDF with sludge-supplemented soils may be greater, leading to quicker SSL breakdown. Nitrogen is required for the synthesis of amino acids and increases the protein content of cereals considerably [61,62]. Yamur et al. [63] stated that SSL application improved protein content from 19.82 to 23.92% in lentils.

3.6. *Sewage Sludge and Fertilizers Influence the Cadmium, Chromium, Nickel and Lead Concentration (mg kg^{-1}) of Rice and Wheat*

The SSL and fertilizer treatments had higher Cd concentration in rice and wheat grain compared to 100% RDF (Table 6) treatment. In III-Rice, the highest Cd concentration was observed in treatment T_3 (100% RDF + 30 Mg ha^{-1} SSL) at 1.27 mg kg^{-1}, followed by T_9 (70% RDF + 30 Mg ha^{-1} SSL) at 1.19 mg kg^{-1}, and T_8 60% (RDF + 30 Mg ha^{-1} SSL) at 1.14 mg kg^{-1}) which showed 184, 166 and 154% increases over T_1 (100% RDF). An almost similar trend was seen in III-Wheat and the highest Cd concentration was observed in treatment T_3 (100% RDF + 30 Mg ha^{-1} SSL) at 1.36 mg kg^{-1}, followed by T_9 (70% RDF + 30 Mg ha^{-1} SSL) at 1.26 mg kg^{-1}, and T_8 (60% RDF + 30 Mg ha^{-1} SSL) at 1.25 mg kg^{-1}, which were 167, 146 and 144% higher that T_1 (0.51 mg kg^{-1}). The lowest Cd concentration was seen in T_0 (0.33 mg kg^{-1}) which was statistically lower than T_1 (100% RDF).

During the 2018–2019 (IV-rice and IV-Wheat) season of the experiment, treatments receiving SSL with CF had significantly higher Cd concentrations then T_1 in grains of IV-rice, and a similar trend was noticed in IV-wheat. In IV-Rice, the Cd concentration was in the order of: T_3 (1.23 mg kg^{-1}) > T_9 (1.16 mg kg^{-1}) > T_8 (1.11 mg kg^{-1}) > T_2 (1.10 mg kg^{-1}) > T_7 (1.05 mg kg^{-1}) > T_6 (0.83 mg kg^{-1}) = T_5 (0.83 mg kg^{-1}) > T_4 (0.80 mg kg^{-1}) > T_1 (0.42 mg kg^{-1}) > T_0 (0.25 mg kg^{-1}), with respective increases of 195, 178, 166, 163, 153, 99, 99 and 93% over T_1 (0.42 mg kg^{-1}). In IV-Wheat, the order was: T_3 (167%) > T_9 (150%) > T_8 (148%) > T_2 (145%) > T_7 (141%) > T_6 (88%) > T_5 (85%) >T_4 (74% increase) over T_1 (100% RDF). Overall, the Cd concentration in grain increased with joint application of sewage sludge, i.e., 20 and 30 Mg ha^{-1}, and full or reduced dose of CF, compared to 100% RDF. This study revealed that Cd concentration was highest (1.36 mg kg^{-1}) in wheat grain but below the permissible limits of Cd, which are 4 mg kg^{-1} and 12 mg kg^{-1} in Poland and the Czech Republic, respectively [64]. Furthermore, the study also found that Cd accumulation in grain did not exceed the phytotoxic threshold level of 5–30 mg kg^{-1} [65]. The Cd concentration was low in rice compared to wheat in the grain. In comparison to wheat, the lower Cd content in rice could be explained due to the submergence condition. Under flooded condition, Cd forms insoluble compounds such as Cd sulfide and/or Cd carbonate which reduce its availability to the rice plants [66,67]. Greger [68], reported that magnification of heavy metals levels in plants is influenced by soil properties, SSL composition, the application rate of SSL, and elemental speciation [69].

Table 6. Impact of sewage sludge use with fertilizers on cadmium and chromium concentration in rice and wheat.

Treatments	Grain			
	Cadmium (mg kg^{-1})			
	2017–2018		2018–2019	
	III-Rice	III-Wheat	IV-Rice	IV-Wheat
T_0 (WF)	0.29 ± 0.04 c	0.33 ± 0.03 d	0.25 ± 0.04 e	0.29 ± 0.05 d
T_1 (RDF 100)	0.45 ± 0.06 c	0.51 ± 0.03 c	0.42 ± 0.05 d	0.49 ± 0.02 c
T_2 (RDF 100 + SSL 20)	1.12 ± 0.07 a	1.24 ± 0.03 a	1.10 ± 0.06 ab	1.20 ± 0.05 a
T_3 (RDF 100 + SSL 30)	1.27 ± 0.12 a	1.36 ± 0.04 a	1.23 ± 0.05 a	1.31 ± 0.01 a
T_4 (RDF 50 + SSL 20)	0.81 ± 0.07 b	0.89 ± 0.06 b	0.80 ± 0.02 c	0.85 ± 0.05 b
T_5 (RDF 60 + SSL 20)	0.85 ± 0.09 b	0.94 ± 0.03 b	0.83 ± 0.05 c	0.91 ± 0.06 b
T_6 (RDF 70 + SSL 20)	0.87 ± 0.07 b	0.95 ± 0.04 b	0.83 ± 0.02 c	0.92 ± 0.05 b
T_7 (RDF 50 + SSL 30)	1.10 ± 0.06 a	1.24 ± 0.07 a	1.05 ± 0.06 ab	1.18 ± 0.04 a
T_8 (RDF 60 + SSL 30)	1.14 ± 0.05 a	1.25 ± 0.06 a	1.11 ± 0.02 ab	1.21 ± 0.06 a
T_9 (RDF 70 + SSL 30)	1.19 ± 0.04 a	1.26 ± 0.08 a	1.16 ± 0.04 ab	1.23 ± 0.01 a
Significance level	**	**	**	**
	Chromium (mg kg^{-1})			
T_0 (WF)	1.37 ± 0.033 b	1.21 ± 0.25 c	1.17 ± 0.092 e	1.04 ± 0.10 g
T_1 (RDF 100)	1.97 ± 0.18 b	1.75 ± 0.09 c	1.46 ± 0.36 e	1.27 ± 0.14 g
T_2 (RDF 100 + SSL 20)	4.78 ± 0.67 a	4.22 ± 0.17 a	3.55 ± 0.058 b	3.24 ± 0.14 bc
T_3 (RDF 100 + SSL 30)	4.90 ± 0.67 a	4.49 ± 0.36 a	4.11 ± 0.51 a	3.55 ± 0.03 a
T_4 (RDF 50 + SSL 20)	4.25 ± 0.79 a	3.54 ± 0.34 b	2.92 ± 0.14 d	2.50 ± 0.08 f
T_5 (RDF 60 + SSL 20)	4.41 ± 0.59 a	3.68 ± 0.26 ab	3.01 ± 0.06 d	2.67 ± 0.10 ef
T_6 (RDF 70 + SSL 20)	4.49 ± 0.29 a	3.79 ± 0.22 ab	3.04 ± 0.03 cd	2.68 ± 0.01 ef
T_7 (RDF 50 + SSL 30)	4.64 ± 0.63 a	3.91 ± 0.16 ab	3.29 ± 0.10 bcd	2.90 ± 0.06 de
T_8 (RDF 60 + SSL 30)	4.79 ± 0.82 a	4.05 ± 0.18 ab	3.50 ± 0.11 bc	3.02 ± 0.08 cd
T_9 (RDF 70 + SSL 30)	4.85 ± 0.69 a	4.34 ± 0.43 ab	4.04 ± 0.10 a	3.47 ± 0.11 ab
Significance level	**	**	**	**

Mean values within the same column having the same letter differ non-significantly ($p \leq 0.01$), while different letters indicate a significant difference ($p \leq 0.01$). Mean (±SE) was taken from 3 replicates for each treatment. **, indicates significant at 1% level of probability.

The data pertaining to Cr concentration in grain (Table 6) showed significantly higher Cr concentration with all combined treatments of SSL with CF compared to RDF. During 2017–2018, the maximum Cr concentration in III-rice was recorded in treatment T_3 (4.90) followed by T_9 (4.85) and T_8 (4.79), whereas in III-Wheat it was in treatment T_3 (4.49) followed by T_9 (4.34) and T_2 (4.22), with a respective significant increase of 148, 146 and 143% in III-Rice and 156, 148 and 141% in III-wheat compared to 100% RDF (T_1). During 2018–2019 (IV-rice and IV-Wheat), the maximum Cr concentration in IV-Rice (4.11 mg kg^{-1}) and IV-wheat (3.55 mg kg^{-1}) was recorded in T_3 (30 Mg ha^{-1} SSL). The application of 20 and 30 Mg ha^{-1} SSL with full or reduced doses of CF (T_2, T_3, T_4, T_5, T_6, T_7, T_8 and T_9) significantly increased the Cr concentration in grain by 2.43, 2.86, 2.00, 2.06, 2.11, 2.37, 2.46 and 2.76 times in IV-Rice and 2.55, 2.79, 1.96, 1.99, 2.11, 2.24, 2.37 and 2.73 times in IV-Wheat, respectively, over 100% RDF (T_1). The phytotoxic limit of Cr is 5–30 mg kg^{-1} [65]. This result indicates that Cr concentration in grain was within the acceptable limit and did not cross the phytotoxic threshold level.

The results show that the Ni concentration in III-rice varied from 7.65 to 19.00 mg kg^{-1}, whereas, in III-wheat it ranged between 7.93 to 17.31 mg kg^{-1} (Table 7). During 2017–2018, the maximum Ni concentration in III-rice (19.00 mg kg^{-1}) and III-wheat (17.31 mg kg^{-1}) was recorded in treatment T_3 (100% RDF + 30 Mg ha^{-1} SSL). The application of 20 and 30 Mg ha^{-1} SSL along with a full dose or reduced dose of CF significantly increased the Ni concentration in grains of III-rice and III-wheat compared to 100% RDF (T_1). During 2018–2019, the Ni concentration in grain ranged between 7.35 to 5.48, and 6.28 to 14.35 mg kg^{-1}, in IV-rice and IV-wheat, respectively. The greatest Ni concentration in IV-Rice (15.48 mg kg^{-1}) and IV-wheat (14.35 mg kg^{-1}) was recorded in T_3 (100% RDF + 30 Mg ha^{-1} SSL). The application of 30 Mg ha^{-1} SSL along with a reduced dose of CF (50, 60, 70% of RDF), i.e., T_7, T_8 and T_9 resulted in significant increases in Ni concentration in grains by 45, 48 and 61% in IV-rice, respectively, over 100% RDF (T_1), whereas in IV-wheat, these treatments were statistically at par with T_1 (100% RDF). All the treatments amended with 20 Mg ha^{-1} SSL along with a reduced dose of CF (50, 60, 70 of RDF), i.e., T_4, T_5 and T_6, did not show a significant increase of Ni in grain compared to 100% RDF in both crops (IV-rice and IV-wheat). The Ni concentration was below the phytotoxic limit of 10–100 mg kg^{-1} as in [65]. Out of the various treatments (Table 7), T_3 (100% RDF + 30 Mg ha^{-1} SSL) had the highest Pb concentration in III-rice grain (1.99) and III-wheat grain (1.94). The joint application of 20 and 30 Mg ha^{-1} SSL and full dose or reduced doses CF (T_2, T_3, T_4, T_5, T_6, T_7, T_8 and T_9) in III-rice significantly increased the Pb concentration in grain by 1.89, 2.34, 1.52, 1.57, 1.65, 1.77, 1.95 and 2.25 times, respectively, compared to 100% RDF. The corresponding increases of Pb concentration in III-wheat were 2.53, 2.88, 1.83, 1.91, 2.04, 2.08, 2.14 and 2.77 times.

Table 7. Impact of sewage sludge use with fertilizers on nickel and lead concentration in rice and wheat.

Treatments	Grain			
	Nickel (mg kg^{-1})			
	2017–2018		2018–2019	
	III-Rice	III-Wheat	IV-Rice	IV-Wheat
T_0 (WF)	7.65 ± 0.67 c	7.93 ± 0.36 d	7.35 ± 0.52 c	6.28 ± 0.87 c
T_1 (RDF 100)	9.70 ± 0.95 c	9.12 ± 0.48 d	9.05 ± 0.83 bc	8.04 ± 1.29 b
T_2 (RDF 100 + SSL 20)	16.35 ± 0.76 ab	14.62 ± 0.37 bc	14.05 ± 1.10 a	13.32 ± 0.45 a
T_3 (RDF 100 + SSL 30)	19.00 ± 1.75 a	17.31 ± 1.12 a	15.48 ± 1.36 a	14.35 ± 1.35 a
T_4 (RDF 50 + SSL 20)	13.96 ± 1.26 b	12.72 ± 0.79 b	12.08 ± 0.96 bc	9.20 ± 1.11 b
T_5 (RDF 60 + SSL 20)	14.63 ± 0.65 b	12.90 ± 0.38 b	12.12 ± 1.67 bc	9.61 ± 0.36 b
T_6 (RDF 70 + SSL 20)	14.94 ± 0.76 b	13.68 ± 0.81 bc	12.20 ±1.10 bc	9.26 ± 0.93 b
T_7 (RDF 50 + SSL 30)	15.06 ± 1.76 b	14.73 ± 0.92 bc	13.23 ± 0.41 a	11.09 ± 0.71 ab
T_8 (RDF 60 + SSL 30)	16.00 ± 0.79 ab	14.42 ± 0.41 bc	13.47 ± 0.52 a	10.48 ± 0.96 ab
T_9 (RDF 70 + SSL 30)	17.67 ± 1.46 ab	15.83 ± 1.32 ab	14.58 ± 1.69 a	12.04 ± 1.19 ab
Significance level	**	**	**	**
Treatments	Lead (mg kg^{-1})			
T_0 (WF)	0.49 ± 0.073 e	0.40 ± 0.08 e	0.40 ± 0.12 c	0.37 ± 0.08 d
T_1 (RDF 100)	0.85 ± 0.04 d	0.67 ± 0.17 d	0.59 ± 0.03 c	0.59 ± 0.03 d
T_2 (RDF 100 + SSL 20)	1.63 ± 0.087 b	1.51 ± 0.04 b	1.31 ± 0.7 b	1.16 ± 0.08 bc
T_3 (RDF 100 + SSL 30)	1.99 ± 0.15 a	1.93 ± 0.04 a	1.81 ± 0.10 a	1.53 ± 0.25 a
T_4 (RDF 50 + SSL 20)	1.30 ± 0.10 c	1.23 ± 0.08 c	1.13 ± 0.05 b	0.94 ± 0.03 c
T_5 (RDF 60 + SSL 20)	1.32 ± 0.03 c	1.28 ± 0.06 bc	1.13 ± 0.03 b	0.99 ± 0.04 c
T_6 (RDF 70 + SSL 20)	1.41 ± 0.04 bc	1.37 ± 0.07 bc	1.19 ± 0.05 b	0.99 ± 0.03 c
T_7 (RDF 50 + SSL 30)	1.51 ± 0.04 bc	1.40 ± 0.04 bc	1.26 ± 0.10 b	1.15 b ± 0.02 c
T_8 (RDF 60 + SSL 30)	1.53 ± 0.14 bc	1.44 ± 0.06 bc	1.27 ± 0.06 b	1.16 ± 0.07 bc
T_9 (RDF 70 + SSL 30)	1.92 ± 0.07 a	1.86 ± 0.07 a	1.69 ± 0.01 a	1.39 ± 0.13 ab
Significance level	**	**	**	**

Mean values within the same column with the same letter differ non-significantly ($p \leq 0.01$), while different letters indicate a significant difference ($p \leq 0.01$). Mean (±SE) was taken from 3 replicates for each treatment. **, indicates significant at 1% level of probability.

During 2018–2019, (IV-rice and IV-wheat), all the treatments of SSL (T_3–T_9) were significantly higher in Pb concentration compared to 100% RDF. The Pb concentration in grain ranged from 0.40 to 1.81 and 0.37 to 1.53 mg kg^{-1} in IV-rice and IV-wheat, respectively. The highest Pb concentrations in IV-Rice (1.81 mg kg^{-1}) and IV-Wheat (1.53 mg kg^{-1}) were recorded in T_3 (30 Mg ha^{-1} SSL). Treatments T_2, T_3, T_4, T_5, T_6, T_7, T_8 and T_9 when compared with RDF had corresponding increases of 2.24, 2.87, 1.87, 1.89, 2.03, 2.13 and 2.75 times in IV-rice and 1.95, 2.57, 1.58, 1.67, 1.67, 1.94, 1.96 and 2 times in IV-wheat. The lowest Pb concentration was recorded in T_0 during the course of the experiment. The present investigation revealed that Pb (highest, 1.99 mg kg^{-1} in III-rice) in rice and wheat grain were below phytotoxic limits (30–300 mg kg^{-1}) as outlined by [65]. Singh and Agrawal [70] and Eid et al. [71] reported an increase in heavy metal concentration in the areal parts of barley by application of SSL. Zoubi et al. [72] reported similar findings.

3.7. Total Heavy Metal Content in Post-Harvest Soil after Completion of the Experiment

The total Cd, Cr, Ni and Pb content in soils after the termination of the experiment ranged from 0.44 to 2.53, 2.25 to 9.26, 7.98 to 21.23 and 4.45 to 27.36 mg kg^{-1}, respectively (Figure 6). Among all the treatments, T_3 (100% RDF + 30 Mg ha^{-1} SSL) had the highest total Cd, Cr, Ni and Pb after harvest of IV-wheat. The treatments that received 30 SSL Mg ha^{-1} (T_3, T_7, T_8 and T_9) increased the total Cd content by 5.75, 3.2 3.3 and 3.6 times, respectively over T_1, whereas, treatments T_2, T_4, T_5 and T_6 that received a lower dose (20 SSL Mg ha^{-1}) increased by 3.2, 2.9, 2.9 and 3.0 times, respectively. However, total Cr content soil after harvest of the IV-wheat crop showed a declining trend in the order of T_3 (9.26) > T_9 (9.15 mg kg^{-1}) > T_8 (8.88 mg kg^{-1}) > T_7 (8.45 mg kg^{-1}) > T_2 (7.77 mg kg^{-1}) > T_6 (7.89 mg kg^{-1}) > T_5 (7.69 mg kg^{-1}) > T_4 (7.22 mg kg^{-1}).

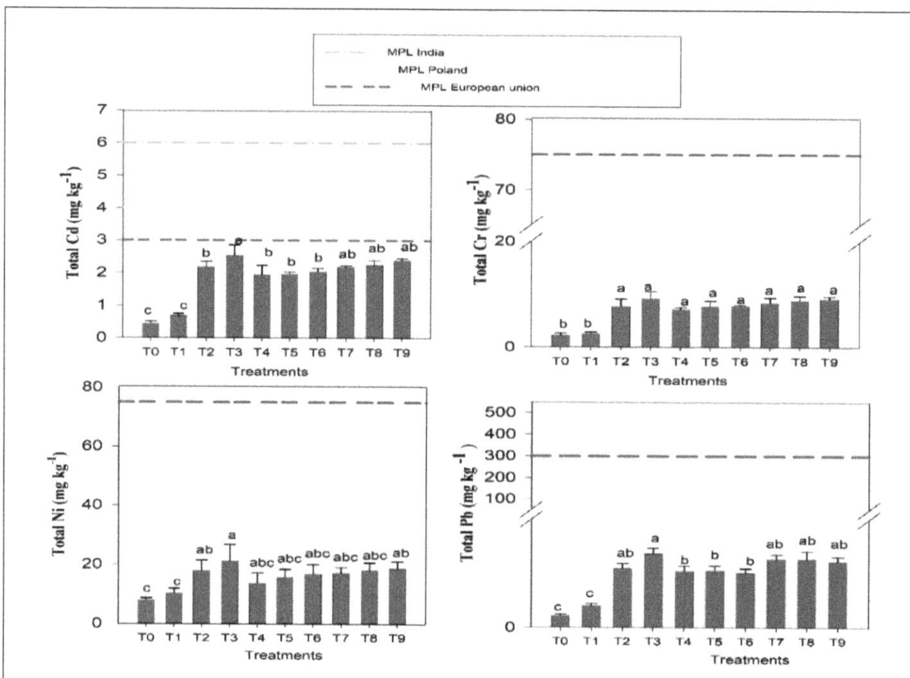

Figure 6. Impact of sewage sludge use with fertilizers on total heavy metals buildup in soils after completion of the experiment. Data (mean ± SE) followed by the same letter differ non-significantly ($p \leq 0.05$), while different letters indicate a significant difference ($p \leq 0.05$).

These data were statistically similar to one another but significantly greater T_1 (2.59 mg kg^{-1}). The total Ni content in post-harvest soils significantly increased due to the application of SSL. The 100% RDF treatment, i.e., T_1 (10.25 mg kg^{-1}) was statistically at par with all other treatments except T_2 (18.0 mg kg^{-1}), T_3 (21.23 mg kg^{-1}) and T_9 (18.65 mg kg^{-1}) which had the highest content of Ni in the post-harvest soil. Treatments T_3 (100% RDF + 30 Mg ha^{-1} SSL), T_7 (50% RDF + 30 Mg ha^{-1} SSL), T_8 (60% RDF + 30 Mg ha^{-1} SSL) and T_9 (70% RDF + 30 Mg ha^{-1} SSL) had increased Pb content by 3.32, 3.09 3.08 and 2.98 times, respectively, over T_1 (100% RDF), whereas in treatments T_2 (100% RDF + 20 Mg ha^{-1} SSL), T_4 (50% RDF + 20 Mg ha^{-1} SSL), T_5 (60% RDF + 20 Mg ha^{-1} SSL) and T_6 (70% RDF + 20 Mg ha^{-1} SSL) with lower dose of SSL (20 Mg ha^{-1}), the respective increases were 2.66, 2.55, 2.58 and 2.45 times. The total heavy metal contents in soil were within the maximum permissible limits (MPL) with respect to the limits [64] for India and for Poland and the European Union [28] (Figure 6). The results indicate that total heavy metal contents in soil were within MPL even after applying the highest dose of SSL (30 Mg ha^{-1}) with 100% RDF.

4. Conclusions

Application of 20 or 30 Mg ha^{-1} SSL in conjunction with 100% RDF significantly improved the productivity of the rice-wheat system compared to 100% RDF. It was observed that 20 or 30 Mg ha^{-1} SSL along with recommended or reduced doses of fertilizer significantly increased the heavy metal content in grains and experimental soil compared to 100% RDF, or absolute control plots, but this buildup was within permissible limits. Therefore, to obtain optimum grain yield (at par with RDF 100%), it is recommended to apply only a 50% dose of RDF in the first two crops and a 70% dose of RDF in the next two crops with a one-time application of 20 t ha^{-1} SSL. Regular monitoring of heavy metal buildup in soil, and its subsequent increase in the edible part of rice and wheat, needs to be strictly tracked to avoid risks related to soil and human health. However, this study needs to be confirmed by long-term experiments before recommendations for other agro-climatic regions.

Author Contributions: Conceptualization: S.S.J. and S.K.S.; methodology: S.S.J. and S.K.S.; formal analysis: S.S.J., M.P. and S.K.S.; data curation: S.S.J. and S.K.S.; statistical expertise: S.S.J. and S.K.S.; writing—original draft preparation: S.S.J. and S.K.S.; writing—review and editing: M.P., A.G., A.M.A. and A.H.; visualization: A.H., S.S.J. and S.K.S.; supervision: S.K.S.; funding acquisition: A.M.A., A.G. and A.H. All authors have read and agreed to the published version of the manuscript.

Funding: The work was financially supported by BHU-UGC, India. The research programme was also partially supported by the Princess Nourah bint Abdulrahman University Researchers Supporting Project number (PNURSP2022R65), Princess Nourah bint Abdulrahman University, Riyadh, Saudi Arabia.

Institutional Review Board Statement: Not applicable.

Informed Consent Statement: Not applicable.

Data Availability Statement: Data are available after request.

Acknowledgments: The authors wish to acknowledge the BHU-UGC, India and the Princess Nourah bint Abdulrahman University Researchers Supporting Project number (PNURSP2022R65), Princess Nourah bint Abdulrahman University, Riyadh, Saudi Arabia for providing financial support for this research. The anonymous reviewers are also gratefully acknowledged for their efforts in helping us to improve the presentation of our work in this manuscript.

Conflicts of Interest: The authors have no conflict of interest to disclose.

Ethical Statement: No living organism (human or animal) was involved in conducting the present experiments.

References

1. Ladha, J.K.; Dawe, D.; Pathak, H.; Padre, A.T.; Yadav, R.L.; Singh, B.; Singh, Y.; Singh, Y.; Singh, P.; Kundu, A.L. How extensive are yield declines in long-term rice-wheat experiments in Asia? *Field Crops Res.* **2003**, *81*, 159–180. [CrossRef]
2. Singh, V.K.; Sharma, B.B. Productivity of rice (*Oryza sativa*) as influenced by crop diversification in wheat (*Triticum aestivum*)-rice cropping system on Mollisols of foothills of Himalayas. *Indian J. Agric. Sci.* **2001**, *71*, 5–8.
3. Bijarniya, D.; Parihar, C.M.; Jat, R.K.; Kalvania, K.; Kakraliya, S.K.; Jat, M.L. Portfolios of climate smart agriculture practices in smallholder rice-wheat system of Eastern Indo-Gangetic plains—Crop productivity, resource use efficiency and environmental foot prints. *Agronomy* **2020**, *10*, 1561. [CrossRef]
4. Sanyal, S.K.; Majumdar, K.; Singh, V.K. Nutrient management in Indian agriculture with special reference to nutrient mining—A relook. *J. Indian Soc. Soil Sci.* **2014**, *62*, 307–325.
5. Timsina, J. Can organic sources of nutrients increase crop yields to meet global food demand? *Agronomy* **2018**, *8*, 214. [CrossRef]
6. Latare, A.M.; Singh, S.K.; Kumar, O. Impact of sewage sludge application on soil fertility, microbial population and enzyme activities in soil under rice-wheat system. *J. Indian Soc. Soil Sci.* **2018**, *66*, 300–309. [CrossRef]
7. Jatav, H.S.; Singh, S.K.; Yadav, J.S. Cumulative Effect of Sewage Sludge and Fertilizers Application on Enhancing Soil Microbial Population Under Rice-Wheat Cropping System. *J. Exp. Biol. Agric. Sci.* **2018**, *6*, 538–543. [CrossRef]
8. Qiu, S.; McComb, A.J.; Bell, R.W.; Davis, J.A. Response of soil microbial activity to temperature, moisture, and litter leaching on a wetland transect during seasonal refilling. *Wetl. Ecol. Manag.* **2005**, *13*, 43–54. [CrossRef]
9. Thomas, C.L.; Acquah, G.E.; Whitmore, A.P.; McGrath, S.P.; Haefele, S.M. The effect of different organic fertilizers on yield and soil and crop nutrient concentrations. *Agronomy* **2019**, *9*, 776. [CrossRef]
10. Swain, A.; Singh, S.K.; Mohapatra, K.K.; Patra, A. Sewage sludge amendment affects spinach yield, heavy metal bioaccumulation, and soil pollution indexes. *Arab. J. Geosci.* **2021**, *14*, 717. [CrossRef]
11. Fijalkowski, K.; Rorat, A.; Grobelak, A.; Kacprzak, M.J. The presence of contaminations in sewage sludge—The current situation. *J. Environ. Manag.* **2017**, *203*, 1126–1136. [CrossRef] [PubMed]
12. Roy, T.; Biswas, D.R.; Ghosh, A.; Patra, A.K.; Singh, R.D.; Sarkar, A.; Biswas, S.S. Dynamics of culturable microbial fraction in an Inceptisol under short-term amendment with municipal sludge from different sources. *Appl. Soil Ecol.* **2019**, *136*, 116–121. [CrossRef]
13. Kayikcioglu, H.H.; Yener, H.; Ongun, A.R.; Okur, B. Evaluation of soil and plant health associated with successive three-year sewage sludge field applications under semi-arid biodegradation condition. *Arch. Agron. Soil Sci.* **2019**, *65*, 1659–1676. [CrossRef]
14. Ociepa-Kubicka, A.; Pachura, P. The use of sewage sludge and compost for fertilization of energy crops on the example of *Miscanthus* and *Virginia mallow*. *Rocznik Ochrona Środowiska* **2013**, *15*, 2267–2278.
15. Seleiman, M.F.; Santanen, A.; Jaakkola, S.; Ekholm, P.; Hartikainen, H.; Stoddard, F.L.; Mäkelä, P.S. Biomass yield and quality of bioenergy crops grown with synthetic and organic fertilizers. *Biomass Bioenergy* **2013**, *59*, 477–485. [CrossRef]
16. Singh, R.P.; Agrawal, M. Potential benefits and risks of land application of sewage sludge. *Waste Manag.* **2008**, *28*, 347–358. [CrossRef]
17. Albiach, R.; Canet, R.; Pomares, F.; Ingelmo, F. Organic matter components, aggregate stability and biological activity in a horticultural soil fertilized with different rates of two sewage sludges during ten years. *Bioresour. Technol.* **2001**, *77*, 109–114. [CrossRef]
18. García-Gil, J.C.; Plaza, C.; Senesi, N.; Brunetti, G.; Polo, A. Effects of sewage sludge amendment on humic acids and microbiological properties of a semiarid Mediterranean soil. *Biol. Fertil. Soils* **2004**, *39*, 320–328.
19. Mondal, S.; Singh, R.D.; Patra, A.K.; Dwivedi, B.S. Changes in soil quality in response to short-term application of municipal sewage sludge in a typic haplustept under cowpea-wheat cropping system. *Environ. Nanotechnol. Monit. Manag.* **2015**, *4*, 37–41. [CrossRef]
20. Sharma, S.; Dhaliwal, S.S. Effect of sewage sludge and rice straw compost on yield, micronutrient availability and soil quality under rice–wheat system. *Commun. Soil Sci. Plant Anal.* **2019**, *50*, 1943–1954. [CrossRef]
21. Latare, A.M.; Kumar, O.; Singh, S.K.; Gupta, A. Direct and residual effect of sewage sludge on yield, heavy metals content and soil fertility under rice–wheat system. *Ecol. Eng.* **2014**, *69*, 17–24. [CrossRef]
22. Kumar, M.; Singh, S.K.; Patra, A. Effect of different nutrient sources on yield and biochemical properties of soil under rice–wheat cropping sequence in middle *Gangetic alluvial* plain. *J. Plant Nutr.* **2021**, *44*, 2310–2330. [CrossRef]
23. Gupta, V.; Sharma, R.S.; Vishwakarma, S.K. Long-term effect of integrated nutrient management on yield sustainability and soil fertility of rice (*Oryza sativa*)–wheat (*Triticum aestivum*) cropping system. *Indian J. Agron.* **2006**, *51*, 160–164.
24. Mahajan, A.; Bhagat, R.M.; Gupta, R.D. Integrated nutrient management in sustainable rice-wheat cropping system for food security in India. *SAARC J. Agric.* **2008**, *6*, 29–32.
25. Roy, T.; Singh, R.D.; Biswas, D.R.; Patra, A.K. Effect of sewage sludge and inorganic fertilizers on productivity and micronutrients accumulation by Palak (*Beta vulgaris*) and their availability in a Typic Haplustept. *J. Indian Soc. Soil Sci.* **2013**, *61*, 207–218.
26. Aggelides, S.M.; Londra, P.A. Effects of compost produced from town wastes and sewage sludge on the physical properties of a loamy and a clay soil. *Bioresour. Technol.* **2000**, *71*, 253–259. [CrossRef]
27. Rashid, I.; Murtaza, G.; Zahir, Z.A.; Farooq, M. Effect of humic and fulvic acid transformation on cadmium availability to wheat cultivars in sewage sludge amended soil. *Environ. Sci. Pollut. Res.* **2018**, *25*, 16071–16079. [CrossRef]

28. European Commission. Council Directive 86/278/EEC of 12 June 1986 on the protection of the environment, and in particular of the soil, when sewage sludge is used in agriculture. *Off. J. L* **1986**, *181*, 6–12.
29. Jackson, M.L. *Soil Chemical Analysis*; Prentic-Hall of India Pvt. & Ltd.: New Delhi, India, 1973.
30. Tandon, H.L.S.; Tandon, H.L.S. *Methods of Analysis of Soils, Plants, Waters, and Fertilisers*; Fertiliser Development and Consultation Organisation: New Delhi, India, 1993; Volume 63.
31. Sparks, D.L.; Page, A.; Helmke, P.; Loeppert, R.H.; Soltanpour, P.N.; Tabatabai, M.A.; Johnston, C.T.; Sumne, M.E. *Methods of Soil Analysis, Part 3: Chemical Methods*; Soil Science Society of America; American Society of Agronomy: Madison, WI, USA, 1996; Volume 14.
32. Walkley, A.; Black, I.A. An examination of the Degtjareff method for determining soil organic matter, and a proposed modification of the chromic acid titration method. *Soil Sci.* **1934**, *37*, 29–38. [CrossRef]
33. Subbiah, B.; Asija, G.L. Alkaline permanganate method of available nitrogen determination. *Curr. Sci.* **1956**, *25*, 259.
34. Olsen, S.R. *Estimation of Available Phosphorus in Soils by Extraction with Sodium Bicarbonate*; US Department of Agriculture: Washington, DC, USA, 1954.
35. Hanway, J.J.; Heidel, H. Soil analysis methods as used in Iowa state college soil testing laboratory. *Iowa Agric.* **1952**, *57*, 1–31.
36. Chesnin, L.; Yien, C.H. Turbidimetric determination of available sulfates. *Soil Sci. Soc. Am. J.* **1951**, *15*, 149–151. [CrossRef]
37. Lindsay, W.L.; Norvell, W.A. Development of a DTPA soil test for zinc, iron, manganese, and copper. *Soil Sci. Soc. Am. J.* **1978**, *42*, 421–428. [CrossRef]
38. Peters, J.; Combs, S.; Hoskins, B.; Jarman, J.; Kovar, J.; Watson, M.; Wolf, A.; Wolf, N. *Recommended Methods of Manure Analysis*; University of Wisconsin Cooperative Extension Publishing: Madison, WI, USA, 2003.
39. Nieuwenhuize, J.; Poley-Vos, C.H.; van den Akker, A.H.; van Delft, W. Comparison of microwave and conventional extraction techniques for the determination of metals in soils, sediment and sludge samples by atomic spectrometry. *Analyst* **1991**, *116*, 347–351. [CrossRef]
40. Sparks, D.L.; Fendorf, S.E.; Toner IV, C.V.; Carski, T.H. Kinetic Methods and Measurements. In *Methods of Soil Analysis, Part 3: Chemical Methods*; Soil Science Society of America; American Society of Agronomy: Madison, WI, USA, 1996; Volume 5, pp. 1275–1307.
41. Yamaguchi, M. Determination of the Nitrogen-to-Protein Conversion Factor in Cereals. In *Seed Analysis*; Springer: Berlin/Heidelberg, Germany, 1992; pp. 95–107.
42. Hossain, M.K.; Strezov, V.; Chan, K.Y.; Nelson, P.F. Agronomic properties of wastewater sludge biochar and bioavailability of metals in production of cherry tomato (*Lycopersicon esculentum*). *Chemosphere* **2010**, *78*, 1167–1171. [CrossRef]
43. Latare, A.M.; Singh, S.K.; Kumar, O. Yield and Profitability of Rice-Wheat Sequence with Conjunctive Application of Sewage Sludge and Chemical Yield and Profitability of Rice-Wheat Sequence with Conjunctive Application of Sewage Sludge and Chemical Fertilisers. *Indian J. Fertil.* **2017**, *13*, 50–61.
44. Zhang, Y.; Chen, T.; Liao, Y.; Reid, B.J.; Chi, H.; Hou, Y.; Cai, C. Modest amendment of sewage sludge biochar to reduce the accumulation of cadmium into rice (*Oryza sativa* L.): A field study. *Environ. Pollut.* **2016**, *216*, 819–825. [CrossRef]
45. Rehman, R.A.; Qayyum, M.F. Co-composts of sewage sludge, farm manure and rock phosphate can substitute phosphorus fertilizers in rice-wheat cropping system. *J. Environ. Manag.* **2020**, *259*, 109700. [CrossRef]
46. Romani, M.; Beltarre, G. Good results in rice fields with sewage sludge. *Inf. Agrar.* **2009**, *65*, 39–42.
47. Bahmanyar, M.A.; Pirdashti, H. Responses of biomass, chlorophyll and macro-and micronutrient uptake of rice (*Oryza sativa* L.) to organic and chemical fertilizers. In Proceedings of the 14th Australian Agronomy Conference, Adelaide, SA, Australia, 21–25 September 2008.
48. Burducea, M.; Lobiuc, A.; Asandulesa, M.; Zaltariov, M.-F.; Burducea, I.; Popescu, S.M.; Zheljazkov, V.D. Effects of sewage sludge amendments on the growth and physiology of sweet basil. *Agronomy* **2019**, *9*, 548. [CrossRef]
49. Jamil, M.; Qasim, M.; Umar, M.; Rehman, K. Impact of organic wastes (sewage sludge) on the yield of wheat (*Triticum aestivum* L.) in a calcareous soil. *Int. J. Agric. Biol.* **2004**, *6*, 465–467.
50. El-Shakweer, M.H.A.; El-Sayad, E.A.; Ewees, M.S.A. Soil and plant analysis as a guide for interpretation of the improvement efficiency of organic conditioners added to different soils in Egypt. *Commun. Soil Sci. Plant Anal.* **1998**, *29*, 2067–2088. [CrossRef]
51. Zhang, M.; Heaney, D.; Solberg, E.; Heriquez, B. The effect of MSW compost on metal uptake and yield of wheat, barley and canola in less productive farming soils of Alberta. *Compost. Sci. Util.* **2000**, *8*, 224–235. [CrossRef]
52. Tamrabet, L.; Bouzerzour, H.; Kribaa, M.; Makhlouf, M. The effect of sewage sludge application on durum wheat (*Triticum durum*). *Int. J. Agric. Biol.* **2009**, *11*, 741–745.
53. Jatav, H.S.; Singh, S.K.; Jatav, S.S.; Rajput, V.D.; Sushkova, S. Feasibility of sewage sludge application in rice-wheat cropping system. *Eurasian J. Soil Sci.* **2021**, *10*, 207–214. [CrossRef]
54. Al-Mustafa, W.A.; El-Shall, A.A.; Abdallah, A.E.; Modaihsh, A.S. Response of wheat to sewage sludge applied under two different moisture regimes. *Exp. Agric.* **1995**, *31*, 355–360. [CrossRef]
55. Yadav, G.S.; Lal, R.; Meena, R.S. Vehicular traffic effects on hydraulic properties of a Crosby silt loam under a long-term no-till farming in Central Ohio, USA. *Soil Tillage Res.* **2020**, *202*, 104654. [CrossRef]
56. Gami, S.; Ladha, J.; Pathak, H.; Shah, M.; Pasuquin, E.; Pandey, S.; Hobbs, P.; Joshy, D.; Mishra, R. Long-term changes in yield and soil fertility in a twenty-year rice-wheat experiment in Nepal. *Biol. Fertil. Soils* **2001**, *34*, 73–78.

57. Bhatt, B.; Chandra, R.; Ram, S.; Pareek, N. Long-term effects of fertilization and manuring on productivity and soil biological properties under rice (*Oryza sativa*)–wheat (*Triticum aestivum*) sequence in Mollisols. *Arch. Agron. Soil Sci.* **2016**, *62*, 1109–1122. [CrossRef]
58. Barbarick, K.A.; Ippolito, J.A.; Westfall, D.G. Extractable trace elements in the soil profile after years of biosolids application. *J. Environ. Qual.* **1998**, *27*, 801–805. [CrossRef]
59. Elsokkary, I.; Salam, A.A. Bioavailability and DTPA-extractability of soil heavy metals from successive sewage sludge treated calcareous soil. *Alex. J. Agric. Res.* **1998**, *43*, 349–366.
60. Warman, P.R.; Termeer, W.C. Evaluation of sewage sludge, septic waste and sludge compost applications to corn and forage: Yields and N, P and K content of crops and soils. *Bioresour. Technol.* **2005**, *96*, 955–961. [CrossRef] [PubMed]
61. Asthir, B.; Jain, D.; Kaur, B.; Bain, N.S. Effect of nitrogen on starch and protein content in grain influence of nitrogen doses on grain starch and protein accumulation in diversified wheat genotypes. *J. Environ. Biol.* **2017**, *38*, 427. [CrossRef]
62. Song, Y.-J.; Choi, I.-Y.; Sharma, P.K.; Kang, C.-H. Effect of different nitrogen doses on the storage proteins and palatability of rice grains of primary and secondary rachis branches. *Plant Prod. Sci.* **2012**, *15*, 253–257. [CrossRef]
63. Yamur, M.; Kaydan, D.; Arvas, O. Effects of sewage biosolid application on seed protein ratios, seed NP contents, some morphological and yield characters in lentil (*Lens culinaris* Medic.). *Res. J. Agric. Biol. Sci.* **2005**, *1*, 308–314.
64. Hussain, B.; Umer, M.J.; Li, J.; Ma, Y.; Abbas, Y.; Ashraf, M.N.; Tahir, N.; Ullah, A.; Gogoi, N.; Farooq, M. Strategies for reducing cadmium accumulation in rice grains. *J. Clean. Prod.* **2020**, *286*, 125557. [CrossRef]
65. Kabata-Pendias, A. *Trace Metals in Soils and Plants*; CRC Press: Boca Raton, FL, USA, 2011.
66. De Livera, J.; McLaughlin, M.J.; Hettiarachchi, G.M.; Kirby, J.K.; Beak, D.G. Cadmium solubility in paddy soils: Effects of soil oxidation, metal sulfides and competitive ions. *Sci. Total Environ.* **2011**, *409*, 1489–1497. [CrossRef]
67. Kosolsaksakul, P.; Farmer, J.G.; Oliver, I.W.; Graham, M.C. Geochemical associations and availability of cadmium (Cd) in a paddy field system, northwestern Thailand. *Environ. Pollut.* **2014**, *187*, 153–161. [CrossRef]
68. Greger, M. Metal availability and bioconcentration in plants. In *Heavy Metal Stress in Plants*; Springer: Berlin/Heidelberg, Germany, 1999; pp. 1–27.
69. Mahdy, A.M.; Elkhatib, E.A.; Fathi, N.O. Cadmium, copper, nickel, and lead availability in biosolids-amended alkaline soils. *Aust. J. Basic Appl. Sci.* **2007**, *1*, 354–363.
70. Singh, R.P.; Agrawal, M. Variations in heavy metal accumulation, growth and yield of rice plants grown at different sewage sludge amendment rates. *Ecotoxicol. Environ. Saf.* **2010**, *73*, 632–641. [CrossRef]
71. Eid, E.M.; Alamri, S.A.M.; Shaltout, K.H.; Galal, T.M.; Ahmed, M.T.; Brima, E.I.; Sewelam, N. A sustainable food security approach: Controlled land application of sewage sludge recirculates nutrients to agricultural soils and enhances crop productivity. *Food Energy Secur.* **2020**, *9*, e197. [CrossRef]
72. Al Zoubi, M.M.; Arslan, A.; Abdelgawad, G.; Pejon, N.; Tabbaa, M.; Jouzdan, O. The effect of sewage sludge on productivity of a crop rotation of wheat, maize and vetch and heavy metals accumulation in soil and plant in Aleppo Governorate. *Agric. Environ. Sci.* **2008**, *3*, 618–625.

life

MDPI

Article

The Anti-Genotoxic Activity of Wastewaters Produced after Water-Steam Distillation of Bulgarian *Rosa damascena* Mill. and *Rosa alba* L. Essential Oils

Svetla Gateva [1], Gabriele Jovtchev [1], Tsveta Angelova [1], Ana Dobreva [2,*] and Milka Mileva [3,*]

[1] Institute of Biodiversity and Ecosystem Research, Bulgarian Academy of Sciences, 1113 Sofia, Bulgaria; spetkova2002@yahoo.co.uk (S.G.); gjovtchev@yahoo.de (G.J.); angelova_ts@abv.bg (T.A.)
[2] Institute for Roses and Aromatic Plants, Agricultural Academy, 49 Osvobojdenie Blvd., 6100 Kazanlak, Bulgaria
[3] The Stephan Angeloff Institute of Microbiology, Bulgarian Academy of Sciences, 26 Acad. G. Bonchev Str., 1113 Sofia, Bulgaria
* Correspondence: anadobreva@abv.bg (A.D.); mileva@microbio.bas.bg (M.M.); Tel.: +359-29793185 (M.M.)

Abstract: The steam distillation of valuable rose essential oil from *R. damascena* Mill. and *R. alba* L. generates large volumes of wastewaters. Although such wastewaters are bio-pollutants, they contain valuable bioactive compounds. In this study we investigated the cytotoxic/genotoxic and anti-cytotoxic/anti-genotoxic potential of these products. We used cytogenetic methods for induction of chromosome aberrations and micronuclei in two different experimental test-systems: ahigher plant and human lymphocyte cultures. Different experimental schemes of treatment with the waste products showed that the genotoxic activity of wastewater from the distillation of oils from *R. alba* and *R. damascena* was low in both test–systems. Human lymphocytes showed a higher sensitivity to the products than plant cells. Both types of waste products manifested anti-genotoxic effect against N-methyl-N′-nitro-N-nitrosoguanidine, a direct mutagen. The wastewaters obtained from steam distillation of rose essential oil have cytoprotective/genoprotective effect and could decrease DNA damage. Data are promising for further use of these products in pharmacy and other areas of human life.

Keywords: rose essential oil by-products; ecofriendly products; test-systems; chromosome aberrations; micronuclei

Citation: Gateva, S.; Jovtchev, G.; Angelova, T.; Dobreva, A.; Mileva, M. The Anti-Genotoxic Activity of Wastewaters Produced after Water-Steam Distillation of Bulgarian *Rosa damascena* Mill. and *Rosa alba* L. Essential Oils. *Life* **2022**, *12*, 455. https://doi.org/10.3390/life12030455

Academic Editor: Othmane Merah

Received: 16 February 2022
Accepted: 17 March 2022
Published: 19 March 2022

1. Introduction

Essential oil of *Rosa damascena* Mill. f. trigintipetala Dieck has brought Bulgaria world fame. The Bulgarian rose oil is of extremely high quality, due to the specific climatic conditions. Since 2014, it is in the European registry of protected geographical indications. The locality where the production of essential oils takes place in Bulgaria is the so-called 'Valley of Roses' near Kazanlak. Together with *Rosa damascena* Mill. there grows one more representative of the Rosaceae family—*Rosa alba* L., with delicate and fine fragrance. The two roses species used to grow together until a few years ago, so the essential oil was distilled together too. The plantations have been separated for 10 years now, and the two oils undergo distillation separately.

The rose essential oil is an attractive ingredient not only in the highest-class perfumery and cosmetics because of its unbelievable fragrance [1]. It is also of interest in pharmacy and medicine because of its valuable bioactive components. Many studies have reported that rose oil has interesting biological activities: antiviral, antibacterial, anticancer, antidepressant, antioxidant, anti-inflammatory activities. *R. damascena* essential oil also possesses relaxant, hypnotic, antisclerotic, hepatoprotective, and antispasmodic effects [2–5].

The biological activity of rose oil is due to its chemical composition. Many studies exist about the chemical content of rose essential oil of *R. damascena* from different origin [6–11]

and of *R. alba* [12,13]. More than 300 different compounds are present in the rose oil from *R. damascena* [11] as well as about 214 in the rose oil from *R. alba* [12]. The main compounds are terpene alcohols citronellol, geraniol, nerol, phenethyl alcohol, saturated and unsaturated aliphatic hydrocarbons.

Industrial production of rose essential oil is mainly based on water steam distillation from fresh rose flowers followed by cohobating or concentration in which the initial distillate undergoes multiple redistillation [8]. More than 3000 kg of rose flowers produce 1 kg of rose oil by water distillation [14]. In the last 18 years the annual production of rose oil in Bulgaria has varied between 750 and 1900 kg [15].

Along with the production of the valuable rose oil, the steam distillation procedure generates two types of by-products in large amounts: wastewaters and solid residual plant materials. Discarding these wastes into the environment causes a serious problem. One kilogram of fresh raw flowers generates about 2 kg of wet waste [16]. The by-products discarded annually in the production of 1500 kg rose oil in Bulgaria amount to approximately 9,000,000 kg distilled petals [17]. In Turkey, the other main rose essential oil producer, the total amount spent flowers (distillation residues) after hydro-distillation of the fresh oil-bearing rose flowers per year is 30 thousand tonnes [18].

These waste materials are considered as bio pollutants, but at the same time they are natural products rich in biologically active compounds such as polyphenols. Researchers seek ways to utilize them and use their bioactivity in pharmacy and cosmetics. However, there are limited data on the chemical characterization of the debris, and the detailed polyphenols profile of the waste water is still unknown. Distillation extracts only the volatile constituents of the essential oil, but the non-volatile valuable components that remain in the waste are a value-added biological material.

Rusanov et al. reported some fractions of rose oil distillation wastewater (RODW) rich in phenolic compounds (ellagic acid, 2-phenylethyl-O-β-glucopyranoside, and several kaempferol and quercetin glycosides) [19]. Other authors have isolated a polyphenol-rich fraction from *R. damascena* wastewaters by Amberlite column as the stationary phase [20]. Schieber et al. obtained that kaempferol 3-O-glucoside and quercetin glycoside are the major components of flavonoids extracted from distilled rose petals of *R. damascene*,after industrial distillation for essential oil [14]. Waste rose petals of *R. damascena* also contain pectic polysaccharides [21]. The waste could be a potential source of high-value products that would have health beneficial properties [3,8,18–20,22–25]. Polyphenol depleted water fraction RF20-(SP-207) of rose oil distillation wastewater showed antiproliferative and antimigratory effects on Human non-tumorigenic HaCaT keratinocytes. It could serve as a basis for supportive, therapy against hyperproliferation-involved in skin diseases [26]. Polyphenol fraction from rose oil WW from *R. damascena* showed strong anti-tyrosinase activities [23].

The presence of such bioactive phytochemicals in rose waste is a potential for their application in the food industry. Antioxidant activity of rose by-products allows use as natural antioxidants in meat and sausages [27]. Extracts from rose petals manifested well-expressed antibacterial activity [3]. They could possibly find application in probiotic lactic acid bacteria dairy products [28]. Rose wastes could be additives for functionalization and aromatization of bakery products [29] and in confectionery to replace the wheat flour in pasty biscuits [30]. Mollov et al. reported that the addition of polyphenolic copigments extracted from distilled rose petals of *Rosa damascena* reduces the degradation of anthocyanins in the thermally treated strawberries beverage [31]. Some studies detected growth capabilities of young chickens depend on the intake of feed enriched with natural bioactive compounds from distilled rose petals (*Rosa damascena* Mill.) [17]. Studies have explored distilled rose waste as natural dyes, for bio-sorption of pollutants such as heavy metals, biogas production, dye-sensitized solar cells, etc. [32–36].

In our previous study we analyzed the chromatographic profile of the waste waters obtained from the water steam distillation of the essential oils of *R. damascena* Mill. and *R. alba* [37]. There we identified 52 chemical compounds in the waste from *R. damascena*

Mill. and respectively 49 in wastewater from *R. alba L*. The obtained data showed that 37 compounds are the same for both waste products.

The most non-volatile polyphenolic compounds, including flavonoids remain in the waste. According to [37] the total polyphenolic content in the wastewater of *R. alba* L. oil production was 7.6 ± 0.3 mg GAE/mL, and in *R. damascena*, it was 7.2 ± 0.2 mg GAE/mL, respectively. The total flavonoid content in the waste from *R. alba* L. was 1.00 ± 0.01 mg/mL and in *R. damascena* wastewater it was 1.14 ± 0.01 mg/mL, respectively. The detected flavonoids are quercetin and kaempferol, and their various derivatives, such as isoquercitrin, catechin and epicatechin. The polyphenolic acids are gallic acid, chlorogenic acid, and ellagic acid. The content of tannins in the wastewater from *R. alba* L. was 2.16 ± 0.35 mg/mL, while in the waste from *R. damascene*, it was 1.61 ± 0.05 mg/mL.

All these data call for re-evaluation of the biological properties of the by-products of hydro distillation of rose flowers. Although there are many studies addressing the valorization of the wastes and especially of the wastewaters, limited data exist about their cytotoxic/genotoxic and anti-genotoxic activities [20]. The authors investigated antioxidant activity, xanthine oxidase inhibition and a significant DNA protection ability against H_2O_2 of polyphenol-enriched fraction from *R. damascena* wastewaters (in concentrations of 25–100 μg/mL) on human lymphocytes.

These data, as well as the available information about the phytochemical properties of the distilled rose wastes and their well-expressed antioxidant activities, served as a background for the present study. Its aim was to investigate the cytotoxic/genotoxic and anticytotoxic/anti-genotoxic activities of concentrations of wastewaters produced after water-steam distillation of essential oils of *R. damascena* and *R. alba* in different test-systems using tests for genotoxicity. This assessment would contribute to elucidating the effectiveness and safety of these ecofriendly by-products.

2. Materials and Methods

2.1. Preparation of Rose Wastewaters

Rose wastewaters (ww)—a liquid aqueous residue used in the present study were generated because of water steam distillation done in Institute for Roses and Aromatic Plants in Kazanlak (IRAP), Bulgaria, at semi-industrial processing line.

Both roses were grown in the experimental field of the institute. The rose flowers used for distillation were from the 2019 harvest. The processing of distillation was performed at the following parameters: raw material was 8–10 kg; hydro module 1:4; the flow rate was 16–20 mL/min. The duration was 150 min. The wastewater was collected and stored at 4 °C for the next stage of the study.

2.2. Cytogenetic Analysis

2.2.1. Chemicals

Most of the chemicals, used in the cytogenetic analysis were provided by Sigma–Aldrich Chemie GmbH, Merck (Darmstadt, Germany). N-methyl-N′-nitro-N-nitrosoguanidine (MNNG) was purchased from Fluka—AG (Buchs, Switzerland).

Two types of test-systems were used to study cytotoxic/genotoxic and anti-cytotoxic/anti-genotoxic activities of both rose wastewaters applying two of the most reliable tests for genotoxicity-induction of chromosome aberrations (CA) and induction of micronuclei (MN). Alkylating direct mutagen N-methyl -N′-nitro-N-nitrosoguanidine (MNNG) 50 μg/mL was used as a positive control for both test-systems. Untreated cells were used as a negative control.

2.2.2. Plant Test-System

Seeds of the reconstructed karyotype MK 14/2034 of *Hordeum vulgare* were-presoaked for 1 h in tap water and germinated for 17 h in Petri dishes on moist filter paper at 24 °C. The barley root meristems were exposed for 1 h and/or 4 h to *R. damascene* Mill. and *R. alba* L. wastewaters at concentrations of 6, 14 and 20%. To test anti-cytotoxic/anti-genotoxic activity of rose wastewaters meristem cells were treated as follows. Part of meristems was conditioning affected (60 min) with non-toxic concentration of 20% wastewater, followed by a challenge with 50 μg/mL of MNNG (60 min) with 4 h inter-treatment time. Another part of the meristems was treated with wastewater (60 min) followed by MNNG 50 μg/mL without any inter-treatment time. After each treatment, the barley roots were washed with distilled water. After the treatment and recovery times of 18, 21, 24, 27 and 30 h, the root tips were treated for 2 h with 0.025% colchicine in a saturated a solution of α-bromonaphthalene, fixed in a solution of ethanol: acetic acid (3:1), hydrolyzed in 1 N HCl at 60 °C for 9 min, Feulgen-stained, macerated in 4% pectinase in distilled water for 12 min and at the end squashed on clean slides for scoring of metaphases with chromosome aberrations [38].

For scoring of micronuclei (MN), the barley root tips were fixed after 30 h recovery time without colchicine treatment.

2.2.3. Human Lymphocytes In Vitro

Peripheral venous blood of healthy nonsmoking/nondrinking donors (men and women) aged between 33 to 40 years was used for the experiments to prepare lymphocyte cultures. All procedures were conducted corresponding to the Declaration of Helsinki and all donors signed written informed consent forms. Each lymphocyte cultures contained RPMI 1640 medium (Sigma-Aldrich, Steinheim, Germany), 12% calf serum (Sigma-Aldrich, Buchs, Switzerland), 40 mg/mL gentamycin (Sopharmacy, Sofia, Bulgaria), and 0.1% phytohemagglutinin PHA (Sigma-Aldrich, Mannheim, Germany). The experiments for chromosome aberrations were performed according to the method of Evans, [39]. The lymphocytes were treated with each rose wastewater in concentrations of 3, 6, 11, 14 and 20% for 1 h and 4 h. To test anti-cytotoxic/anti-genotoxic activity of rose wastewaters some cultures were conditioning treated with wastewater (60 min) with non-toxic concentration (6%) followed by 4 h of inter-treatment time and after that challenged with 50 μg /mL of MNNG (60 min). Another part of the lymphocytes was affected with 6% of wastewater (60 min), followed immediately by MNNG 50 μg /mL (60 min) without any inter-treatment time. After each treatment, the lymphocytes were washed in fresh medium and cultured at 37 °C. At the 72nd hour of cultivation 0.02% colchicine was added to each sample, followed by 0.56 % KCl, fixation in methanol: acetic acid (3:1, *v/v*), and stained in 2% Giemsa.

For analysis of micronuclei (MN) cytochalasin-B (6 μg/mL) was added to each culture at the 44th hour after PHA stimulation to arrest cytokinesis according to cytokinesis-block micronucleus (CBMN) assay [40]. At the 24th hour after adding Cyt-B the lymphocyte cultures were centrifuged, hypotonized with 0.56% KCl and fixed in methanol: acetic acid (3:1). After centrifugation the suspension was dropped onto clean slides and stained in 2% Giemsa.

2.2.4. Endpoints

The percentage of metaphases with chromosome aberrations (MwA% ± SD) was calculated using test for chromosome aberrations, to assess the genotoxic effect of the wastewaters in both test systems. Chromatid, isochromatid breaks, chromatid translocations and intercalary deletions were determined (Figures 1 and 2). The cell division that gives information about cytotoxicity was assessed by value of mitotic index (MI) that represents the number of metaphases per 1000 observed cells per each experimental variant. "Aberration hot spots" in barley chromosomes (reconstructed barley karyotype MK14/2034) were determined.

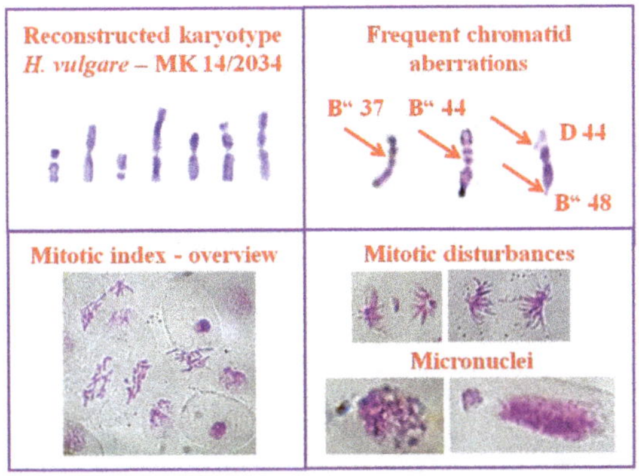

Figure 1. Chromosome aberrations, micronuclei, and mitotic disturbances observed in *H. vulgare* reconstructed karyotype after treatment with wastewaters produced after water steam distillation of essential oils of Bulgarian *R. damascena* and *R. alba*.

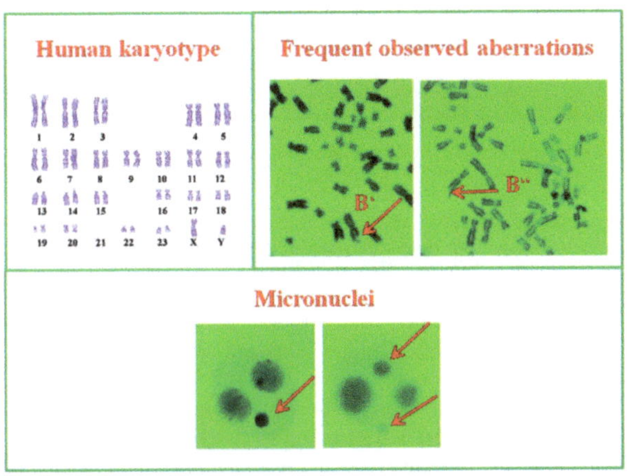

Figure 2. Chromosome aberrations and micronuclei observed in human lymphocyte cells after treatment with wastewaters produced following water steam distillation of essential oils of Bulgarian *R. damascena* and *R. alba*.

Using a test for micronuclei, the percentage of micronuclei (MN% ± SD) was calculated for both barley and human lymphocyte cells based on 5000 cells per each experimental variant (Figures 1 and 2). In human lymphocytes, the nuclear division index (NDI) was calculated, which gives additional information about cytotoxicity of the tested substances. The following formula was used: $(N1 + 2N2 + 3N3 + 4N4)/N$, where N1–N4 represents the number of cells with 1–4 nuclei and N is the total number of scored cells.

2.2.5. Statistics

Each experiment was repeated three times. The two-tailed Fisher's exact test was used for statistical analysis of the different treatment variants.

3. Results

3.1. Cytotoxic and Genotoxic Effects of Wastewater from Distillation of Oils of R. alba *and* R. damascena

Both wastewaters were tested for cytotoxicity using value of mitotic index (MI) as endpoint. No cytotoxic effect was observed both for wastewater from *R. alba* and *R.*

damascena applied at concentrations of 6–20% for 1 and 4 h in *H. vulgare*, compared to the untreated root tip meristems (Figure 3A). Human lymphocyte cultures were more sensitive to wastes than barley. The values of mitotic index decreesed with increasing the wastewater concentration (3, 6, 11, 14 and 20%), irrespectively of the duration of treatment (Figure 3B).

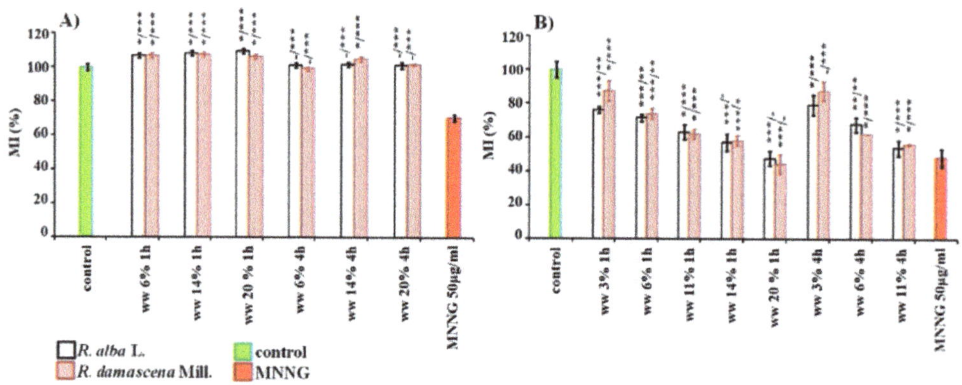

Figure 3. Cytotoxic activity of wastewaters (ww) obtained from distillation of essential oils from *R. alba*, and *R. damascena* assessed by the value of MI in: *H. vulgare* (**A**) and in human lymphocytes (**B**). Mitotic activity was assessed as a percent of negative control. * $p < 0.05$, ** $p < 0.01$, *** $p < 0.001$, and—non-significantly versus negative control (before slash) versus positive control MNNG (after slash).

In genotoxic analysis conducted by induction of chromosome aberrations of variants treated with wastewater from *R. alba*, a low but statistically significant ($p < 0.05$; $p < 0.001$) clastogenic effect was found for all tested concentrations (3–20%), compared to the untreated control both in plant test-system and in human lymphocytes (Figure 4). It should be noted that lymphocyte cultures are more sensitive than barley cells to the wastewater of *R. alba*, as the frequency of chromosome damage is higher. In *H. vulgare*, no clear concentration dependence was observed for both 1 h and 4 h treatments (Figure 4A). The frequency of induced chromosome aberrations ranged from 4.7% ± 0.35 (for 6%) to 5.6% ± 0.48 (for 20%) when the cells were treated for 1 h and from 3.5% ± 0.36 (for 6%) to 4.4% ± 0.49 (for 20%) when the treatment was for 4 h. Human lymphocytes showed a clear concentration dependence with an increase in the frequency of chromosome aberrations after treatment with the waste product for 1 h and for 4 h (Figure 4B). The highest frequency of aberrations was reported in the cultures affected with a concentration of 20% (9.50% ± 0.70 for 1 h). The lowest was reported for a concentration of 3% for 1 h and 4 h (2.00% ± 1.70).

The wastewater from *R. damascena* showed a low but statistically significant clastogenic effect ($p < 0.05$; $p < 0.001$) with all tested concentrations compared to the negative, untreated control in both test-systems (Figure 4). In *H. vulgare* the frequency of aberrations after 1 h treatment was in the range of 5.9% ± 0.41 (with 6%) to 6.8% ± 0.47 (with 20%), and treatment for 4 h induced from 6.1 % ± 0.37 (with 6%) to 7.7% ± 0.38 (with 20%). No dependence on the duration of treatment and applied concentration was found with respect to the values of induced aberration in human lymphocytes, as well as in *H. vulgare* (Figure 4). The frequency of observed chromosomal abnormalities ranged from 4.40% ± 1.00 (for 3%/1 h) to 7.60% ± 2.60 (11%/4 h). The wastewater from *R. damascena* induced higher frequencies of chromosome damages than the wastewater from *R. alba* in both test-systems after treatment for 1 h and 4 h, except for the concentrations of 14% and 20% (1 h) in lymphocyte cultures (Figure 4).

The genotoxic activity of both wastewaters at all studied concentrations was much lower ($p < 0.001$) than that of the alkylating mutagen MNNG in both plant and lymphocyte test- systems (Figure 4).

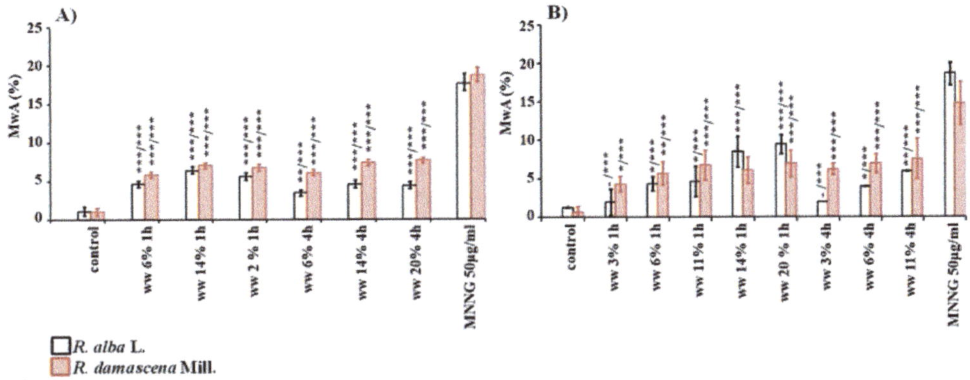

Figure 4. Genotoxic effect of wastewaters (ww) obtained from distillation of essential oils from *R. alba*, and *R. damascena* assessed by induction of chromosome aberrations (CA) in: *H. vulgare* (**A**) and human lymphocyte cultures (**B**). * $p < 0.05$, ** $p < 0.01$, *** $p < 0.001$, and—non-significantly versus negative control (before slash) versus positive control MNNG (after slash).

The spectrum of observed chromosome aberrations in *H. vulgare* induced by different concentrations of wastewater of *R. alba*, is composed mainly of isochromatid breaks and a small number of chromatid breaks. In human lymphocyte cultures predominantly isochromatid breaks followed by chromatid breaks, with no structural chromosomal changes such as translocations were observed (Figure 5).

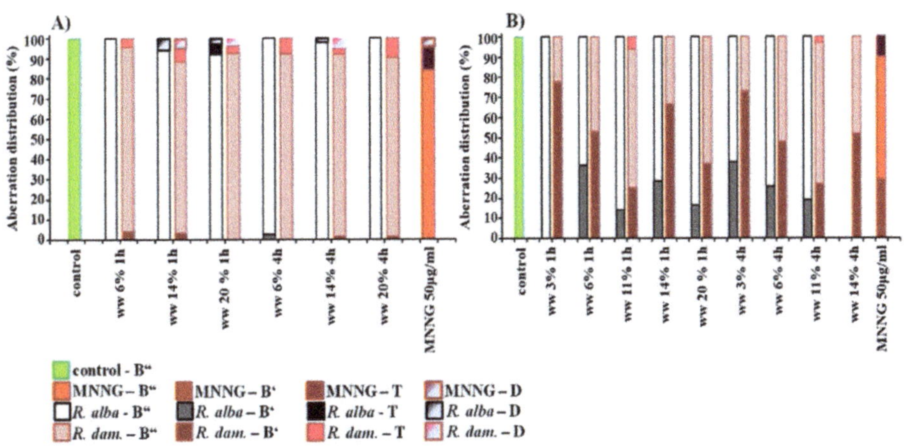

Figure 5. Distribution of aberrations observed after treatment with wastewaters (ww) of *R. alba* and *R. damascena* in *H. vulgare* (**A**) and human lymphocytes (**B**), B″—isochromatid breaks, B′—chromatid breaks, T—translocations, D—intercalary deletions.

Mainly isochromatid breaks followed by translocations, intercalary deletions and chromatid cleavages in *H. vulgare* were induced by wastewater from *R. damascena* in *H. vulgare*. More chromatid breaks were detected in human lymphocytes in variants treated with *R. damascena* waste compared to that of with *R. alba* wastewater. A small percentage of translocations were also observed.

The aberration hot spots in *H. vulgare* in reconstructed karyotype showed that the wastewater from *R. alba* has the potential to reduce their occurrence. In the variants with conditioning treatments with 20% wastewater, aberration hot spots decreased by 60% compared to those induced by the MNNG treatment alone (Table 1).

Table 1. Aberration "hot spots" observed in the chromosomes of root meristem cells of barley—the reconstructed karyotype MK14/2034 after different experimental schemes of treatment with wastewater from the production of essential oil from *R. alba* and *R. damascena* and MNNG.

	Treatment Variants			Hot	Spot		Segments	
		Non-Hot Spot Segments	**Chr.2 Segm.12** 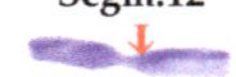	**Chr.4 [3] * Segm.21** 	**Chr.6 Segm.37** 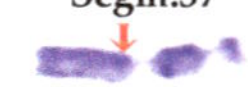	**Chr.6 Segm.39** 	**Chr.7 [1] * Segm.44** 	**Chr.7 [1] * Segm.48** 
1	Untreated control	100%	-	-	-	-	-	-
2	*R. alba* ww 6% 1 h	100%	-	-	-	-	-	-
	R. damascena ww 6% 1 h	85.7%	-	-	-	-	-	14.3%
3	*R. alba* ww 14% 1 h	100%	-	-	-	-	-	-
	R. damascena ww 14% 1 h	100%	-	-	-	-	-	-
4	*R. alba* ww 20% 1 h	73.7%	-	-	15.8%	10.5%	-	-
	R. damascena ww 20% 1 h	87.3%	-	-	-	-	-	12.7%
5	*R. alba* ww 6% 4 h	100%	-	-	-	-	-	-
	R. damascena ww 6% 4 h	100%	-	-	-	-	-	-
6	*R. alba* ww 14% 4 h	100%	-	-	-	-	-	-
	R. damascena ww 14% 4 h	87.1%	-	-	-	-	-	12.9%
7	*R. alba* ww 20% 4 h	86.4%	-	-	13.6%	-	-	-
	R. damascena ww 20% 4 h	100%	-	-	-	-	-	-
8	MNNG 50 μg/mL	52.2%	6.8%	5.3%	7.8%	7.1%	9.9%	10.6%

* Reconstructed karyotype of *Hordeum vulgare* (MK 14/2034) is a result of the combination of two simple reciprocal translocations between parts of chromosomes 1 and 7 and chromosomes 3 and 4.

The cytotoxic and genotoxic effects of both types of wastewaters were assessed with one more endpoint for genotoxicity—induction of micronuclei (MN) (Figure 6). *H. vulgare* meristem cells were treated with concentrations from 6 to 20% and human lymphocytes treated for concentrations from 3 to 11% for 1 h and 4 h (*R. alba*) and from 3% to 20% with *R. damascena*. A low but significant increase in the value of micronuclei induced by waste of *R. alba* while increasing the concentration was found in barley compared with the negative control. The value of the induced MN varied from 0.10% ± 0.01 (6%) to 0.23 ± 0.06 (20%) in the case of 1 h treatment and from 0.13% ± 0.06 (6%) to 0.57% ± 0.06 (20%) in the case of the longer period of treatment—4 h (Figure 6A). In lymphocyte cultures, all wastewater concentrations of *R. alba* induced close values of micronuclei (0.80% ± 0.10 at 3%, 6 % and 11% for 1 h and 0.50% ± 0.10 at 6% and 11% for 4 h), which is higher ($p < 0.001$) than that of the negative control 0.20% ± 0.10 (Figure 6B). It is interestingly to note, that with the extension of the treatment time to 4 h, the values of the induced micronuclei were lower than those observed after 1 h, probably due to the induction of defense mechanisms in the cells.

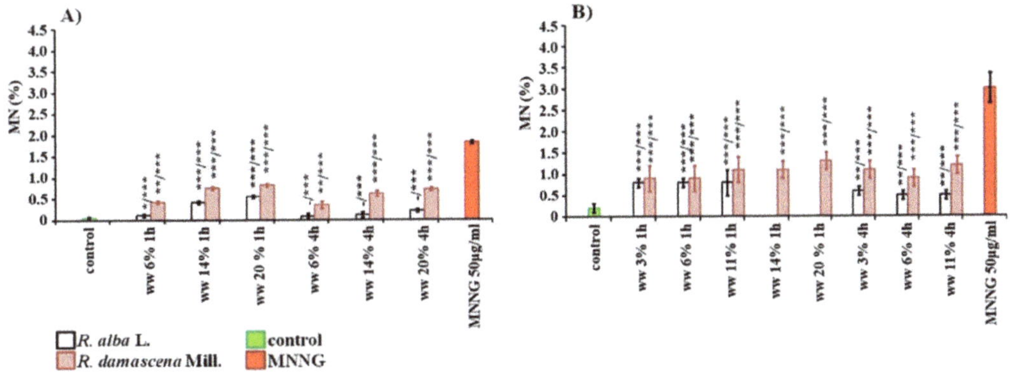

Figure 6. Genotoxic effect of wastewaters (ww) obtained from distillation of essential oil from *R. alba* and *R. damascena* assessed by the induction of micronuclei (MN) in: *H. vulgare* (**A**) and human lymphocyte cultures (**B**). * $p < 0.05$, ** $p < 0.01$, *** $p < 0.001$, and—non-significantly versus negative control (before slash), versus positive control MNNG (after slash).

The observed micronuclei after treatment with *R. damascena* wastewater ranged from 0.90% ± 0.30 (after treatment with 3%/1 h) to 1.30% ± 0.20 (with 20%/1 h) in lymphocyte cultures (Figure 6B). No concentration dependence and treatment duration dependence were observed after treatment with the waste product of *R. damascena* essential oil distillation. In barley the lowest concentration of 6% induced the lowest number MN, whereas close micronucleus frequencies were reported for 14 and 20%, and no significant difference was found between them.

The genotoxicity of both wastewaters applied in all used by us concentrations was much lower ($p < 0.001$) than that of the positive control MNNG in plant as well in lymphocyte test-system (Figure 6).

Nuclear division index (NDI), used as another indicator for assessment of the cytotoxic activity of the wastewaters, was calculated for lymphocyte cultures. Its value was reduced ($p < 0.01$) by treatment with the higher concentrations of wastewater from *R. alba* 6% and 11% compared to the negative control. Calculating the NDI in the variants treated with wastewater from *R. damascena*, a slight, but in some treatment variants a significant decrease in the index was found compared to the negative control (1.35% ± 0.02) (Figure 7). It ranged from 1.29% ± 0.03 (for 3.03%/1 h) to 1.22% ± 0.04 (for 14%, 20% /1 h and for 14%/4 h).

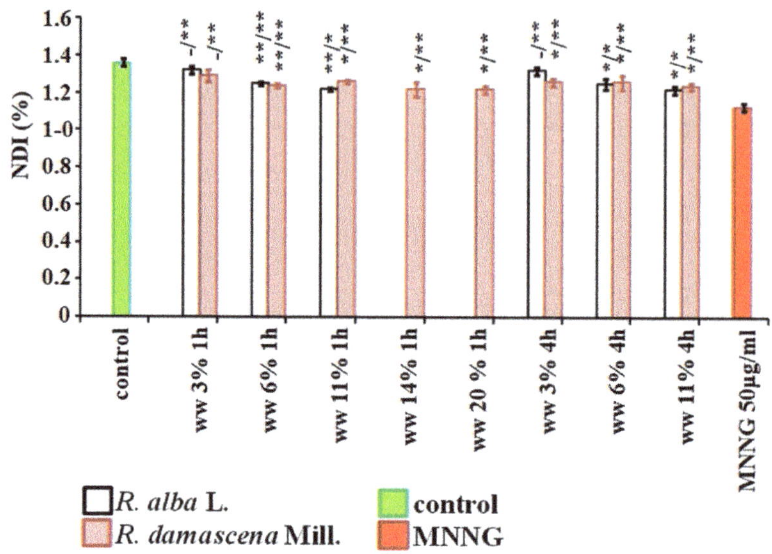

Figure 7. Cytotoxic potential of wastewaters (ww) obtained from distillation of essential oil from *R. alba* and *R. damascena*, assessed by the value nuclear division index (NDI) in human lymphocyte cultures after application of different schemes of treatment. * $p < 0.05$, ** $p < 0.01$, and—non-significantly versus negative control (before slash), versus positive control MNNG (after slash).

3.2. Anti-Cytotoxic and Anti-Genotoxic Activities of Wastewaters from Distillation of Oils of R. alba and R. damascena

The waste products of *R. alba* demonstrate a well pronounced anti-cytotoxic potential ($p < 0.001$, $p < 0.01$) against direct mutagen MNNG when the schemes with combined treatment was applied in both test-systems. This was obtained as in variants with conditioning treatment with concentration of 20% (for barley) and with 6% (for human lymphocytes), followed by a challenge with MNNG (50 µg/mL) and 4 h inter-treatment time, as well as in variants without any time between treatments (Figure 8). The values of MI in these combined variants were higher than in the variants with the mutagen, and reach the values calculated for samples with the wastewater only.

The similar anti-cytotoxic effect was detected also for wastewater of *R. damascena* assessed by MI applying the same experimental schemes with combined treatment as that with waste of *R. alba* (Figure 8). The cytotoxic activity was significantly lower ($p < 0.001$) compared with that of direct mutagen in both plant and lymphocyte test-systems both in variants with conditioning treatment with non-toxic concentrations of wastewater from *R. damascena* (20% for barley and 6% for human lymphocytes, respectively), followed by the challenge with MNNG (50 µg/ml) with 4 h inter-treatment time, and without any inter-treatment time.

Assessing the anti-genotoxic potential of wastewater from *R. alba* by endpoint for genotoxicity CA, applying both schemes with combined treatments with non-toxic concentrations of wastewater and harmful effects of MNNG in both test-systems, a clear anti-genotoxic effect was observed (Figure 9).

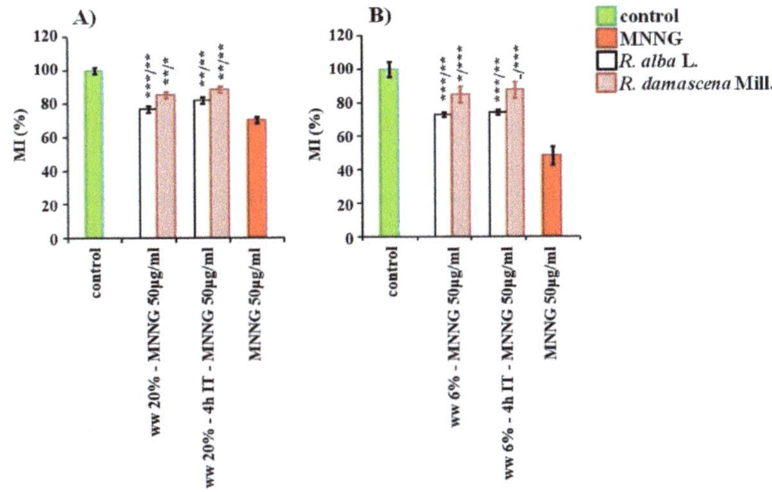

Figure 8. Anti-cytotoxic activity of wastewater (ww) from *R. alba*, and *R. damascena* assessed by the value of mitotic index (MI) after application of different experimental schemes of treatment with wastewater conditioning prior to MNNG challenge (50 µg/mL) with 4 h inter-treatment time and without any inter-treatment time in: *H. vulgare* (**A**) and in human lymphocytes (**B**). Mitotic activity was assessed as a percent of negative control. * $p < 0.05$, ** $p < 0.01$, *** $p < 0.001$, and—non-significantly versus negative control (before slash), versus positive control MNNG (after slash).

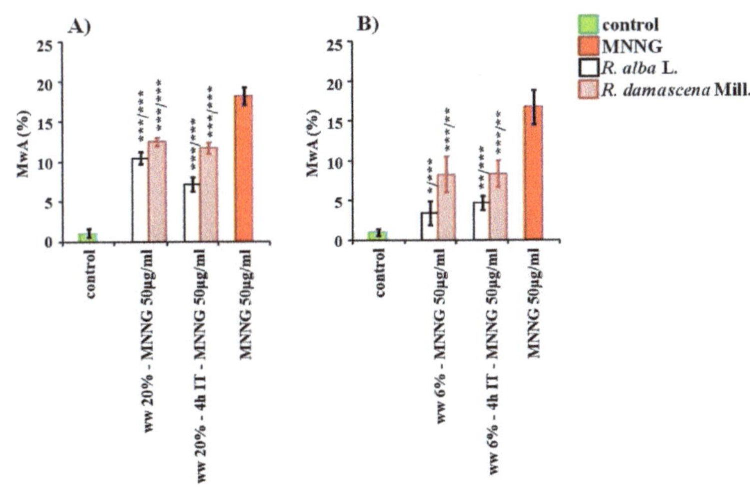

Figure 9. Anti-genotoxic effect of wastewaters (ww) from *R. alba*, and *R. damascena* assessed by induction of chromosome aberrations (CA) after application of different experimental schemes of treatment with: -wastewater conditioning prior to MNNG challenge (50 µg/mL) with 4 h inter-treatment time and—without any inter-treatment time in: *H. vulgare* (**A**) and in human lymphocytes (**B**). * $p < 0.05$, ** $p < 0.01$, *** $p < 0.001$, and—non-significantly versus negative control (before slash), versus positive control MNNG (after slash).

A significant decrease ($p < 0.001$) of the frequency of induced chromosome aberrations (4.70% ± 0.90 and 3.40% ± 1.50 in lymphocytes and 7.2% ± 0.89 in barley) compared to those observed after treatment with MNNG alone (50 µg/mL) in barley (17.6% ± 1.28) and lymphocyte cultures (18.60% ± 1.50), was obtained. The values of aberrations in barley were twice as low and in lymphocyte cultures—five times as low (Figure 9). No significant difference was obtained between the frequencies of aberrations induced after different variants of combined treatment in human lymphocytes, whereas in *H. vulgare* the conditioning treatment with waste from *R. alba* followed by challenge with mutagen and 4 h inter-treatment time reduced the chromosome aberrations to a higher extent (ww *R. alba*→ MNNG10.5% ± 0.72 versus ww *R. alba*→ 4 h → MNNG7.2% ± 0.89 which indicates 31% stronger protection against MNNG).

Wastewater from *R.damascena* also exhibited anti-clastogenic activity after application of both schemes with combined treatment, with 4 h inter-treatment time and without any inter-treatment time. The frequencies of chromosome aberrations were decreased ($p < 0.01$, $p < 0.001$) compared to those of the alkylating agent MNNG in cells of both test-systems (Figure 9). Close values of aberrations were obtained for both variants with combined treatment in barley as well in human lymphocytes.

As seen from Figure 9, the waste water from *R. alba* has more pronounced defense potential against MNNG than waste water from *R. damascena*, as the frequency of chromosome aberrations was decreased to a higher extent for both test-systems ($p < 0.001$).

Analyzing the spectrum of induced chromosome aberrations in the variants with combined treatment with waste from *R. alba* and MNNG in barley, along with the isochromatid breaks and a low number of chromatid breaks, translocations and intercalary deletions was also detected (Figure 10). In human lymphocytes mainly isochromatid breaks followed by chromatid breaks were observed. The spectrum of aberrations observed after combined treatment with wastewater from *R. damascena* essential oil production and MNNG in *H. vulgare* was consist mainly of isochromatid breaks followed by translocations, intercalary deletions and chromatid breaks whereas in human lymphocytes the chromatid breaks were more than that in variants with *R. alba* wastewater.

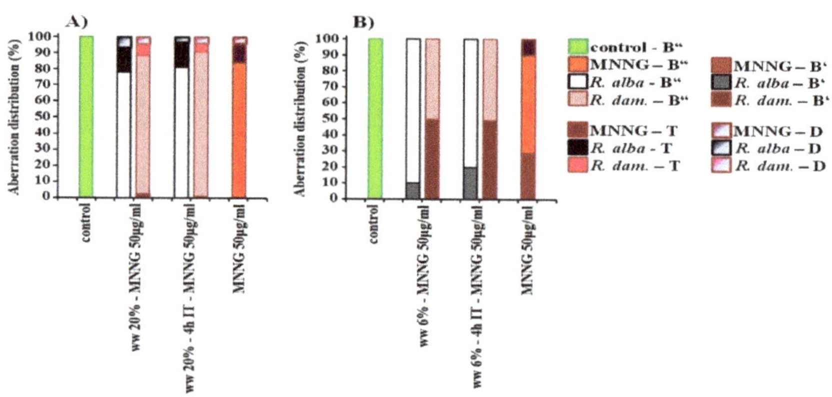

Figure 10. Distribution of aberrations observed after different experimental schemes of treatment with rose wastewaters (ww) and MNNG (50 µg/mL) in: *H. vulgare* (**A**) and in human lymphocytes (**B**), B″—isochromatid breaks, B′—chromatid breaks, T—translocations, D—intercalary deletions.

Analysis of the aberration hot spots in *H. vulgare* meristem cells treated with non-toxic conditioning treatment with 20% wastewater from *R. alba* showed that they decrease by 60% compared to those induced by MNNG alone. Aberration hot spots detected after conditioning treatment with 20% waste product from *R. damascena* and subsequently affected with MNNG were 33% less compared with those induced with MNNG alone and 67% less when the experimental scheme of treatment was wastewater 20% → 4 h → MNNG (Table 2).

Table 2. Aberration "hot spots" observed in the chromosomes of root meristem cells of barley—the reconstructed karyotype MK14/2034 after different variants of conditioning treatment with wastewater from the production of essential oil from *R. alba* and *R. damascena*.

Treatment variants				Hot	Spot		Segments		
		Non-hot spot segments	Chr.2 segm.12 	Chr.4 [3] * segm.21 	Chr.4 [3] * segm.24 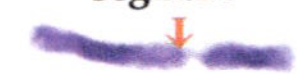	Chr.6 segm.37 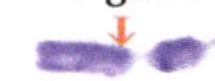	Chr.6 segm.39 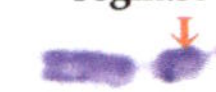	Chr.7 [1] * segm.44 	Chr.7 [1] * segm.48 
1	Untreated control	100%	-	-	-	-	-	-	-
2	*R. alba* ww 20%-MNNG50	80.4%	-	-	7.6%	-	-	12.0%	-
	R. damacsena ww 20%-MNNG50	77.2%	-	-	-	-	-	12.3%	10.5%
3	*R. alba* ww 20%→4 h→MNNG50	77.1%	-	-	10.9%	-	-	12.0%	-
	R. damascena ww 20%→4 h→MNNG50	86.0%	-	6.5%	-	-	-	-	7.5%
4	MNNG 50 μg/ml	52.2%	6.8%	5.3%	-	7.8%	7.1%	9.9%	10.6%

* Reconstructed karyotype of *Hordeum vulgare* (MK 14/2034) is a result of the combination of two simple reciprocal translocations between parts of chromosomes 1 and 7 and chromosomes 3 and 4.

Well-defined anti-genotoxic potential ($p < 0.001$) of the wastewater from *R. alba*, was observed also using MN as endpoint for genotoxicity, applying the experimental schemes with combined treatment in both test-systems. The induction of MN was decreased more than five times in lymphocyte cells and twice in barley compared to those of the alkylating direct mutagen (Figure 11). The wastewater from *R. damascena* also exhibits a very pronounced defense potential against damages induced by alkylating direct mutagen MNNG assessed by MN induction. A significant decrease of the frequencies of MN ($p < 0.001$) compared with that induced by MNNG (2.50% ± 0.30) was obtained using the experimental schemes with combined treatment (with time between treatments and without any) both in the plant test system and in the lymphocyte cultures. The frequency of MN observed in lymphocyte cultures after these treatments was even lower (0.84% ± 0.11) than that of the wastewater alone (0.90% ± 0.30).

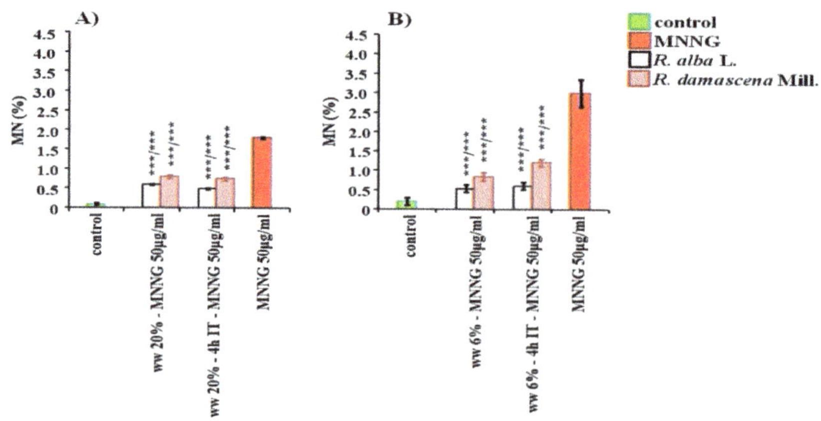

Figure 11. Anti-genotoxic effect of wastewaters (ww) from *R. alba*, and *R. damascena* assessed by induction of micronuclei (MN) after application of different experimental schemes of treatment with: -wastewater conditioning prior to MNNG challenge (50 μg/mL) with 4 h inter-treatment time and,— without any inter-treatment time in: *H. vulgare* (**A**) and in human lymphocytes (**B**). *** $p < 0.001$, and non-significantly versus negative control (before slash) versus positive control MNNG (after slash).

NDI calculated for variants where the experimental schemes with conditioning treatment with wastewater of *R. alba* with 4-h inter-treatment time between treatments and without any time were applied, was significantly higher compared to that of the positive MNNG control (Figure 12). The variants with combined treatment with waste from *R. damascena* had higher level of NDI 1.29% ± 0.04 compared with that of MNNG and close to that of the single treatment with wastewater. This data support the results obtained both with MN and CA endpoints.

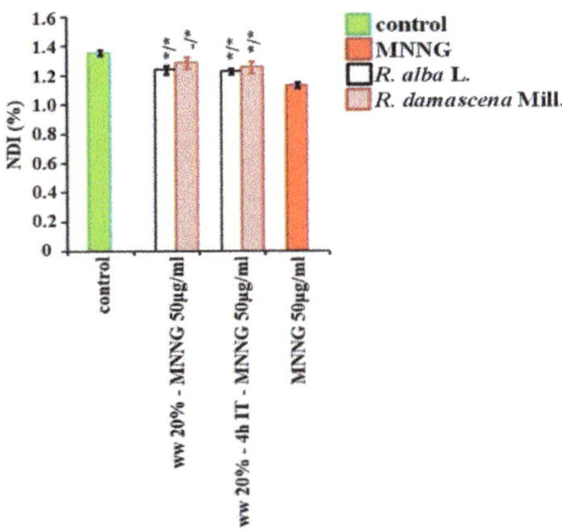

Figure 12. Anti-cytotoxic activity of wastewater (ww) from *R. alba*, and *R. damascena* assessed by the value of nuclear division index (NDI) in human lymphocyte cultures after application of different schemes of treatment. * $p < 0.05$, and non-significantly versus negative control (before slash), versus MNNG (after slash).

4. Discussion

Rose oil is an economically important product for Bulgaria. Along with the valuable essential oil, however, the production generates many waste products that pollute the environment. There has been a growing interest in the study of their utilization in recent years The main components of the rose wastewaters are polyphenols and polysaccharides. The by-product from *R. damascena* oil distillation contains kaempferol, quercetin, ellagic acid and their glycoside derivatives [19,22,23]. The chromatographic analysis of the present wastewater from water steam distillation of essential oils of Bulgarian *R. damascena* and *R. alba* [37], detected various mono-, di-, and acylated glycosides of polyphenolic compounds quercetin and kaempferol, ellagic acid, as well as many of their derivatives, gallic acid and its derivatives—catechin and epicatechin. (Figure 13).

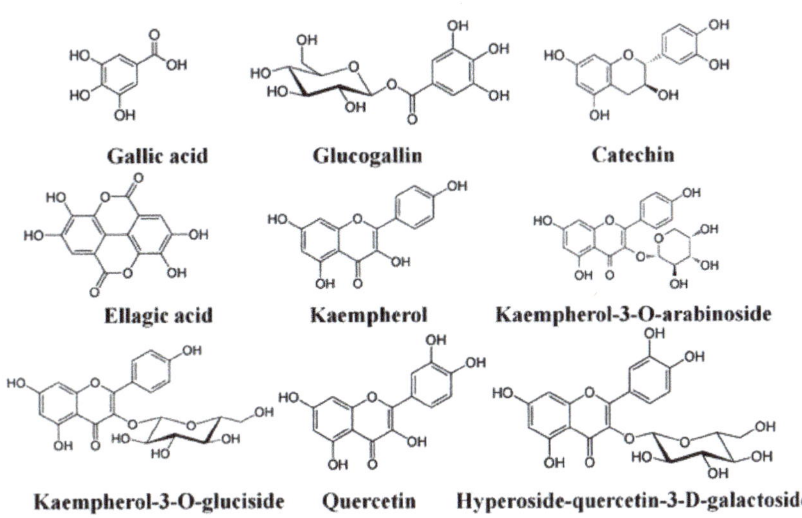

Figure 13. Basic chemical constituents in wastewater from *R. damascena*, and *R. alba* by UHPLC-HRMS/MS analysis (according to Georgieva et al. [37]).

Their quantity and quality varied depending on the rose species that generated the wastewater. Gallic acid is about two times more in *R. alba* L. than in *R. damascena*; ellagic acid is also present in a larger amount in *R. alba* than in *R. damascena*. Hyperoside and isoquercitrin are present more than twice higher in *R. damascena* than in *R. alba* The quantities of kaempherol-3-O-glucoside, kaempherol-3-O-arabinoside and kaempherol-p-coumarouyl-hexoside are more than 2.5 times higher in *R. damascena* than in *R. alba* [37]. This suggests some differences in the biological activity of the wastewater. Hence, it was interesting to investigate and compare the cytotoxic/genotoxic activities and anti-cytotoxic/anti-genotoxic potential of concentrations of wastewater derived from the production of essential oils of both rose species, *R. damascena* and *R. alba* using two different types of experimental test-systems.

At concentrations from 6% to 20% both types of wastewaters did not induce any cytotoxic effect in *H. vulgare* test-system. The human lymphocytes were more sensitive than barley and the cell's viability and mitotic activity clearly depended on the concentration (3% to 20%) and treatment time duration (1 h and 4 h). This was so both in the MI and NDI assay. Wedler et al. [26] also reported a dose dependent antiproliferative activity of rose wastewater fraction in non-tumorigenic HaCaT cells. Some authors reported that a fraction of wastewater from *R. damascena* is toxic at concentrations of 100 μg/mL and higher in HepG2 cells in the MTT assay [20]. Georgieva et al. have found that wastewater from *R. damascena* Mill. and of *R. alba* L. did not exert very strong cytotoxic effect on some human cancer and normal cell lines in the MTT assay, but different type of cells showed different sensitivity to the tested substances [37]. The cells differ in their sensitivity to the tested substances.

In our study both kinds of rose wastewater applied for 1 h and 4 h at concentrations in the range of 3–20% demonstrated clastogenic activity to DNA in both test-systems compared to the negative control. The hereditary material of the human lymphocytes was more susceptible to damage than that of barley. Conversely, in a study by other authors [20], treatment for 15 min with wastewater fraction from *R. damascena* (25, 50, 100 μg/mL) did not show any genotoxic effect in the comet assay in human lymphocytes. This was probably because of the shorter treatment time duration in [20], as both lower concentrations are similar to those used by us.

Here, we established that the value of induced chromosome aberrations depends on the type of the wastewater. *R. alba* wastewater showed lower clastogenic activity than that of *R. damascena* in both test-systems. The frequency of damage was lower after treatment for a longer time of 4 h than for 1 h. The frequency of chromosome aberrations induced by the by-products from *R. damascena* was not time-dependent. Interestingly, the treatment with waste from *R. alba* essential oil distillation induced predominantly chromatid breaks and less isochromatid breaks in lymphocyte cells, whereas both types of breaks were with almost equal values after treatment with waste from *R. damascena* These results probably reflect some differences in the chemical composition and the quantitative ratio of the components in the two kinds of wastewater. More substances were found in the waste of *R. damascena* than in that of *R. alba*. The observed cytotoxic and genotoxic effects of the wastes is probably due on the presence of quercetin and kaempferol. The information available in the literature reports such activities of quercetin [41] and kaempherol [42] and weaker ones of ellagic acid [41] in various cancer cells. These effects depend on a complex of conditions. The concentrations used, target organisms or cell lines are important [43]. The treatment scheme, including possible combinations with other drugs, is also of great importance.

We tested the anti-cytotoxic and anti-genotoxic activities of both wastewaters against the direct mutagen MNNG by applying two experimental schemes using tests for genotoxicity. Our results showed that both kinds of wastewater possessed anti-cytotoxic and anti-genotoxic potential. The frequencies of chromosome aberrations and micronuclei decreased following conditioning treatment with the wastewater in a non-toxic concentration before MNNG challenge with 4 h inter-treatment time, compared to the samples treated only with MNNG. The defense potential of the waste products manifests itself independently of the experimental conditions in both test-system. The damage induced by the alkylating agent also decreased both in barley and in lymphocytes when non-toxic concentrations (6% for lymphocytes and 20% for barley) of rose wastewater were applied before the mutagen without any inter-treatment time. The protective effect demonstrated by both kinds of wastewater in our study is probably due to the presence of polyphenolic compounds, which are the major components of both kinds of rose water by-products. Our results are in accordance with the study of Georgieva et al. [37], who obtained a well-expressed redox-modulating capacity of wastewater from four Bulgarian rose species including *R. alba* and *R. damascena* The fact that the defense potential of the waste from *R. damascena* manifested itself to a lesser extent than that of *R. alba* is probably due to a difference between the chemical composition of the wastewater. Georgieva et al. [37] showed a slightly higher value of total polyphenolic content in the waste of *R. damascena* than in *R. alba*. The chromatographic profile previously reported [37] showed that both kinds of wastewater contain hyperoside, mono-, di-, and acylated glycosides of kaempherol and quercitin and their derivatives, as well as ellagic acid and gallic acid and their derivatives. These compounds are known for their biological activities- antioxidant, antiradical activities, and other pharmacological properties [44–47]. The presence of these main chemical compounds in the tested by-products would explain the defence potential in our study. The flavonoid kaempferol can protect DNA, proteins and lipids against damage induced by oxidative stress. It has manifested anticancer and anti-inflammatory activities [48,49]. Our previous study showed its anti-cytotoxic and anti-genotoxic potential against the radiomimetic zeocin [50]. The kaempferol glycoside derivatives can decrease DNA damage induced by etoposide in human peripheral blood mononuclear cells [51]. Ellagic acid has anti-genotoxic potential in human sperm [52] and in Zebrafish Blood Cells exposed to benzene [53], It also has anticarcinogenic and hepatoprotective properties [45,54]. Both ellagic acid and quercetin significantly increase the value of GSH and decrease NADPH and ascorbate dependent lipid peroxidation in mice [55]. Quercetin is well known by its anticancerogenic effect [41,56]. Quercetin has anti-genotoxic activity against DNA damage induced by the mutagen methyl methane sulfonate, aflatoxin B1, and doxorubicin in human hepatoma HepG2 cells [44]. The authors suggest that quercetin acts both as a desmutagenic and bioantimutagenic agent, repairing DNA damage. Tannins, which are secondary metabolites derived from phenolic acids also show antimutagenic effects in terms of DNA-breaking activity [57] and possess anticancerogenic activity [58].

The well-known monofunctional alkylating mutagen N-methyl-N′-nitro-N-nitrosoguanidine used in the present study, directly alkylates the nitrogen and oxygen end of the DNA bases and can induce intra-strand, inter-strand crosslinks and double-strand breaks in DNA [59]. Base excision reparation(BER) and nucleotide excision reparation (NER) are the main processes for the repair of N-alkylated bases [60,61]. The well expressed anti-cytotoxic and anti-genotoxic effects of the wastewater from water steam distillation of rose oil from Bulgarian *R. damascena* and *R. alba* against MNNG, in our study, suggest that the chemical compounds in the waste could also activate repair functions. Such repair pathways could act in addition to the anti-oxidant and scavenging activities, of the main polyphenolic compounds. Our previous study showed that *R. alba* essential oil demonstrates good anti-cytotoxic and anti-genotoxic potential against MNNG in plant and lymphocyte test-systems [13]. Based on our present results, we suggest that the tested rose wastewater shows good defense potential and is comparable to that of fresh rose oil samples. However, more studies are necessary to confirm this.

5. Conclusions

This study assessed the cytotoxic/genotoxic and anti-cytotoxic/anti-genotoxic effects of wastewater produced after water steam distillation of essential oils of Bulgarian *R. damascena* and *R. alba*. The waste products did not show strong cytotoxic/genotoxic effects. This depends on the type of rose (their polyphenolic content) from which the waste originated, the concentrations applied and the sensitivity of the test-system used. Both kinds of rose wastewater showed good biological effect, i.e., cytoprotective/genoprotective activity against the direct alkylating mutagen MNNG. The level of DNA damage decreased, regardless of the experimental conditions in two different types of test-systems. The obtained data are promising for further utilization and successful use of these by-products in appropriate concentrations in pharmacy and other areas of human life, just as the rose essential oils themselves. On the other hand, the valorisation of these products would contribute to the reduction of waste from rose oils distillation in the environment.

Author Contributions: Conceptualization, S.G. and G.J.; performed cytogenetic experiments, S.G., G.J. and T.A.; provided rose wastewater materials, A.D.; wrote—original draft, S.G.; writing—review and editing, G.J. and M.M.; visualization, G.J.; supervision, M.M.; project administration, M.M.; funding acquisition, M.M. All authors have read and agreed to the published version of the manuscript.

Funding: This research was funded by Bulgarian National Science Fund, project N KP-06-H36/17 "Biological activities of Bulgarian rose oils and approach to valorization of waste obtained from their production" from 17 December 2019, granted to M. Mileva.

Institutional Review Board Statement: The study was conducted in accordance with the Declaration of Helsinki and approved by Commission on Ethics and Academic Unity Institute of biodiversity and ecosystem research (Number: No 1, Data: 5 March 2021).

Informed Consent Statement: Written informed consent was obtained from all subjects involved in the study.

Data Availability Statement: All the obtained data of this research are presented in the manuscript.

Acknowledgments: The authors would like to thank to all volunteer donors who participated in the experiments. Thanks to Stefan Tsonchev for his contributions in proofreading and editing the article.

Conflicts of Interest: The authors declare no conflict of interest.

References

1. Widrlechner, M.P. History and utilization of *Rosa damascena*. *Econ. Bot.* **1981**, *35*, 42–58. [CrossRef]
2. Schulz, H. Fragrance and pigments. In *Odoriferous Substances and Pigments*; Roberts, A.V., Debener, T., Gudin, S., Eds.; Elsevier: Amsterdam, The Netherlands, 2003; pp. 231–240.
3. Özkan, G.; Sagdiç, O.; Göktürk-Baydar, N.; Baydar, H. Anti-oxidant and antibacterial activities of *Rosa damascena* flower extracts. *Food Sci. Technol. Int.* **2004**, *10*, 277–281. [CrossRef]
4. Gochev, V.; Wlcek, K.; Buchbauer, G.; Stoyanova, A.; Dobreva, A.; Schmidt, E.; Jirovetz, L. Comparative Evaluation of Antimicrobial Activity and Composition of Rose Oils from Various Geographic Origins, in Particular Bulgarian Rose Oil. *Nat. Prod. Commun.* **2008**, *3*, 1063–1068. [CrossRef]
5. Mahboubi, M. *Rosa damascena* as holy ancient herb with novel applications. *J. Tradit. Complement. Med.* **2016**, *6*, 10–16. [CrossRef] [PubMed]
6. Gochev, V.; Jirovetz, L.; Wlcek, K.; Buchbauer, G.; Schmidt, E.; Stoyanova, A.; Dobreva, A. Chemical Composition and Antimicrobial Activity of Historical Rose Oil from Bulgaria. *J. Essent. Oil Bear. Plants* **2009**, *12*, 1–6. [CrossRef]
7. Ulusoy, S.; Boşgelmez-Tınaz, G.; Seçilmiş-Canbay, H. Tocopherol, Carotene, Phenolic Contents and Antibacterial Properties of Rose Essential Oil, Hydrosol and Absolute. *Curr. Microbiol.* **2009**, *59*, 554–558. [CrossRef] [PubMed]
8. Kovacheva, N.; Rusanov, K.; Atanassov, I. Industrial cultivation of oil bearing rose and rose oil production in Bulgaria during 21st century, directions and challenges. *Biotechnol. Biotechnol. Equip.* **2010**, *24*, 1793–1798. [CrossRef]
9. Koksal, N.; Aslancan, H.; Sadighazadi, S.; Kafkas, E. Chemical investigation on *Rose damascena* Mill. volatiles; effects of storage and drying conditions. *Acta Sci. Pol. Hortorum Cultus* **2015**, *14*, 105–114.
10. Nunes, H.S.; Miguel, M.G. *Rosa damascena* essential oils: A brief review about chemical composition and biological properties. *Trends Phytochem. Res.* **2017**, *1*, 111–128.
11. Antonova, D.V.; Medarska, Y.N.; Stoyanova, A.S.; Nenov, N.S.; Slavov, A.M.; Antonov, L.M. Chemical profile and sensory evaluation of Bulgarian rose (*Rosa damascena* Mill.) aroma products, isolated by different techniques. *J. Essent. Oil Res.* **2020**, *33*, 171–181. [CrossRef]
12. Nedkov, N. Bulgarian Rose Oil of White Oil-Bearing Rose. *Bulg. J. Agric. Sci.* **2009**, *15*, 319–323.
13. Gateva, S.; Jovtchev, G.; Chanev, C.; Georgieva, A.; Stankov, A.; Dobreva, A.; Mileva, M. Assessment of anti-cytotoxic, antigenotoxic and antioxidant potential of Bulgarian *Rosa alba* L. essential oil. *Caryologia Int. J. Cytol. Cytosyst. Cytogenet.* **2020**, *73*, 71–88. [CrossRef]

14. Schieber, A.; Mihalev, K.; Berardini, N.; Mollov, P.; Carle, R. Flavonol Glycosides from Distilled Petals of *Rosa damascena* Mill. *Zeitschrift für Naturforschung C* **2005**, *60*, 379–384. [CrossRef] [PubMed]
15. Atanasova, T.; Kakalova, M.; Stefanof, L.; Petkova, M.; Stoyanova, A.; Damyanova, S.; Desyk, M. Chemical composition of essential oil from *Rosa damascena* Mill., growing in new region of Bulgaria. *Ukr. Food J.* **2016**, *5*, 492–498. [CrossRef]
16. Erbas¸, S.; Baydar, H. Variation in scent compounds of oil bearing rose (Rosa damascena Mill.) produced by headspace solid phase microextraction, hydrodistillation and solvent extraction. *Rec. Nat. Prod.* **2016**, *10*, 555–565.
17. Balev, D.; Vlahova, V.D.; Mihalev, K.; Shikov, V.; Dragoev, S.; Nikolov, V. Application of Natural Dietary Antioxidants in Broiler Feeds. *J. Mt. Agric. Balk.* **2015**, *18*, 224–232.
18. Baydar, N.G.; Baydar, H. Phenolic compounds, antiradical activity and antioxidant capacity of oil-bearing rose (*Rosa damascena* Mill.) extracts. *Ind. Crop. Prod.* **2013**, *41*, 375–380. [CrossRef]
19. Rusanov, K.; Garo, E.; Rusanova, M.; Fertig, O.; Hamburger, M.; Atanassov, I.; Butterweck, V. Recovery of Polyphenols from Rose Oil Distillation Wastewater Using Adsorption Resins—A Pilot Study. *Planta Med.* **2014**, *80*, 1657–1664. [CrossRef] [PubMed]
20. Sabahi, Z.; Farmani, F.; Mousavinoor, E.; Moein, M. Valorization of Waste Water of *Rosa damascena* Oil Distillation Process by Ion Exchange Chromatography. *Sci. World J.* **2020**, *2020*, 5409493. [CrossRef] [PubMed]
21. Slavov, A.; Kiyohara, H.; Yamada, H. Immunomodulating pectic polysaccharides from waste rose petals of *Rosa damascena* Mill. *Int. J. Biol. Macromol.* **2013**, *59*, 192–200. [CrossRef] [PubMed]
22. Abdel-Hameed, E.-S.; Bazaid, S.A.; Salih, A.; Sabra, A.N.A. Total phenolic, in vitro antioxidant activity and safety assessment (Acute, sub-chronic and chronic toxicity) of industrial Taif rose water by-product in mice. *Der. Pharm. Lett.* **2015**, *7*, 251–259.
23. Solimine, J.; Garo, E.; Wedler, J.; Rusanov, K.; Fertig, O.; Hamburger, M.; Atanassov, I.; Butterweck, V. Tyrosinase inhibitory constituents from a polyphenol enriched fraction of rose oil distillation wastewater. *Fitoterapia* **2016**, *108*, 13–19. [CrossRef] [PubMed]
24. Vasileva, I.; Krastev, L.; Petkova, N.; Yantcheva, N.; Nenov, N.; Krachmarov, A.; Atanasova, A.; Slavov, A. Valorization of cacao and rose waste for preparation of liqueurs. *Food Sci. Appl. Biotechnol.* **2019**, *2*, 8–17. [CrossRef]
25. Alizadeh, A.; Fattahi, M. The study of active compounds and antioxidant activity of waste residues obtained during Essential oil and Rose water production in 24 Damask Rose Populations of East and West Azerbaijan: By-product. *J. Plant Res. (Iran. J. Biol.)* **2021**, *34*, 945–956.
26. Wedler, J.; Rusanov, K.; Atanassov, I.; Butterweck, V. A Polyphenol-Enriched Fraction of Rose Oil Distillation Wastewater Inhibits Cell Proliferation, Migration and TNF-α-Induced VEGF Secretion in Human Immortalized Keratinocytes. *Planta Medica* **2016**, *82*, 1000–1008. [CrossRef] [PubMed]
27. Oswell, N.J.; Thippareddi, H.; Pegg, R.B. Practical use of natural antioxidants in meat products in the U.S.: A review. *Meat Sci.* **2018**, *145*, 469–479. [CrossRef] [PubMed]
28. Dimitrova, M.; Ivanov, G.; Mihalev, K.; Slavchev, A.; Ivanova, I.; Vlaseva, R. Investigation of antimicrobial activity of polyphenol-enriched extracts against probiotic lactic acid bacteria. *Food Sci. Appl. Biotechnol.* **2019**, *2*, 67–73. [CrossRef]
29. Slavov, A.; Vasileva, I.; Nenov, N.; Stefanov, L.; Shikov, V.; Mihalev, K.; Stoyanova, A. Wastes from the rose oil industry–approaches for valorization and recycling. *Inf. Bull. Bulg. Natl. Assoc. Essent. Oils Perfum. Cosmet.* **2018**, *76*, 39–45.
30. Chochkov, R.; Vasileva, I.; Ivanova, P.; Slavov, A. Effect of industrial *Rosa Damascena* Mill Waste in Pastry Biscuits. In Proceedings of the 87 International Scientific Conference of Young Scientist and Students "Youth Scientific Achievements to the 21st Century Nutrition Problem Solution", Kyiv, Ukraine, 7–8 April 2021.
31. Mollov, P.; Mihalev, K.; Shikov, V.; Yoncheva, N.; Karagyozov, V. Colour stability improvement of strawberry beverage by fortification with polyphenolic copigments naturally occurring in rose petals. *Innov. Food Sci. Emerg. Technol.* **2007**, *8*, 318–321. [CrossRef]
32. Bhatti, H.N.; Khalid, R.; Hanif, M.A. Dynamic biosorption of Zn(II) and Cu(II) using pretreated *Rosa gruss* an teplitz (red rose) distillation sludge. *Chem. Eng. J.* **2009**, *148*, 434–443. [CrossRef]
33. Ansari, T.M.; Hanif, M.A.; Mahmood, A.; Ijaz, U.; Khan, M.A.; Nadeem, R.; Ali, M. Immobilization of rosewaste biomass for uptake of Pb(II) from aqueous solutions. *Biotechnol. Res. Int.* **2011**, *2011*, 685023. [CrossRef] [PubMed]
34. Karaboyaci, M.; Karaboyacı, M. Recycling of rose wastes for use in natural plant dye and industrial applications. *J. Text. Inst.* **2014**, *105*, 1160–1166. [CrossRef]
35. Eren, E.; Gok, E.C.; Seyhan, B.N.; Maslakci, N.N.; Oksuz, A.U. Evaluation of Anthocyanin, a Rose Residue Extract, for Use in Dye-Sensitized Solar Cell. *Asian J. Chem.* **2015**, *27*, 3745–3748. [CrossRef]
36. Rabbani, D.; Mahmoudkas, N.; Mehdizad, F.; Shaterian, M. Green Approach to Wastewater Treatment by Application of *Rosa damascena* Waste as Nano-Biosorbent. *J. Environ. Sci. Technol.* **2016**, *9*, 121–130. [CrossRef]
37. Georgieva, A.; Ilieva, Y.; Kokanova-Nedialkova, Z.; Zaharieva, M.M.; Nedialkov, P.; Dobreva, A.; Kroumov, A.; Najdenski, H.; Mileva, M. Redox-Modulating Capacity and Antineoplastic Activity of Wastewater Obtained from the Distillation of the Essential Oils of Four Bulgarian Oil-Bearing Roses. *Antioxidants* **2021**, *10*, 1615. [CrossRef] [PubMed]
38. Jovtchev, G.; Stergios, M.; Schubert, I. A comparison of N-methyl-N-nitrosourea-induced chromatid aberrations and micronuclei in barley meristems using FISH techniques. *Mutat. Res. Toxicol. Environ. Mutagen.* **2002**, *517*, 47–51. [CrossRef]
39. Evans, H. *Handbook of Mutagenicity Test Procedures*; Kilbey, B., Legator, M., Nicols, W., Ramel, C., Eds.; Elsevier Science Publishers: Amsterdam, The Netherlands, 1984; pp. 405–427.
40. Fenech, M. Cytokinesis-block micronucleus cytome assay. *Nat. Protoc.* **2007**, *2*, 1084–1104. [CrossRef] [PubMed]

41. Srivastava, S.; Somasagara, R.R.; Hegde, M.; Nishana, M.; Tadi, S.K.; Srivastava, M.; Choudhary, B.; Raghavan, S.C. Quercetin, a Natural Flavonoid Interacts with DNA, Arrests Cell Cycle and Causes Tumor Regression by Activating Mitochondrial Pathway of Apoptosis. *Sci. Rep.* **2016**, *6*, 24049. [CrossRef] [PubMed]
42. Kluska, M.; Juszczak, M.; Żuchowski, J.; Stochmal, A.; Woźniak, K. Kaempferol and Its Glycoside Derivatives as Modulators of Etoposide Activity in HL-60 Cells. *Int. J. Mol. Sci.* **2021**, *22*, 3520. [CrossRef]
43. Harwood, M.; Danielewska-Nikiel, B.; Borzelleca, J.F.; Flamm, G.W.; Williams, G.M.; Lines, T.C. A critical review of the data related to the safety of quercetin and lack of evidence of in vivo toxicity, including lack of genotoxic/carcinogenic properties. *Food Chem. Toxicol.* **2007**, *45*, 2179–2205. [CrossRef]
44. Barcelos, G.; Grotto, D.; Angeli, J.P.F.; Serpeloni, J.M.; Rocha, B.; Bastos, J.K.; Barbosa, F., Jr. Evaluation of Antigenotoxic Effects of Plant Flavonoids Quercetin and Rutin on HepG2 Cells. *Phytother. Res.* **2011**, *25*, 1381–1388. [CrossRef] [PubMed]
45. Vanella, L.; Di Giacomo, C.; Acquaviva, R.; Barbagallo, I.; Volti, G.L.; Cardile, V.; Abraham, N.G.; Sorrenti, V. Effects of Ellagic Acid on Angiogenic Factors in Prostate Cancer Cells. *Cancers* **2013**, *5*, 726–738. [CrossRef] [PubMed]
46. Azqueta, A.; Collins, A. Polyphenols and DNA Damage: A Mixed Blessing. *Nutrients* **2016**, *8*, 785. [CrossRef] [PubMed]
47. Bae, J.; Kim, N.; Shin, Y.; Kim, S.-Y.; Kim, Y.-J. Activity of catechins and their applications. *Biomed. Dermatol.* **2020**, *4*, 8. [CrossRef]
48. Rho, H.S.; Ghimeray, A.K.; Yoo, D.S.; Ahn, S.M.; Kwon, S.S.; Lee, K.H.; Cho, D.H.; Cho, J.Y. Kaempferol and Kaempferol Rhamnosides with Depigmenting and Anti-Inflammatory Properties. *Molecules* **2011**, *16*, 3338–3344. [CrossRef] [PubMed]
49. Calderon-Montaño, J.M.; Burgos-Morón, E.; Perez-Guerrero, C.; Lopez-Lazaro, M. A Review on the Dietary Flavonoid Kaempferol. *Mini-Rev. Med. Chem.* **2011**, *11*, 298–344. [CrossRef] [PubMed]
50. Jovtchev, G.; Gateva, S.; Stankov, A. Lilium compounds kaempferol and jatropham can modulate cytotoxic and genotoxic effects of radiomimetic zeocin in plants and human lymphocytesIn vitro. *Environ. Toxicol.* **2016**, *31*, 751–764. [CrossRef] [PubMed]
51. Kluska, M.; Juszczak, M.; Wysokinski, D.; Żuchowski, J.; Stochmal, A.; Wozniak, K. Kaempferol derivatives isolated from Lens culinaris Medik. reduce DNA damage induced by etoposide in peripheral blood mononuclear cells. *Toxicol. Res.* **2019**, *8*, 896–907. [CrossRef] [PubMed]
52. Iovine, C.; Mottola, F.; Santonastaso, M.; Finelli, R.; Agarwal, A.; Rocco, L. In vitro ameliorative effects of ellagic acid on vitality, motility and DNA quality in human spermatozoa. *Mol. Reprod. Dev.* **2021**, *88*, 167–174. [CrossRef] [PubMed]
53. Mottola, F.; Finelli, R.; Iovine, C.; Carannante, M.; Santonastaso, M.; Rocco, L. Anti-Genotoxicity Evaluation of Ellagic Acid and Curcumin—An In Vitro Study on Zebrafish Blood Cells. *Appl. Sci.* **2021**, *11*, 8142. [CrossRef]
54. Girish, C.; Koner, B.C.; Jayanthi, S.; Rao, K.R.; Rajesh, B.; Pradhan, S.C. Hepatoprotective activity of picroliv, curcumin and ellagic acid compared to silymarin on paracetamol induced liver toxicity in mice. *Fundam. Clin. Pharmacol.* **2009**, *23*, 735–745. [CrossRef] [PubMed]
55. Khanduja, K.L.; Gandhi, R.K.; Pathania, V.; Syal, N. Prevention of N-nitrosodiethylamine-induced lung tumourigenesis by ellagic-acid and quercetin in mice. *Food Chem. Toxicol.* **1999**, *37*, 313–318. [CrossRef]
56. Davis, W.L.; Matthew, S.B. Antioxidants and cancer III: Quercetin. *Altern. Med. Rev.* **2000**, *5*, 196–208.
57. Shirahata, S.; Murakami, H.; Nishiyama, K.; Yamada, K.; Nonaka, G.; Nishioka, I.; Omura, H. DNA breakage by flavan-3-ols and procyanidins in the presence of cupric ion. *J. Agric. Food Chem.* **1989**, *37*, 299–303. [CrossRef]
58. Efenberger-Szmechtyk, M.; Nowak, A.; Nowak, A. Cytotoxic and DNA-Damaging Effects of *Aronia melanocarpa*, *Cornus mas*, and *Chaenomeles superba* Leaf Extracts on the Human Colon Adenocarcinoma Cell Line Caco-2. *Antioxidants* **2020**, *9*, 1030. [CrossRef] [PubMed]
59. Black, A.K.; McFarland, R.D.; Grisham, J.W.; Smith, G.J. Cell cycle perturbation and cell death after exposure of a human lymphoblastoid cell strain to N-methyl-N′-nitro-N-nitrosoguanidine. *Am. J. Pathol.* **1989**, *134*, 53–61. [PubMed]
60. Sancar, A.; Lindsey-Boltz, L.A.; Ünsal-Kaçmaz, K.; Linn, S. Molecular Mechanisms of Mammalian DNA Repair and the DNA Damage Checkpoints. *Annu. Rev. Biochem.* **2004**, *73*, 39–85. [CrossRef] [PubMed]
61. Drabløs, F.; Feyzi, E.; Aas, P.A.; Vaagbø, C.B.; Kavli, B.; Bratlie, M.S.; Peña-Diaz, J.; Otterlei, M.; Slupphaug, G.; Krokan, H.E. Alkylation damage in DNA and RNA—Repair mechanisms and medical significance. *DNA Repair* **2004**, *3*, 1389–1407. [CrossRef] [PubMed]

life

MDPI

Article

Evaluation of the Phytochemical and Pharmacological Potential of Taif's Rose (*Rosa damascena* Mill var. *trigintipetala*) for Possible Recycling of Pruning Wastes

Tarek M. Galal [1], Hatim M. Al-Yasi [1], Mustafa A. Fawzy [1], Tharwat G. Abdelkader [1], Reham Z. Hamza [1], Ebrahem M. Eid [2,3] and Esmat F. Ali [1,*]

1 Department of Biology, College of Sciences, Taif University, P.O. Box 11099, Taif 21944, Saudi Arabia; t.aboseree@tu.edu.sa (T.M.G.); h.alyasi@tu.edu.sa (H.M.A.-Y.); mafawzy@tu.edu.sa (M.A.F.); t.abdelkader@tu.edu.sa (T.G.A.); reham.z@tu.edu.sa (R.Z.H.)

2 Biology Department, College of Science, King Khalid University, P.O. Box 9004, Abha 61321, Saudi Arabia; ebrahem.eid@sci.kfs.edu.eg

3 Botany Department, Faculty of Science, Kafrelsheikh University, Kafr El-Sheikh 33516, Egypt

* Correspondence: a.esmat@tu.edu.sa

Abstract: This study investigated the phytochemical contents of Taif's rose pruning wastes and their potential application as phytomedicine, thereby practicing a waste-recycling perspective. In the Al-Shafa highland, four Taif rose farms of various ages were chosen for gathering the pruning wastes (leaves and stems) for phytochemical and pharmacological studies. The leaves and stems included significant amounts of carbohydrates, cardiac glycosides, alkaloids, flavonoids, and other phenolic compounds. The cardiac glycoside and flavonoid contents were higher in Taif rose stems, while the phenolic and alkaloid contents were higher in the plant leaves. Cardiovascular glycosides (2.98–5.69 mg g^{-1}), phenolics (3.14–12.41 mg GAE g^{-1}), flavonoids (5.09–9.33 mg RUE g^{-1}), and alkaloids (3.22–10.96 mg AE g^{-1}) were among the phytoconstituents found in rose tissues. According to the HPLC analysis of the phenolic compounds, Taif's rose contains flavonoid components such as luteolin, apigenin, quercetin, rutin, kaempferol, and chrysoeriol; phenolics such as ellagic acid, catechol, resorcinol, gallic acid, and phloroglucinol; alkaloids such as berbamine, jatrorrhizine, palmatine, reticuline, isocorydine, and boldine. Warm water extract was highly effective against *Bacillus subtilis*, *Escherichia coli*, and *Proteus vulgaris*, whereas methanol and cold water extracts were moderately effective against *Aspergillus fumigatus* and *Candida albicans*. The study's findings suggested that Taif's rose wastes could be used for varied medical purposes.

Keywords: Damask rose; pruning wastes; phytochemicals; biological activity; recycling

Citation: Galal, T.M.; Al-Yasi, H.M.; Fawzy, M.A.; Abdelkader, T.G.; Hamza, R.Z.; Eid, E.M.; Ali, E.F. Evaluation of the Phytochemical and Pharmacological Potential of Taif's Rose (*Rosa damascena* Mill var. *trigintipetala*) for Possible Recycling of Pruning Wastes. *Life* **2022**, *12*, 273. https://doi.org/10.3390/life12020273

Academic Editor: Milka Mileva

Received: 16 January 2022
Accepted: 8 February 2022
Published: 12 February 2022

1. Introduction

Rosa damascena Mill. var. *trigintipetala* (Taif's rose or Damask rose), a Rosaceae plant, is one of the most important commercial crops farmed due to the high value of its essential oils worldwide [1]. It is a tall shrub that can reach 2.5 meters in height and blooms once a year (in May–June), with a fully mature plant producing 500–600 flowers [2]. Taif's rose grows in temperate and subtropical climates at elevations ranging from 300 to 2500 meters [3]. It is commercially grown in Saudi Arabia, Egypt, Turkey, Morocco, Bulgaria, Iran, France, China, and India, among other countries [4]. It is also one of the attractive and aromatic plants grown for use in the perfume, pharmaceutical, and food industries in numerous Taif governorate locations [5]. In contrast to the Bulgarian variety, the Saudi Arabia rose oils do not fully comply with the ISO 9842:2003 standard for rose oil, but they have a high olfactive potential [6]. Taif's rose has been shown to have antioxidant, antidiabetic, anti-HIV, antibacterial, anti-inflammatory, and cardiotonic properties due to the presence of various phytochemical compounds such as alkaloids, phenolic acids, flavonoids, and other phenolic compounds [7–9].

In terms of the current state of the Taif rose, the governorate of Taif has approximately 860 farms ranging in size from large (1.0 ha) to small (0.03 ha) with most of them ranging from 0.3 to 0.7 ha. This variation in size may be due to the cultivation conditions of the Taif rose on mountain ridges and in wadi beds, which differs from the corresponding species worldwide. The wastes generated by these farms (>2700 ton) could be agriculturally produced from rose bush pruning and industrially created from the oil distillation process [10]. A tiny portion is used for vegetative propagation, but the majority is dried and burned, posing environmental issues such as air and soil pollution, as well as health risks to surrounding residents [11]. The output of Taif rose blooms, and thus the highest oil percentage, is closely related to trimming [12]. Pruning is carried out such that the lower branches get sufficient light to create food, changing growth phases to encourage new axillary and bloom buds, and removing disease dependent on the variety [1].

Thousands of plant species are employed in many traditional systems of medicine around the world and are recognized for their contributions to contemporary medicine, with some of them, such as *Brugmansia* and *Rosa* species, being used to treat cancer and cardiovascular problems [13,14]. The use of medicinal plants is a centuries-old tradition, and recent advances in contemporary therapies have boosted the use of natural products for a variety of maladies and disorders around the world [15]. Secondary metabolites have a variety of biological effects, and they provide the scientific foundation for many ancient civilizations' uses of herbs in traditional medicine [16]. Phenolic compounds are widely distributed and the most abundant secondary metabolites in plants; they include flavonoids, alkaloids, and phenolic acids, which are involved in the defense against ultraviolet radiation or aggression by pathogens, parasites, and predators [17]. These secondary plant metabolites were investigated for their activity against cardiovascular and neurodegenerative diseases and cancer [17,18].

Pharmacological investigations have demonstrated that rose blooms of Taif provide a wide range of health benefits due to their high polyphenolic content [19]. According to Karkania et al. [20], *R. damascena* has potential antimicrobial activities against both Gram-negative and Gram-positive bacteria as well as fungi. Moreover, strong antimicrobial activity has been reported against different bacterial and fungal strains such as *Escherichia coli*, *Proteus vulgaris*, *Candida albicans*, and *Staphylococcus aureus* [21,22]. On a scientific level, empirical knowledge from folk medicine is becoming an increasingly important component of in vitro and in vivo studies, including preclinical and clinical trials. These studies investigated and explained the therapeutic efficacy of rose products and their ingredients, such as antidepressant effects, psychological relaxation, sexual dysfunction improvement, antioxidant, antimicrobial, antifungal, probiotic, and antipyretic effects, smooth muscle relaxation, lipid-lowering content, antiulcerogenic effects, and so on [4,23–25].

Because of the associated transportation, storage, and processing requirements, direct burning of agricultural biomass or trash is not cost-effective [20]. Furthermore, inappropriate agricultural waste storage pollutes the environment (soil, air, water, and sight) [26]. Several studies were carried out on the pharmacological activity of the essential oil of Taif's rose [6,27], while, to the authors' knowledge, no studies have been conducted on the recycling of its vegetative wastes. As a result, recycling and reusing these solid wastes for commercial purposes is extremely important and essential. As a result, the current study intends to investigate the phytochemical elements including cardiac glycosides, flavonoid, alkaloids, and other phenolic compounds, of Taif's rose pruning wastes and their pharmacological potential as a phytomedicine. These compounds, in addition to having antioxidant properties, have several other specific biological actions in preventing and/or treating diseases.

2. Material and Methods

2.1. Plant Sampling

During December 2020, four Taif rose farms on the Al-Shafa highland, Taif Province, Saudi Arabia, were chosen to collect pruning wastes for prospective recycling in medical uses. Farms F1, F2, F3, and F4 had ages ranging from 10, 12, 20, and 4 years, respectively. At each farm, ten rose plants of various sizes were chosen to estimate the biomass of their fresh pruning wastes. Shrubs were pruned until they reached a height of 80–90 cm. Fresh wastes were collected and weighed to determine their fresh biomass (kg ha^{-1}) by multiplying the average individual weight by the number of individuals per farm. Then, samples were left for air drying for about two weeks until constant weight.

2.2. *Sample Preparation*

For plant analysis, three composite samples (leaves and stems) of trimmed vegetative wastes were collected from each farm. Plant materials were rinsed in tap and distilled water, then air-dried at room temperature in the shade before being homogenized in a planetary high-energy mill with a hardened chromium steel vial.

2.3. *Quality Analysis*

Approximately 250 g sample of plant powder was shaken in 1000 mL ethanol for 24 hours on an orbital shaker at room temperature, and then the extract was filtered with Whatman No 1 filter paper. The filtrate was concentrated to dryness under reduced pressure at 40 °C through evaporator. The extract was stored between 2 and 8 °C for analysis of alkaloids, phenolic acids, flavonoids, and cardiac glycosides. HCl, $NaCO_3$, ethanol, Baljet's solution, picric acid, NaOH, $AlCl_3$, methanol, Folin reagent, $NaHCO_3$, formic acid, acetonitrile, glacial acetic acid, diethylamine, dimethyl sulfoxide, Ketoconazole, and Gentamicin were the used chemical reagents.

2.3.1. Determination of Soluble Carbohydrates

According to Sadasivam and Manickam [28], the total soluble carbohydrates were calculated using the anthrone method. Approximately 100 mg of Taif's rose powder was hydrolyzed in a boiling water bath for 3 hours with 5 mL of 2.5 N HCl. The acid digested sample was chilled to room temperature before adding sodium carbonate to neutralize it. Using distilled water, the final volume was diluted to 100 mL and centrifuged for 15 min at 5000 rpm. The total soluble carbohydrates were then determined by collecting the supernatant.

2.3.2. Determination of Cardiac Glycosides

Solich et al. [29] and Tofighi et al. [30] used techniques to measure cardiac glycosides. To detect cardiac glycosides, a 10% ethanol extract was mixed with 10 mL newly prepared Baljet's solution (95 mL of 1% picric acid + 5 mL of 10% NaOH). After an hour, the liquid was diluted with 20 mL distilled water, and the absorbance was measured with a spectrophotometer (CECIL CE 1021, Cecil Instruments Limited, Corston, UK) at 495 nm.

2.3.3. Determination of Total Flavonoid Contents (TFC)

Tofighi et al. [30] published procedures for calculating the TFC of vegetative pruning wastes. Ten milligrams of plant leaves were extracted under reflux (80 °C) for 60 min with a 20 mL water–ethanol solution 60% (v/v) (pH = 5.06). After cooling to room temperature, the extract was filtered, and the residue was extracted again under the same conditions. The hydroalcoholic extract and the re-extract were combined, and the volume was raised to 50 mL of water–ethanol solution at 60% (v/v) (stock solution). To bring the stock solution to volume, a part of it was transferred to a 10 mL volumetric flask and mixed with methanol (blank solution). A second aliquot of the stock solution was transferred to a new 10 mL volumetric flask, which was then filled with 2% $AlCl_3$ and brought to volume with methanol (test solution). After 25 min, the absorbance of the test solution was measured at 430 nm against a blank solution.

The rutin content of the TFC herbal material was determined as the average of three determinations. The flavonoid content (mg g^{-1}) of herbal material (adjusted for moisture content) was determined as follows: TFC herbal material = (TFC tested solution × 1.25 × 50)/(w-ld), where TFC test solution is the total concentration of flavonoids in the test solution (mg mL^{-1}), 1.25 corresponds to the dilution factor, 50 is the volume of the stock solution (mL), "w" is the weight of herbal material (g), and "l d" is the loss on drying of herbal material.

2.3.4. Determination of The Total Phenolic Compounds

The concentration of phenolics in the plant ethanol extract was determined using a spectrophotometric method [30,31]. The reaction mixture consisted of 0.5 mL ethanol extract, 2.5 mL 10% Folin–Ciocalteu's reagent mixed in water, and 2.5 mL 7.5% $NaHCO_3$. The blank was made with 0.5 mL methanol, 2.5 mL 10 percent Folin–Ciocalteu's reagent dissolved in water, and 2.5 mL 7.5 percent $NaHCO_3$. The samples were then incubated in a thermostat for 45 min at 45 °C. A spectrophotometer was used to measure the absorbance at 765 nm (CECIL CE 1021, Cecil Instruments Limited, Corston, UK). The samples were made in triplicate for each assay, and the mean absorbance value was computed. The calibration curve for the standard gallic acid solution was created using the same method. The amount of phenolics was measured in milligrams of gallic acid equivalent (GAE) per gram of dry weight (DW).

2.3.5. Estimation of Phenolic and Flavonoid Compounds Using HPLC

High-performance liquid chromatography (HPLC) was used to estimate the flavonoid and phenolic compounds of Taif's rose plants. HPLC-MS techniques are often used for the separation, identification, and quantitation of flavonoids, phenolic acids, and other phenolic compounds in plants. The HPLC-MS system (Agilent 1100: Agilent Corp., Palo Alto, Calif.) is composed of a quaternary pump, a photodiode-array detector, a *Uv*/*v* is detector, and a single quadrupole MS detector with an ion source (ESI). Flavonoids were separated in 70 min using a gradient solvent system of 0.1% formic acid solution with a flow rate of 1.0 mL min^{-1}, detected at 280 nm, and identified by ESI-MS [32]. Phenolic acid was separated in 60 min using a gradient mobile phase of water/acetonitrile/glacial acetic acid (980/20/5, *v*/*v*/*v*, pH 2.68) and acetonitrile/glacial acetic acid (1000/5, *v*/*v*), with a flow rate of 0.8 mL min^{-1} and detection at 325 nm [33]. Moreover, alkaloids were analyzed by HPLC (0 min, 80:20 (A–B); 5 min, 80:20; 20 min, 60:40; 25 min, 0:100) using 0.2% diethylamine and 0.16% formic acid as solvent system A, and 0.2% diethylamine and 0.16% formic acid in acetonitrile as solvent system B. The column used was the GraceSmart RP18 (Grace Vydac, Hesperia, CA, USA), 5 µm, 250 mm × 4.6 mm with a flow rate of 1.0 mL min^{-1}. The peaks were detected at 226 nm.

2.4. Biological Activity

2.4.1. Preparation of Extracts

Approximately 250 grams of plant powder was steeped in 1.5 liters of 95% ethanol and methanol and boiled in cold (approximate room temperature) and warm (50 °C) water at room temperature for 5 days. The combination was blended daily to provide a consistent infusion. The extract was filtered using Whatman filter paper No. 1 after 5 days. A rotary evaporator at 60 °C was used to dry the filtrate. The dried extract was kept at −20 °C in sterile glass vials until use [34].

2.4.2. Microorganisms Used

The following microorganisms were obtained from Al-Azhar University, Faculty of Science: gram-positive bacteria (*Bacillus subtilis*), gram-negative bacteria (*Escherichia coli* and *Proteus vulgaris*), and fungal strains (*Aspergillus fumigatus* and *Candida albicans*). The bacterial and fungal strains were cultured in nutrient agar and malt extract, respectively.

2.4.3. In Vitro Evaluation of the Antimicrobial Activity

An antimicrobial susceptibility test was performed using the agar disc well diffusion method [35] with some modifications. The diameter of inhibitory zones was used to measure antimicrobial activity. Plant extracts were tested against bacterial isolates as antimicrobial agents. On the surface media, inoculum suspensions of all bacterial and fungal isolates were distributed. Using a 6 mm Cork borer, holes (diameter 6 mm) were drilled into the media. The dried plant extracts were treated in dimethyl sulfoxide (DMSO) to make a 10 mg mL^{-1} final extract. Each plate's well was filled with 100 µL of plant extract. The inoculated agar plates were incubated for 24 hours at 37 °C for bacterial growth and 48 hours at 28 °C for fungal growth. After 24–48 hours of incubation, inhibition zones caused by active extract components were measured. The studies were performed in triplicate, and the inhibition zone was assessed using a standard scale [36]. Ketoconazole antibiotic (MIC = 100 µg mL^{-1}) was used as control treatment for fungi, while gentamicin (MIC = 4 µg mL^{-1}) was used for the bacteria.

2.5. Statistical Analysis

The differences in plant's chemical constituents in separate farms were analyzed by one-way analysis of variance (ANOVA I), using SPSS software (version 22), after the data were checked for normality [37]. When there were substantial variations among the farms, a post-hoc test was used (Duncan's test).

3. Results

3.1. Biomass of Pruning Wastes

Taif's rose was pruned in December after the rainy season, where the fresh biomass (FW: fresh weight) of the pruning wastes increased with increasing plant age (Table 1). The highest biomass (5.2 t FW ha^{-1}) was recorded at the oldest farm (F3), while the lowest (1.3 t FW ha^{-1}) was recorded at the youngest farm (F4). The average biomass produced from the different farms was 3.2 t FW ha^{-1}.

Table 1. Fresh biomass (mean: upper line, SD: lower line) of the vegetative wastes produced after pruning of four Taif's rose farms on the Al-Shafa highland.

Farm	Farm 1	Farm 2	Farm 3	Farm 4	F-Value
Mean	2.5c	3.7b	5.2a	1.3d	128.5 ***
SD	0.4	0.5	0.6	0.4	

Means with the same letter in the same row are not significantly different (Duncan's multiple range tests at $p < 0.05$), ***: meam $p < 0.001$).

3.2. Chemical Constituents

The analysis of total soluble carbohydrates (in dry weight) indicated significant variation ($p < 0.001$) in their tissue contents among the different study farms (Table 2). It was found that Taif's rose leaves had higher carbohydrate contents than the stems, where the highest content (3.05%) was recorded in the leaves of F3 plants, while the lowest (0.78%) was found in F1 stems. The phytochemical screening of the ethanolic extract of Taif's rose detected significant variations ($p < 0.001$) in the contents of cardiac glycosides, flavonoids, alkaloids, and phenolic compounds among the study farms (Table 2). Notably, Taif's rose stems had higher cardiac glycoside and flavonoid contents, while leaves had higher phenolic and alkaloid contents. The plant leaves from F3 had the highest phenolic content (12.41 mg GAE g^{-1}) but the lowest cardiac glycoside and flavonoid contents (2.98 mg securiaside g^{-1} and 5.09 mg RUE g^{-1}). In addition, the stems of the F4 plants had the highest cardiac glycosides (5.69 mg securiaside g^{-1}) but the lowest phenolic and alkaloid contents (3.14 mg GAE g^{-1} and 3.22 mg AE g^{-1}). Moreover, the highest flavonoids (9.33 mg RUE g^{-1}) were recorded in the stems of the F2 plants, while the highest alkaloid content (10.96 mg AE g^{-1}) was found in the leaves of the F1 plants.

Table 2. Phytochemical constituents (mean ± SD) of the leaves and stems of Taif's rose collected from different rose farms. Maximum and minimum values are underlined.

Farm		Carbohydrates %	Cardiac Glycosides mg Securiaside g^{-1}	Phenolics mg GAE g^{-1}	Flavonoids mg RUE g^{-1}	Alkaloids mg AE g^{-1}
Farm 1	L	1.26±0.05d	4.33±0.23c	10.26±1.01b	5.33±0.62g	10.96±1.32a
	S	0.78±0.10g	4.33±0.20c	7.22±0.87e	7.11±1.02c	7.12±0.94d
Farm 2	L	1.69±0.04c	3.97±0.22e	8.36±1.32c	6.14±1.04f	8.36±2.08c
	S	0.96±0.10e	5.14±0.19b	6.07±0.71f	9.33±2.13a	8.36±1.62c
Farm 3	L	3.05±0.05a	2.98±0.32g	12.41±2.13a	5.09±0.92h	10.07±2.12b
	S	0.88±0.12f	4.23±0.26d	5.36±1.06g	7.01±0.82d	4.09±0.26e
Farm 4	L	2.36±0.09b	3.14±0.41f	8.19±1.04d	8.76±1.63b	8.36±1.27c
	S	0.79±0.14g	5.69±0.72a	3.14±0.23h	6.39±0.05e	3.22±0.43f
F-value		2346.5 ***	2839.7 ***	1929.4 ***	3980.4 ***	1038.5 ***

L: leaves, S: stem, GAE: gallic acid equivalent, RUE: rutin equivalent, AE: atropine equivalent. Means with the same letter in the same row are not significantly different (Duncan's multiple range tests at $p < 0.05$), ***: meam $p < 0.001$).

3.3. HPLC of Phytochemical Compounds

3.3.1. Phenolic Compounds

Ellagic acid, catechol, resorcinol, gallic acid, and phloroglucinol were the main phenolic compounds, which were separated and identified using the HPLC in Taif's rose extract (Table 3 and Figure 1). Plants collected from F4 had the highest contents of ellagic and gallic acid (23.54, and 37.40 mg g^{-1}, respectively), while those from F2 had the highest resorcinol content (18.74 mg g^{-1}) in their stems. Additionally, F3 and F2 plant leaves had the highest catechol and phloroglucinol contents (21.60 and 6.24 mg g^{-1}, respectively).

3.3.2. Flavonoid Compounds

Using HPLC, the separated and identified flavonoid compounds were apigenin, luteolin, chrysoeriol, rutin, and kaempferol (Table 4 and Figure 2). It was clear that plants had higher contents of the separated compounds (except rutin) in their stems than in their leaves. The highest contents of luteolin and chrysoeriol (30.56 and 66.20 mg g^{-1}) were recorded in the stems, while the highest rutin (25.30 mg g^{-1}) was recorded in the leaves of the F1 plants. In addition, the highest quercetin and apigenin (25.41 and 30.44 mg g^{-1}) and kaempferol (38.74 mg g^{-1}) were recorded in the stems of F4 and F3, respectively.

Table 3. HPLC analysis of the phenolic concentration of the leaves and stems of Taif's rose collected from different rose farms. ND: not detected.

Farm		Phenolics Concentration (mg g^{-1})				
		Gallic Acid	Ellagic Acid	Catechol	Resorcinol	Phloroglucinol
Farm 1	L	5.60	ND	23.54	34.20	ND
	S	9.45	ND	13.44	18.74	0.96
Farm 2	L	19.58	14.50	3.54	0.25	6.24
	S	33.60	19.58	2.66	1.23	ND
Farm 3	L	15.63	ND	21.60	4.12	ND
	S	13.65	1.58	6.28	ND	0.89
Farm 4	L	16.04	1.38	6.11	9.05	ND
	S	37.40	23.54	3.96	2.74	ND

L: leaves, S: stem.

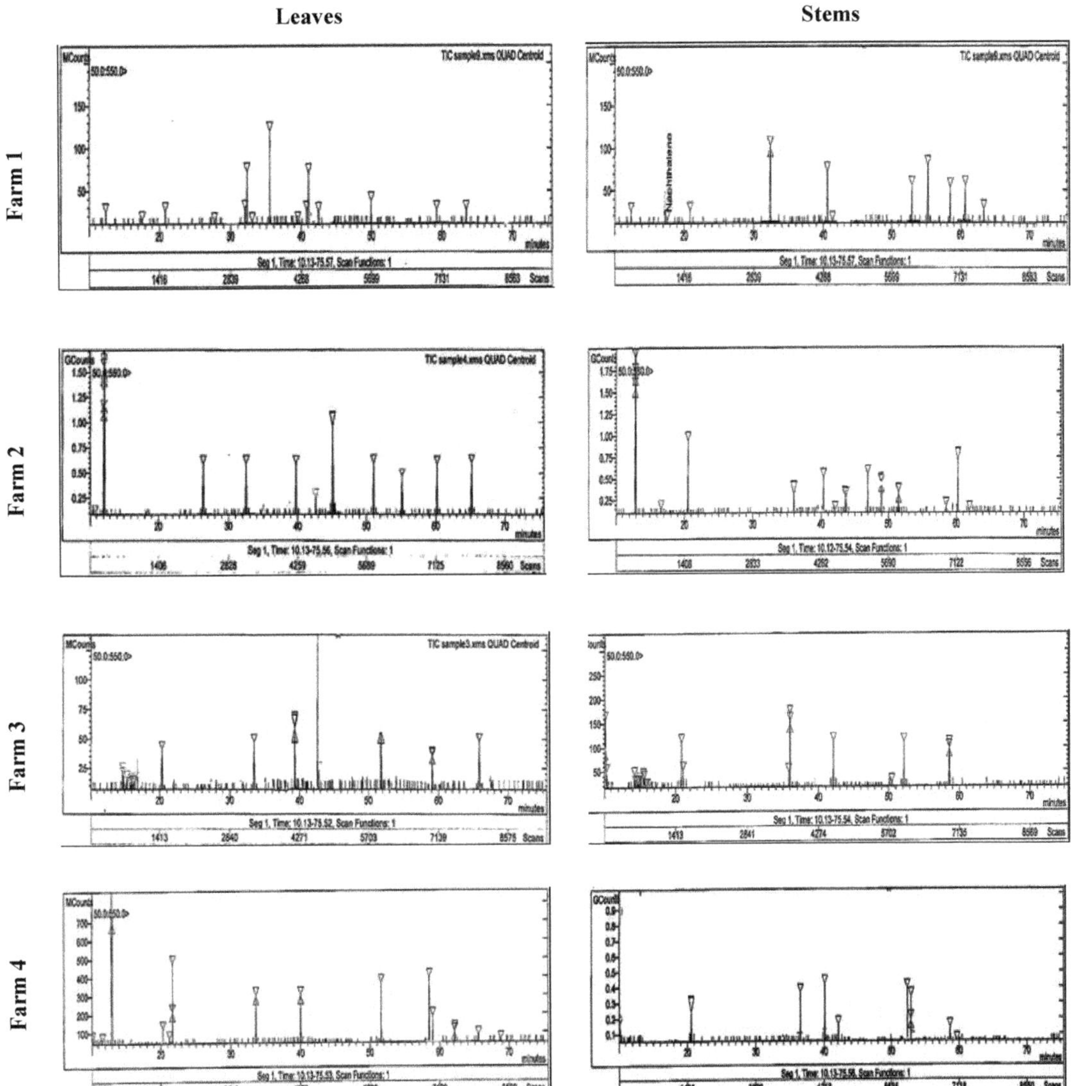

Figure 1. HPLC analysis of the phenolic compounds in the stem and leaves of Taif's rose collected from different farms.

Table 4. HPLC analysis of the flavonoid concentration of the leaves and stems of Taif's rose collected from different rose farms. ND: not detected.

Farm		Flavonoids Concentration (mg g^{-1})					
		Quercetin	Apigenin	Luteolin	Chrysoeriol	Rutin	Kaempferol
Farm 1	L	18.77	2.6	1.99	ND	25.3	ND
	S	0.87	17.22	30.56	66.2	ND	18.32
Farm 2	L	12.87	6.58	10.5	ND	19.85	21.68
	S	17.02	ND	25.6	44.05	3.8	38.74
Farm 3	L	1.63	ND	18.02	19.2	6.74	ND
	S	7.15	ND	22.6	14.8	ND	19.23
Farm 4	L	ND	ND	ND	ND	15.6	21.5
	S	25.41	30.44	5.6	55.69	18.52	33.91

L: leaves, S: stem.

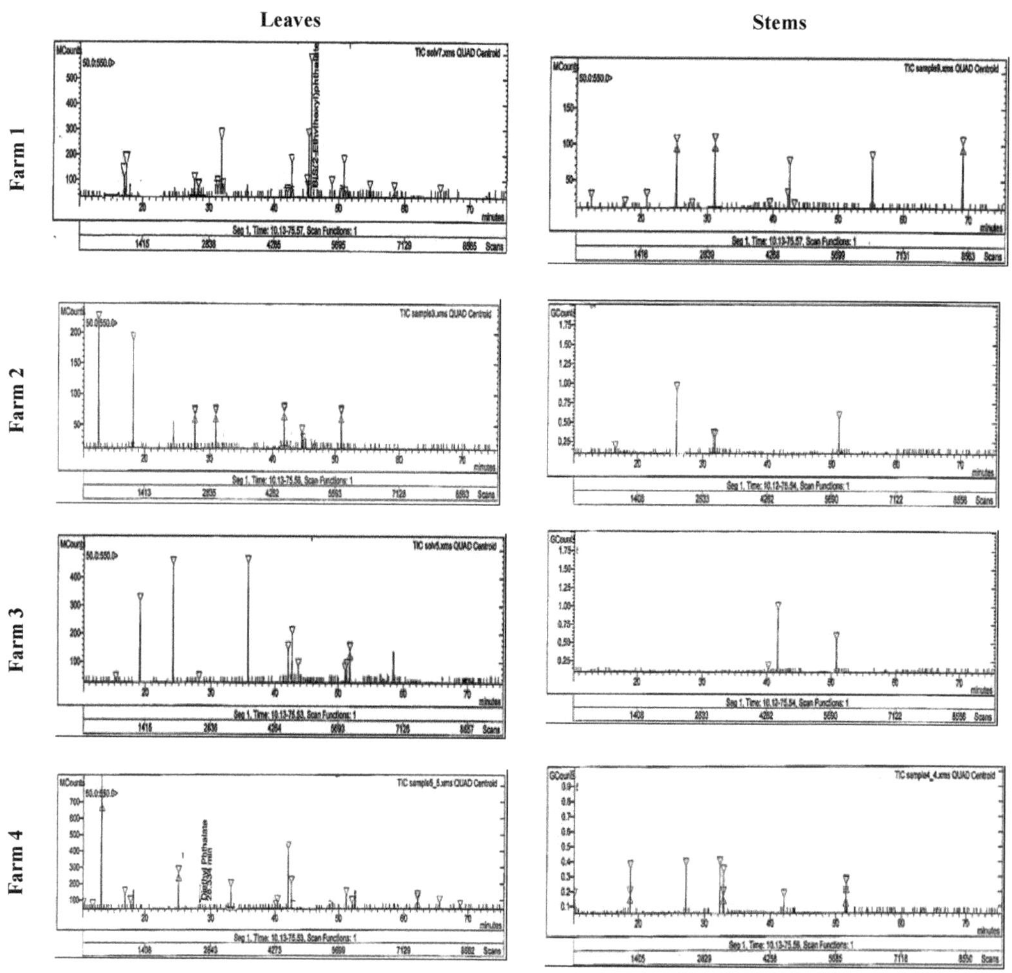

Figure 2. HPLC analysis of the flavonoid compounds in the stem and leaves of Taif's rose collected from different farms.

3.3.3. Alkaloid Compounds

Six alkaloid compounds (berbamine, jatrorrhizine, palmatine, reticuline, isocorydine, and boldine) were separated and identified using HPLC in Taif's rose extract (Table 5 and Figure 3). It was found that the stems of Taif's rose plants had higher contents of the separated compounds (except jatrorrhizine) than the leaves. The stems of the F1 plants had the highest contents of berbamine, palmatine, and isocorydine (5.24, 6.36, and 5.69 mg g^{-1}, respectively), while their leaves had the highest jatrorrhizine content (9.50 mg g^{-1}). In addition, the highest boldine and reticuline contents (0.89 and 8.5 mg g^{-1}) were recorded in the stems of F2 and F4 plants, respectively.

Table 5. HPLC analysis of the alkaloid concentration of the leaves and stems of Taif's rose collected from different rose farms. ND: not detected.

Farm		Alkaloid's Concentration (mg g^{-1})					
		Berbamine	**Jatrorrhizine**	**Palmatine**	**Reticuline**	**Isocorydine**	**Boldine**
Farm 1	L	3.14	0.69	ND	2.36	ND	ND
	S	5.24	7.13	6.36	ND	5.69	ND
Farm 2	L	4.69	9.5	ND	ND	1.6	N.D
	S	ND	ND	ND	ND	3.7	0.89
Farm 3	L	1.25	ND	ND	ND	ND	ND
	S	ND	ND	2.45	ND	ND	0.39
Farm 4	L	ND	ND	ND	1.4	ND	ND
	S	ND	4.44	1.69	8.5	ND	ND

L: leaves, S: stem.

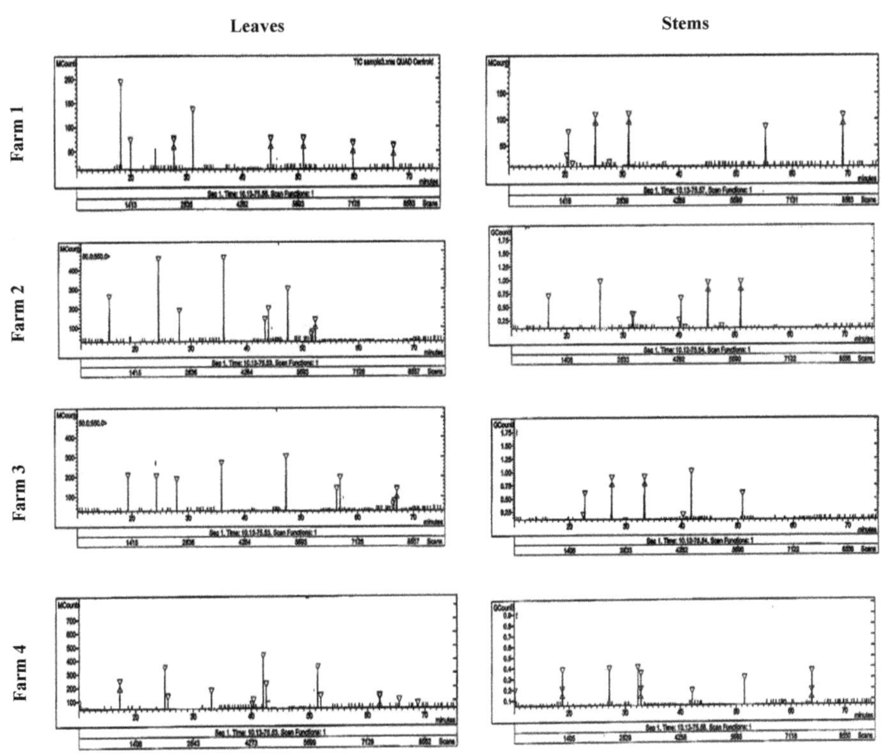

Figure 3. HPLC analysis of the alkaloid compounds in the stem and leaves of Taif's rose collected from different farms.

3.4. *In Vitro Antimicrobial Activity*

3.4.1. Leaf Extracts

The pharmacological properties of Taif's rose leaf extracts showed that the boiling water extract was exclusively active against all the studied bacterial and fungal strains, while the remaining extracts had no antifungal activities (Table 6 and Figure 4). *Bacillus subtilis, Escherichia coli*, and *Proteus vulgaris* were highly sensitive (inhibition zones = 24, 24, and 41 mm, respectively) to warm water extract compared with gentamicin (26, 30, and 17 mm). In addition, the boiling water extract was moderately active against fungal and bacterial strains with activities of 12, 10, 12, 12, and 16 mm for *Aspergillus fumigatus*, *Candida albicans*, *B. subtilis*, *E. coli*, and *P. vulgaris*, respectively. It is worth noting that *P. vulgaris* was exclusively sensitive to all Taif's rose extracts as follows: warm water > ethanol > boiling water > methanol > cold water.

3.4.2. Stem Extracts

The antimicrobial activity data of the stem extracts of Taif's rose showed that the methanol and cold water extracts were moderately active against all studied bacterial and fungal strains, while ethanol, boiling water, and warm water had no activity against fungal strains (Table 7 and Figure 4). It was clear that *P. vulgaris* was highly susceptible to all stem extracts with the highest inhibition zone (24 mm) for cold water and the lowest (13 mm) for methanol extracts. In addition, *B. subtilis* was moderately sensitive against most stem extracts (except warm water) with the highest activity (14 mm) for methanol and the lowest (11 mm) for ethanol extracts. Moreover, *A. fumigatus* and *C. albicans* were moderately susceptible to methanol (11 mm) and cold water (12 and 13 mm) extracts.

Table 6. Antimicrobial activity (mm) of the different extracts of Taif's rose leaves on the pathogenic bacterial and fungal strains. NA: no activity.

Extract	Fungi		Bacteria		
			Gram-Positive	Gram-Negative	
	Aspergillus fumigatus	*Candida albicans*	*Bacillus subtilis*	*Escherichia coli*	*Proteus vulgaris*
Control	Ketoconazole		Gentamicin	Gentamicin	
	17	20	26	30	17
Methanol	NA	NA	NA	NA	12
Ethanol	NA	NA	NA	NA	17
Boiling Water	12	10	12	12	16
Cold Water	NA	NA	10	NA	11
Warm Water	NA	NA	24	24	41

Table 7. Antimicrobial activity (mm) of the different extracts of Taif's rose stem on the pathogenic bacterial and fungal strains. NA: no activity.

Extract	Fungi		Bacteria		
			Gram-Positive	Gram-Negative	
	Aspergillus fumigatus	*Candida albicans*	*Bacillus subtilis*	*Escherichia coli*	*Proteus vulgaris*
Control	Ketoconazole		Gentamicin	Gentamicin	
	17	20	26	30	17
Methanol	11	11	14	13	13
Ethanol	NA	NA	11	NA	18
Boiling Water	NA	NA	12	NA	15
Cold Water	12	13	12	14	24
Warm Water	NA	NA	NA	NA	20

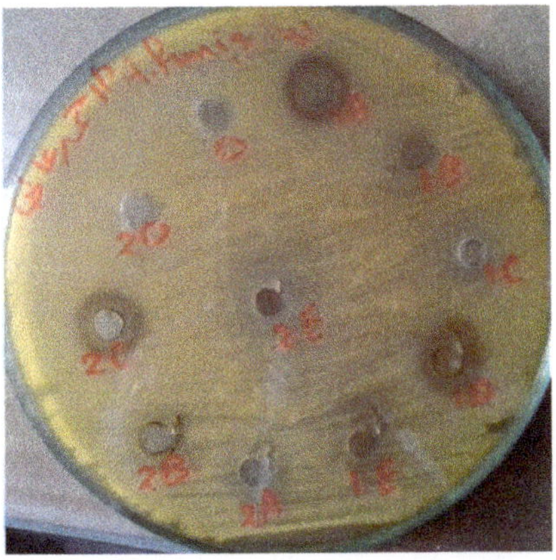

Aspergillus fumigatus

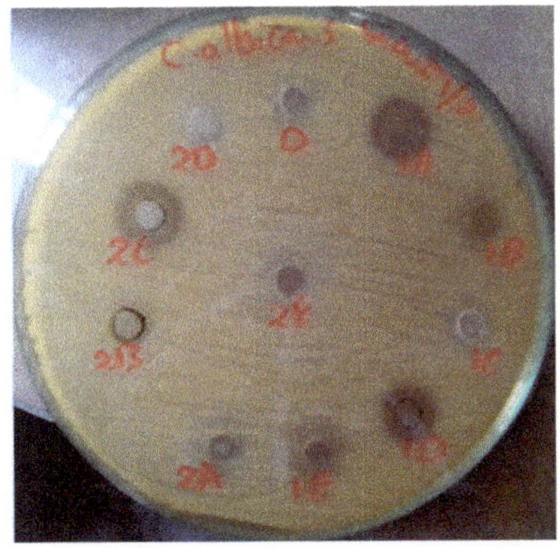

Candida albicans

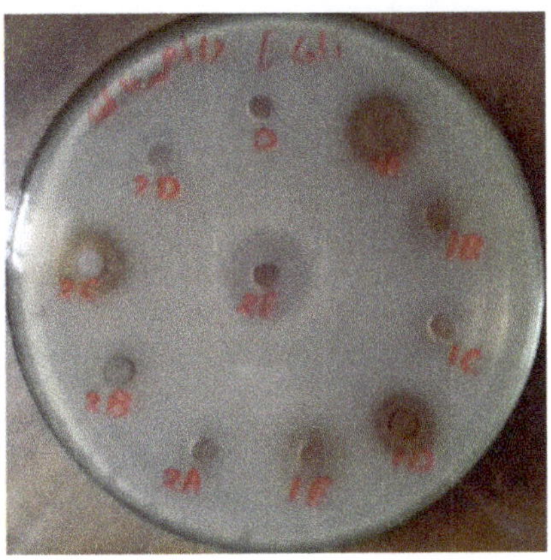

Escherichia coli

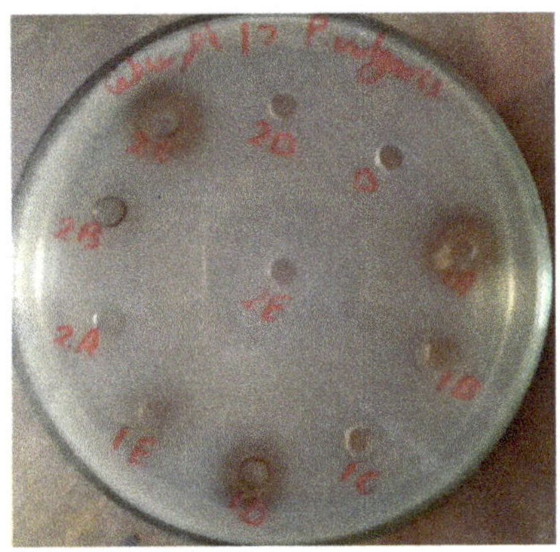

Proteus vulgaris

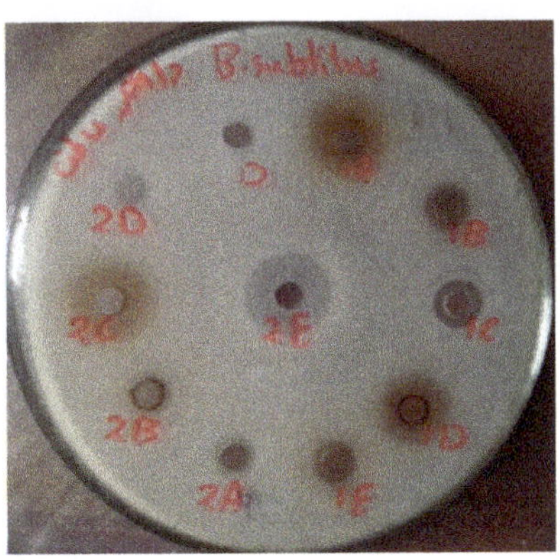

Bacillus subtilis

Figure 4. Antimicrobial activity of the different extracts of Taif's rose. 1: stem, 2: leaf, A: methanol extract, B: ethanol extract, C: boiled water, D: cold water, E: warm water.

4. Discussion

The pruning of Taif's rose is a horticultural art for manipulating plant architecture to force the plant into artificial rest or a dormant period before flowering [11]. Pruning waste disposal by drying and burning or storage causes environmental pollution [28]. In the current study, the biomass of the pruning wastes ranged between 1.3 t FW ha^{-1} and 5.2 t FW ha^{-1} in F4 (youngest farm) and F3 (oldest farm), respectively. According to Al-Yasi et al. [5], approximately 860 farms with different areas in the Taif governorate and its suburbs are cultivated with Taif's rose, which produces approximately 2730 tons of pruning waste and can cause a tremendous environmental problem. Therefore, it is of great importance and urgent need to recycle these agricultural wastes and reuse them for various economic purposes.

Carbohydrates are energy-rich molecules that play an important role in the immune system, pathogenesis, blood clotting, fertilization, and protein folding and placement. Their determination in plants is important for quality control analysis, as they are bioinformative macromolecules [38]. It was found that Taif's rose leaves had higher carbohydrate contents than the stems, where the contents ranged between 0.78% in the stem and 3.05% in the leaves. These values were comparable to 0.25–4.05% in the leaves of the medicinal aloe plants [39], but lower than 9.0–16.3% recorded in the leaves of *Calotropis procera* [40]. The high carbohydrate content in the leaves may be a response to the drought stress of rose plants before and after pruning [5]. In addition, the accumulation of carbohydrates may be due to the reduction in their utilization, either as a source of energy or for the formation of new cells and tissues or as an osmolyte of the cells [41]. According to Chesney and Vasquez [42], the biosynthesis of carbohydrates is influenced by pruning practices, and the stored carbohydrates can be used for plant regrowth [12].

The use of herbal/natural drugs as complementary/alternative medicines is gaining popularity throughout the world and many drugs are directly extracted from plants, whereas others are chemically modified [14]. Phytochemical exploration of the ethanolic extract of the Taif's rose stem and leaves revealed the presence of cardiac glycosides, alkaloids, and phenolic compounds. Similar findings were reported by Fathima and Murthy [43] in the petals of the same species. According to Alzletni et al. [44], the determination of these phytochemical compounds is important to show the nutritional and medicinal value of plants. Notably, the Taif rose stem had higher cardiac glycoside and flavonoid contents, while the plant leaves had higher phenolic and alkaloid contents. According to Baydar and Baydar [7], Taif's rose leaf extracts were rich in phenolic acids, including ferulic and gallic acids, and flavonoids, including catechin, compared to the other extracts. Cardiac glycosides are a type of secondary metabolite that has traditionally been utilized to augment cardiac contractile force in individuals suffering from cardiac arrhythmias or congestive heart failure [45]. Their contents ranged between 2.98 mg g^{-1} in the leaves of the oldest farm and 5.69 mg g^{-1} in the stems of the youngest farm. These values were lower than the 9.5–15.2 mg g^{-1} and 9.07–21.09 mg g^{-1} recorded in *C. procera* [40] and *Aloe* spp. [39], respectively.

Plant-based secondary metabolites are known to represent several structurally diverse classes of polyphenols with potential pharmacological activities, including anticancer, anti-inflammatory, antioxidant, and antipathogenic properties [46]. The plant leaves from the oldest farm had the highest phenolic contents (12.41 mg g^{-1}), while the stems of the youngest farm had the lowest (3.14 mg g^{-1}). These values are lower than the 386.4 mg g^{-1} recorded in the flower residue of Taif's rose [47] and 4.21–25.02 mg g^{-1} recorded in *Aloe* spp. [40]. This means that the pharmacological activities (antioxidant, anti-ageing, whitening, antitumor) of the flowers are greater than those of leaves and stems. According to Nayebi et al. [19], the cardioprotective effect of Taif's rose bioactive phenolics may be attributed to the inhibition of the enzymes related to atherosclerosis and hypertension. In addition, several investigations have demonstrated the antibacterial and disinfectant activity of Taif's rose and indicated the role of large phenolic contents such as flavonoids, terpenoids, and phenyl ethyl alcohol [20]. The highest flavonoid content

(9.33 mg g^{-1}) was recorded in the stems of Farm 2 plants, while the lowest (5.09 mg g^{-1}) was found in the leaves of Farm 3 plants.

Ellagic acid, catechol, resorcinol, gallic acid, and phloroglucinol were the main phenolic compounds separated and identified using HPLC in Taif's rose extract. The content of gallic acid as the main indicator in Taif's rose ranged between 5.6 and 37.4 mg g^{-1}, which is lower than the 50.3 mg g^{-1} recorded in the flower residue of the same species [48], 7.21–40.12 mg g^{-1} recorded in *Aloe* spp. [39]. Investigations have shown that gallic acid possesses a lot of biological activities such as antioxidant properties, antimicrobial activity, anti-inflammatory, antiviral, and antimutagenic activities, and anticancer activity [49,50]. Gallic acid also has anti-biofilm activity versus *Staphylococcus aureus* [51]. Moreover, quercetin is an abundant polyphenolic flavonoid that provides many health-promoting benefits such as being potent vasodilatory agents, cancer-reducing agents, anti-inflammatory, protective against asthma, and many others [52]. Furthermore, ellagic acid is an important compound used as an anticarcinogenic, multifunctional protector against oxidative stress and an anti-inflammatory agent in the treatment of chronic ulcerative colitis [4,53,54].

HPLC analysis showed that Taif rose leaves and stems produced flavonoid compounds including luteolin, apigenin, quercetin, rutin, kaempferol, and chrysoeriol. These compounds have potential antioxidant, anti-inflammatory, and antimicrobial properties [54]. Moreover, according to Dahat et al. [55], quercetin and its glycoside rutin have been reported in extracts displaying nephroprotective properties. In addition, luteolin and apigenin have been shown to inhibit the viability of leukemic cells, colon and ovarian carcinoma cells, and particularly human breast cancer cells, as well as reduce the occurrence of mouth sores and induce mild symptomatic relief [56]. Quercetin also helps protect against certain types of cancers, especially colon cancer [49], reduces the occurrence of mouth sores, and helps to induce mild symptomatic relief [52]. Moreover, alkaloids are biologically active compounds widely used as pharmaceuticals and synthesized as secondary metabolites in plants, and many of these compounds are highly toxic [13,15]. Berbamine, jatrorrhizine, palmatine, reticuline, isocorydine, and boldine were the main alkaloid compounds in Taif's rose. These compounds are common constituents of many Chinese medicinal plants and are known to have antibacterial, anti-inflammatory, anticancer, and choleretic properties as well as promote leukocytosis [57]. According to Duke [58], the long-term consumption of boldine led to color hallucinations, depression, partial motor aphasia, and sound hallucinations. High excitement exaggerates reflexes and respiratory movements, increases diuresis, causes cramps and convulsions, ends in death from centric respiratory paralysis, and heartbeats sometimes fail after respiration.

The antimicrobial activity of a plant depends on the phytogeographical area, the plant part, and the extraction process [59]. The pharmacological properties of Taif's rose leaf extracts showed that the boiling water extract was moderately active against all studied bacterial and fungal strains, and the remaining extracts had no antifungal activities. Gram-positive (*B. subtilis*), and gram-negative (*E. coli*, and *P. vulgaris*) bacteria were highly sensitive to warm water extracts compared with gentamicin antibiotics (control). In a similar study on *Rosa indica* extracts, Saeed et al. [60] found antibacterial activity against *Proteus* sp. and *E. coli*, and antifungal activity against *A. fumigatus* strains. The most effective reason is the presence of various phytochemical compounds such as alkaloids, phenolic acids, flavonoids, tannins, and other phenolic compounds. According to Baydar and Baydar [7], the total phenolics were higher in the cold and hot extractions of the leaf. In addition, the total phenolic and flavonoid contents have a good correlation with antioxidant activity [61], which is an important factor in assessing the biological activity of medicinal plants in the rose species [62]. Moreover, Samuelsen [63] and Abd Razik et al. [64] attributed the antimicrobial activity to the presence of some intermediately polar or nonpolar substances of relatively low molecular weight in the plant extract.

The methanol and cold water extracts of the stem were active against all studied bacterial and fungal strains; however, the remaining extracts had no activity against fungal strains. Similar findings reported that the methanol extract of *P. major* showed higher antimicrobial activity than the ethanol extract [55,65]. According to Norziah et al. [66], the use of water as the extracting solvent is more desirable than the use of organic solvents due to its environmentally friendly and non-toxic characteristics. Moreover, water is a good solvent in extracting a considerable quantity of phenolic and flavonoid compounds with high activities that can safely be exploited in numerous food applications. In one study, the intraperitoneal administration of 10 mg kg^{-1} of *R. damascena* Mill. methanolic extract in infected mice significantly reduced the parasitemia of *Plasmodium berghei* [19]. It was clear that *P. vulgaris* was highly susceptible, while *B. subtilis* was moderately sensitive against all stem extracts. Similar results were reported by Halawani [67] on the different extracts of *R. damascena*. Conversely, the aqueous extract of the medicinal plant *P. major* has no antimicrobial activity [55]. Therefore, pharmaceutical studies are required to separate, purify, and identify the phytochemical compounds in the ethanolic, methanolic, and water extracts of the pruning wastes of Taif's rose. In addition, the antibacterial activity of each compound was investigated to determine the compound/s that has antibacterial activity against pathogenic microorganisms.

5. Conclusions

Processing waste materials (such as rose wastes) and converting them into useful and efficient materials is an important issue that needs more consideration. The current study revealed that the pruning wastes of Taif's rose could be recycled due to their biologically active compounds including alkaloids, flavonoids, and phenolic compounds. More than 2700 tons of pruning waste is produced annually from approximately 860 rose farms in Taif Province. The phytochemical screening of Taif's rose indicated the presence of a considerable content of cardiac glycosides, which have traditionally been utilized to augment cardiac contractile force in individuals suffering from cardiac arrhythmias or congestive heart failure. HPLC analysis showed that Taif's rose contains flavonoid compounds including luteolin, apigenin, quercetin, rutin, kaempferol, and chrysoeriol; phenolics including ellagic acid, catechol, resorcinol, gallic acid, and phloroglucinol; alkaloids including berbamine, jatrorrhizine, palmatine, reticuline, isocorydine, and boldine. These compounds have several pharmacological properties including antimicrobial activities. Ethanol and methanol extracts of Taif's rose showed antimicrobial activity, but the highest was found in the water extracts. Further studies on the phytochemical constituents and pharmacological activity of the distillation wastes of Taif rose are currently underway.

Author Contributions: Conceptualization, T.M.G., E.F.A. and H.M.A.-Y.; data curation, M.A F., T.G.A.; R.Z.H., T.M.G., E.M.E. and E.F.A.; funding acquisition, T.M.G., M.A.F. and H.M.A.-Y.; project administration, T.M.G., E.M.E. and E.F.A.; formal analysis, H.M.A.-Y., E.F.A., T.M.G., E.M.E., E.M.E., and T.G.A.; investigation, E.F.A., T.M.G. and R.Z.H.; methodology, T.M.G., M.A.F., H.M.A.-Y.; E.F.A. and E.M.E.; resources, T.M.G. and E.F.A.; software, H.M.A.-Y., R.Z.H. and T.M.G., supervision, T.M.G. and E.F.A.; validation, R.Z.H., E.M.E. and T.G.A.; visualization, E.F.A. and T.M.G.; writing, T.M.G., H.M.A.-Y., T.G.A., E.F.A., E.M.E. and M.A.F., writing, review and editing, T.M.G. and E.F.A. All authors have read and agreed to the published version of the manuscript.

Funding: The authors extend their appreciation to the Deputyship for Research & Innovation, Ministry of Education in Saudi Arabia for funding this research work through the project number 1-441-129.

Informed Consent Statement: Not applicable.

Data Availability Statement: Not applicable.

Conflicts of Interest: There were no conflicts of interest from the authors.

References

1. Thakur, M.; Sharma, S.; Sharma, U.; Kumar, R. Study on effect of pruning time on growth, yield and quality of scented rose (*Rosa Damascena* Mill.) varieties under acidic conditions of western Himalayas. *J. Appl. Res. Med. Aromat. Plants* **2019**, *13*, 100202. [CrossRef]
2. Nunes, H.; Miguel, M.G. *Rosa damascena* essential oils: A brief review about chemical composition and biological properties. *Trends Pharmacol. Sci* **2017**, *1*, 111–128.
3. Najem, W.; Beyrouthy, M.E.; Wakim, L.H.; Neema, C.; Ouaini, N. Essential oil composition of *Rosa damascene* Mill. from different localities in Lebanon. *Acta Bot. Gall.* **2011**, *158*, 365–373. [CrossRef]
4. Mileva, M.; Ilieva, Y.; Jovtchev, G.; Gateva, S.; Zaharieva, M.M.; Georgieva, A.; Dimitrova, L.; Dobreva, A.; Angelova, T.; Vilhelmova-Ilieva, N.; et al. Rose Flowers—A Delicate Perfume or a Natural Healer? *Biomolecules* **2021**, *11*, 127. [CrossRef]
5. Al-Yasi, H.; Attia, H.; Alamer, K.; Hassan, F.; Ali, E.; Elshazly, S. Impact of drought on growth, photosynthesis, osmotic adjustment, and cell wall elasticity in damask rose. *Plant Physiol. Biochem.* **2020**, *150*, 133–139. [CrossRef] [PubMed]
6. Dobreva, A.; Getchovska, K.; Nedeltcheva-Antonova, D. A comparative study of Saudi Arabia and Bulgarian Rose oil chemical profile: The effect of the technology and geographic origin. *Flavour Fragr J.* **2020**, *35*, 584–596. [CrossRef]
7. Baydar, N.G.; Baydar, H. Phenolic compounds, antiradical activity and antioxidant capacity of oil-bearing rose (*Rosa damascena* Mill.) extracts. *Ind. Crop. Prod.* **2013**, *41*, 375–380. [CrossRef]
8. Baniasad, A.; Khajavirad, A.; Hosseini, M.; Shafei, M.N.; Aminzadah, S.; Ghavi, M. Effect of hydro-alcoholic extract of *Rosa damascena* on cardiovascular responses in normotensive rat. *Avicenna J. Phytomed.* **2015**, *5*, 319–324. [PubMed]
9. Labban, L.; Thallaj, N. The medicinal and pharmacological properties of Damascene Rose (*Rosa damascena*): A review. *Int. J. Herb. Med.* **2020**, *8*, 33–37.
10. Shawl, A.S.; Adams, R. Rose oil in Kashmiri India, an emerging cash crop benefiting industry and local agribusiness. *Perfum. Flavorist* **2009**, *34*, 2–5.
11. Pal, P.K. Evaluation, genetic diversity, recent development of distillation method, challenges and opportunities of *Rosa damascena:* a review. *J. Essent. Oil Bear. Plants* **2013**, *16*, 1–10. [CrossRef]
12. Pal, P.K.; Mahajan, M. Pruning system and foliar application of $MgSO_4$ alter yield and secondary metabolite profile of *Rosa damascena* under rainfed acidic conditions. *Front. Plant Sci.* **2017**, *8*, 507. [CrossRef] [PubMed]
13. Koul, B.; Taak, P.; Kumar, A.; Kumar, A.; Sanyal, I. Genus Psoralea: a review of the traditional and modern uses, phytochemistry and pharmacology. *J. Ethnopharmacol.* **2019**, *232*, 201–226. [CrossRef]
14. Algradi, A.M.; Liu, Y.; Yang, B.; Kuang, H. Review on the genus Brugmansia: traditional usage, phytochemistry, pharmacology, and toxicity. *J. Ethnopharmacol.* **2021**, *279*, 113910. [CrossRef] [PubMed]
15. Malhotra, B.; Kulkarni, G.T.; Dhiman, N.; Joshi, D.D.; Chander, S.; Kharkwal, A.; Sharma, A.K.; Kharkwal, H. Recent advances on *Berberis aristata* emphasizing berberine alkaloid including phytochemistry, pharmacology and drug delivery system. *J. Herb. Medicine* **2021**, *27*, 100433. [CrossRef]
16. Kaur, L.; Singh, D.; Cooper, R.; Kaur, M.; Singh, H.; Mutreja, V.; Sharma, A. Comprehensive review on ethnobotanical uses, phytochemistry, biological potential and toxicology of *Parthenium Hysterophorus* L.: A journey from noxious weed to a therapeutic medicinal plant. *J. Ethnopharmacol.* **2021**, *281*, 114525. [CrossRef]
17. Dai, J.; Mumper, R.J. Plant phenolics: extraction, analysis and their antioxidant and anticancer properties. *Molecules* **2010**, *15*, 7313–7352. [CrossRef]
18. Rasmussen, S.E.; Frederiksen, H.; Struntze Krogholm, K.; Poulsen, L. Dietary proanthocyanidins: occurrence, dietary intake, bioavailability, and protection against cardiovascular disease. *Mol. Nutr. Food Res.* **2005**, *49*, 159–174. [CrossRef]
19. Nayebi, N.; Khalili, N.; Kamalinejad, M.; Emtiazy, M. A systematic review of the efficacy and safety of *Rosa Damascena* Mill. with an overview on its phytopharmacological properties. *Complementary Ther. Med.* **2017**, *34*, 129–140. [CrossRef]
20. Karkania, V.; Fanara, E.; Zabaniotou, A. Review of sustainable biomass pellets production—A study for agricultural residues pellets' market in Greece. *Renew. Sustain. Energy Rev.* **2012**, *16*, 1426–1436. [CrossRef]
21. Shohayeb, M.; El-Sayed, S.; Abdel-Hameed, S.; Bazaid, A.; Maghrabi, I. Antibacterial and antifungal activity of *Rosa damascena* Mill. essential oil, diferent extracts of rose petals. *Glob. J. Pharm.* **2014**, *1*, 1–7.
22. Ghavam, M.; Afzali, A.; Manconi, M.; Bacchetta, G.; Manca, M.L. Variability in chemical composition and antimicrobial activity of essential oil of *Rosa* × *damascena* Herrm. from mountainous regions of Iran. *Chem. Biol. Technol. Agric.* **2021**, *8*, 22. [CrossRef]
23. Mahboubi, M. Rosa damascena as holy ancient herb with novel applications. *J. Tradit. Complement. Med.* **2016**, *6*, 10–16. [CrossRef]
24. Mileva, M.; Kusovski, V.K.; Krastev, D.S.; Dobreva, A.M.; Galabov, A.S. Chemical composition, in vitro antiradical and antimicrobial activities of Bulgarian *Rosa alba* L. essential oil against some oral pathogens. *Int. J. Curr. Microbiol. Appl. Sci.* **2014**, *3*, 11–20.
25. Mileva, M.; Krumova, E.; Miteva-Staleva, J.; Kostadinova, N.; Dobreva, A.; Galabov, A.S. Chemical compounds, in vitro antioxidant and antifungal activities of some plant essentia l oils belonging to Rosaceae family. *Compt. Rend. Acad. Bulg. Sci.* **2014**, *67*, 1363–1368.
26. Atay, O.A.; Ekinci, K. Characterization of pellets made from rose oil processing solid wastes /coal powder/pine bark. *Renew. Energy* **2020**, *149*, 933–939. [CrossRef]
27. Rusanov, K.; Garo, E.; Rusanova, M.; Fertig, O.; Hamburger, M.; Atanassov, I.; Butterweck, V. Recovery of Polyphenols from Rose Oil Distillation Wastewater Using Adsorption Resins—A Pilot Study. *Planta Med.* **2014**, *80*, 1657–1664. [CrossRef]

28. Sadasivam, S.; Manickam, A. *Biochemical Methods*, 3rd ed; New Age International Publishers: New Delhi, India, 2008.
29. Solich, P.; Sedliaková, V.; Karlíček, R. Spectrophotometric determination of cardiac glycosides by flow-injection analysis. *Anal. Chim. Acta* **1992**, *269*, 199–203. [CrossRef]
30. Tofighi, Z.; Ghazi, N.; Hadjiakhoondi, A.; Yassa, N. Determination of cardiac glycosides and total phenols in different generations of Securigera Securidaca suspension culture. *Res. J. Pharmacogn.* **2016**, *3*, 25–31.
31. Singleton, V.L.; Orthofer, R.; Rosa, M.B.T.; Raventós, L. Analysis of total phenols and other oxidation substrates and antioxidants by means of Folin-Ciocalteu reagent. *Methods Enzymol.* **1999**, *299*, 152–178.
32. Schütz, K.; Kammerer, D.R.; Carle, R.; Schieber, A. Characterization of phenolic acids and flavonoids in dandelion (*Taraxacum officinale* WEB. Ex WIGG) root and herb by high performance liquid chromatography/electrospray ionization mass spectrometry. *Rapid Commun. Mass Spectrom.* **2005**, *19*, 179–186. [CrossRef] [PubMed]
33. Zheng, W.; Clifford, M.N. Profiling the chlorogenic acids of sweet potato (*Ipomea batatas*) from China. *Food Chem.* **2008**, *106*, 147–152. [CrossRef]
34. Kandil, O.; Radwan, N.M.; Hassan, A.B.; Amer, A.M.; El-Banna, H.A.; Amer, W.M. Extracts and fractions of *Thymus capitatus* exhibit antimicrobial activities. *J. Ethnopharmacol.* **1994**, *44*, 19–24. [CrossRef]
35. Das, M.M.; Deka, D.C. Evaluation of anticancer and antimicrobial activity of arborinine from *Glycosmis pentaphylla*. *J. Biol. Act. Prod. Nat.* **2017**, *7*, 131–139. [CrossRef]
36. Abd El-Kader, H.A.; Sedde, S.R.; El-Shanawany, A.A. In vitro study of the effect of some medicinal plants on the growth of some dermatophytes. *Assiut Vet. Med. J.* **1995**, *34*, 36–42.
37. SPSS. *SPSS Base 15.0 User's Guide*; SPSS inc.: Chicago, IL, USA, 2016.
38. Campa, C.; Coslovi, A.; Flamigni, A.; Rossi, M. Overview on advances in capillary electrophoresis–mass spectrometry of carbohydrates: A tabulated review. *Electrophoresis* **2006**, *27*, 2027–2050. [CrossRef] [PubMed]
39. Aseeri, S.A.; Al-Yasi, H.M.; Galal, T.M. *Aloe Species in the Kingdom of Saudi Arabia: Morphological, Phytochemical and Molecular Characterization*; Lambert Academic Publishing Gmbh & Co.KG.: Saarbrücken, Germany, 2020; ISBN 978-620-2-66802-6.
40. El-Bakry, A.A.; Hammad, I.A.; Galal, T.M.; Ghazi, S.M.; Rafat, F.A. Polymorphism in *Calotropis procera*: variation of metabolites in populations from different phytogeographical regions of Egypt. *Rendiconti Rend. Fis. Acc. Lincei* **2014**, *25*, 461–469. [CrossRef]
41. Harish, S.R.; Murugan, K. Biochemical and genetical variation in the mangrove associate *Clerodendron inerme* (L.) Gaertn. under different habitats of Kerala Asian. *J. Exp. Biol. Sci.* **2011**, *2*, 553–561.
42. Chesney, P.; Vasquez, N. Dynamics of non-structural carbohydrate reserves in pruned *Erythrina poeppigiana* and *Gliricidia sepium* trees. *Agrofor. Syst.* **2007**, *69*, 89–105. [CrossRef]
43. Fathima, S.N.; Murthy, S.V. Cardioprotective effects to chronic administration of *Rosa damascena* petals in isoproterenol induced myocardial infarction: biochemical, histopathological and ultrastructural studies. *Biomed. Pharmacol. J.* **2019**, *12*, 1155–1166. [CrossRef]
44. Alzletni, H.; Galal, T.; Khalafallah, A. *The Arable Weed* Malva parviflora *L.: Ecophysiology and Phytochemistry*; Lambert Academic Publishing Gmbh & Co.KG.: Saarbrücken, Germany, 2020; 220p, ISBN 978-3-613-97570-9.
45. Abarquez, R.F. The old but reliable digitalis: persistent concerns and expanded indications. *Int. J. Clin. Pract.* **2001**, *55*, 108–114. [PubMed]
46. Rajendran, A.; Narayanan, V.; Gnanavel, I. Evaluation of therapeutic efficacy of *Aloe vera* sap in diabetes and treating wounds and inflammation in animals. *J. Appl. Sci. Res.* **2007**, *3*, 1434–1436.
47. Liu, W.; Chen, L.; Huang, Y.; Fu, L.; Song, L.; Wang, Y.; Bai, Z.; Meng, F.; Bi, Y. Antioxidation and active constituents analysis of flower residue of *Rosa damascena*. *Chin. Herb. Med.* **2020**, *12*, 336–341. [CrossRef]
48. Marlene, R.P.; Camila, K.P.; Cesar, M.B.; Evelyn, W.; Tânia, B.C.; Claudriana, L. Gallic acid and dodecyl gallate prevents carbon tetrachloride-induced acute and chronic hepatotoxicity by enhancing hepatic antioxidant status and increasing p53 expression. *Biol. Pharmaceut. Bull. J.* **2017**, *40*, 425–434.
49. Yanni, Y.; Mengyao, W.; Yingjie, H.; Chuankai, L.; Xin, P.; Wen, Z.; Youyi, H. Appropriately raising fermentation temperature beneficial to the increase of antioxidant activity and gallic acid content in *Eurotium cristatum*-fermented loose tea. *LWT-Food Sci. Technol.* **2017**, *82*, 248–254.
50. Murakami, M.; Yamaguchi, T.; Takamura, H.; Matoba, T. Effects of thermal treatment on radical-scavenging activity of single and mixed polyphenolics compounds. *Food Chem. Toxicol. J.* **2004**, *69*, 7–10. [CrossRef]
51. Liu, M.; Wu, X.; Li, J.; Liu, L.; Zhang, R.; Shao, D.; Du, X. The Specific Anti-Biofilm Effect of Gallic Acid on Staphylococcus Aureus by Regulating the Expression of the Ica Operon. *Food Control* **2016**, *73*, 1–6. [CrossRef]
52. Sharma, A.; Gupta, H. Quercetin-a flavonoid. *Chron. Young Sci.* **2010**, *1*, 10–15.
53. Galano, A.; Francisco Marquez, M.; Pérez-González, A. Ellagic acid: An unusually versatile protector against oxidative stress. *Chem. Res. Toxicol.* **2014**, *27*, 904–918.
54. Mirsane, S.A.; Mirsane, S.M. Benefits of ellagic acid from grapes and pomegranates against colorectal cancer. *Caspian J. Int. Med.* **2017**, *8*, 133–134.
55. Dahat, Y.; Saha, P.; Mathew, J.T.; Chaudhary, S.K.; Srivastava, A.K.; Kumar, D. Traditional uses, phytochemistry and pharmacological attributes of Pterocarpus santalinus and future directions: A review. *J. Ethnopharmacol.* **2021**, *276*, 114127. [CrossRef]
56. Adom, M.B.; Taher, M.; Mutalabisin, M.F.; Amri, M.S.; Abdul Kudos, M.B.; Sulaiman, M.W.; Sengupta, P.; Susanti, D. Chemical constituents and medical benefits of *Plantago major*. *Biomed. Pharmacother.* **2017**, *96*, 348–360. [CrossRef]

57. Petruczynik, A. Analysis of alkaloids from different chemical groups by different liquid chromatography methods. *Cent. Eur. J. Chem.* **2012**, *2012 10*, 802–835. [CrossRef]
58. Duke, J.A. *Duke's Handbook of Medicinal Plants of Latin America*; CRC Press: Boca Raton, FL, USA; Taylor & Francis Group, LLC: Abingdon, UK, 2009; 962p.
59. Metiner, K.; Özkan, O.; AK, S. Antibacterial effects of ethanol and acetone extract of *Plantago major* L. on gram positive and gram-negative bacteria. *Kafkas Univ. Vet. Fak. Derg.* **2012**, *18*, 503–505.
60. Saeed, R.; Ali, S.; Ullah, H.; Ullah, M.; Hassan, S.; Ahmed, S.; Akhwan, S. Phytochemical analysis and anti-microbial activities of *Rosa indica* collected from Kohat Pakistan. *Am. J. Phytomed. Clin. Ther.* **2014**, *2*, 1370–1377.
61. Alizadeh, Z.; Fattahi, M. Essential oil, total phenolic, flavonoids, anthocyanins, carotenoids and antioxidant activity of cultivated damask rose (*Rosa damascena*) from Iran: with chemotyping approach concerning morphology and composition. *Sci. Hortic.* **2021**, *288*, 110341. [CrossRef]
62. Shameh, S.; Alirezalu, A.; Hosseini, B.; Maleki, R. Fruit phytochemical composition and color parameters of 21 accessions of five Rosa species grown in North West Iran. *J. Sci. Food. Agric.* **2019**, *99*, 5740–5751. [CrossRef]
63. Samuelsen, A.B. The traditional uses, chemical constituents and biological activities of Plantago major L.: A review. *J. Ethnopharmacol.* **2000**, *71*, 1–21. [CrossRef]
64. Abd Razik, B.M.; Hasan, H.A.; Murtadha, M.K. The study of antibacterial activity of *Plantago major* and *Ceratonia siliqua*. *Iraqi Postgrad. Med. J.* **2012**, *11*, 130–135.
65. Sharifa, A.A.; Neoh, Y.L.; Iswadi, M.I.; Khairul, O.; Abdul Halim, M.; Jamaludin, M.; Mohamed, A.B.; Hing, H.L. Effects of methanol, ethanol and aqueous extract of *Plantago major* on gram positive bacteria, gram negative bacteria and yeast. *Ann. Microsc.* **2008**, *8*, 42–44.
66. rziah, M.H.; Fezea, F.A.; Bhat, R.; Ahmad, M. Effect of extraction solvents on antioxidant and antimicrobial properties of fenugreek seeds (*Trigonella foenum-graecum* L.). *Int. Food Res. J.* **2015**, *22*, 1261–1271.
67. Halawani, E.M. Antimicrobial activity of *Rosa damascena* petals extracts and chemical composition by gas chromatography-mass spectrometry (GC/MS) analysis. *Afr. J. Microbiol. Res.* **2014**, *8*, 2359–2367.

MDPI
St. Alban-Anlage 66
4052 Basel
Switzerland
www.mdpi.com

Life Editorial Office
E-mail: life@mdpi.com
www.mdpi.com/journal/life

www.ingramcontent.com/pod-product-compliance
Lightning Source LLC
Chambersburg PA
CBHW061546210926
51850CB00009B/2012